Lecture Notes in Artificial I

Edited by R. Goebel, J. Siekmann, anc

Subseries of Lecture Notes in Computer Science

António Teixeira
Vera Lúcia Strube de Lima
Luís Caldas de Oliveira   Paulo Quaresma (Eds.)

Computational Processing of the Portuguese Language

8th International Conference, PROPOR 2008
Aveiro, Portugal, September 8-10, 2008
Proceedings

 Springer

Series Editors

Randy Goebel, University of Alberta, Edmonton, Canada
Jörg Siekmann, University of Saarland, Saarbrücken, Germany
Wolfgang Wahlster, DFKI and University of Saarland, Saarbrücken, Germany

Volume Editors

António Teixeira
Universidade de Aveiro, Dep. de Electrónica, Telecomunicações e Informática, and
Instituto de Engenharia Electrónica e Telemática de Aveiro (IEETA)
3810-193 Aveiro, Portugal
E-mail: ajst@ua.pt

Vera Lúcia Strube de Lima
Pontifícia Universidade Católica do Rio Grande do Sul
Faculdade de Informática, Grupo PLN
90619-900 Porto Alegre, RS, Brazil
E-mail: vera.strube@pucrs.br

Luís Caldas de Oliveira
Universidade Técnica de Lisboa, and
INESC-ID, L2F
1000 Lisboa, Portugal
E-mail: lco@inesc-id.pt

Paulo Quaresma
Universidade de Évora, Departamento de Informática
7000-671 Évora, Portugal
E-mail: pq@di.uevora.pt

Library of Congress Control Number: 2008933855

CR Subject Classification (1998): H.3.1, H.5.2, I.2.1, I.2.7

LNCS Sublibrary: SL 7 – Artificial Intelligence

ISSN   0302-9743
ISBN-10   3-540-85979-9 Springer Berlin Heidelberg New York
ISBN-13   978-3-540-85979-6 Springer Berlin Heidelberg New York

Springer is a part of Springer Science+Business Media

springer.com

© Springer-Verlag Berlin Heidelberg 2008
Printed in Germany

Typesetting: Camera-ready by author, data conversion by Scientific Publishing Services, Chennai, India
Printed on acid-free paper   SPIN: 12513574   06/3180   5 4 3 2 1 0

Preface

The International Conference on Computational Processing on Portuguese, formerly the Workshop on Computational Processing of the Portuguese Language – PROPOR – is the main event in the area of Natural Language Processing that focuses on Portuguese and the theoretical and technological issues related to this specific language. The meeting has been a very rich forum for the interchange of ideas and partnerships for the research communities dedicated to the automated processing of the Portuguese language.

This year's PROPOR, the first one to adopt the International Conference label, followed workshops held in Lisbon, Portugal (1993), Curitiba, Brazil (1996), Porto Alegre, Brazil (1998), Évora, Portugal (1999), Atibaia, Brazil (2000), Faro, Portugal (2003) and Itatiaia, Brazil (2006).

The constitution of a steering committee (PROPOR Committee), an international program committee, the adoption of high-standard refereing procedures and the support of the prestigious ACL and ISCA international associations demonstrate the steady development of the field and of its scientific community.

A total of 63 papers were submitted to PROPOR 2008. Each submitted paper received a careful, triple-blind review by the program committee or by their commitment. All those who contributed are mentioned on the following pages. The reviewing process led to the selection of 21 regular papers for oral presentation and 16 short papers for poster sessions.

The workshop and this book were structured around the following main topics: Speech Analysis; Ontologies, Semantics and Anaphora Resolution; Speech Synthesis; Machine Learning Applied to Natural Language Processing; Speech Recognition and Natural Language Processing Tools and Applications. Short papers and related posters were organized according to the two main areas of PROPOR: Natural Language Processing and Speech Technology.

This year's PROPOR had two important novelties: one was the fact that the two main areas of the conference were more equally represented and the other was the inclusion of a special session dedicated to Applications of Portuguese Speech and Language Technologies. The special session, promoted by the Microsoft Language Development Center (MLDC), provided an opportunity for university and industrial communities working on portuguese natural language processing and speech technology to report their most recent products, systems, resources or tools for Portuguese. Two satellite events were also organized in association with PROPOR: the Second HAREM Workshop, Named Entity Recognition in Portuguese, and the workshop "Ten years of Linguateca".

We would like to express here our thanks to all members of our technical program committee and additional reviewers, as listed on the following pages.

We are especially grateful to our invited speakers, Tanja Schultz (University of Karlsruhe and CMU) and Chris Quirk (Microsoft), for their invaluable contribution, which undoubtedly increased the interest in the conference and its quality.

We are indebted to the PROPOR 2008 secretary, Anabela Viegas, for all her support.

We would like to publicly acknowledge the institutions and companies without which this conference would not have been possible: Universidade de Aveiro, Institute of Electronics and Telematics Engineering of Aveiro (IEETA), Association for Computational Linguistics (ACL), International Speech Communication Association (ISCA), ISCA Special Interest Group on Iberian Language (SIG-IL), Fundação para a Ciência e a Tecnologia (FCT), Microsoft, Springer, !UZ Technologies, DESIGNEED and Grande Hotel da Curia.

June 2008

António Teixeira
Vera Lúcia Strube de Lima
Luís Caldas de Oliveira
Paulo Quaresma

Organization

Conference Chair

António Teixeira   DETI/IEETA, Universidade de Aveiro, Portugal

Program Co-chairs

Vera Lúcia Strube
de Lima   Pontifícia Universidade Católica do Rio Grande do Sul, Brazil
Luís Caldas de Oliveira   L2F/INESC-ID, IST, Portugal

Publication Chair

Paulo Quaresma   Universidade de Évora, Portugal

Program Committee

Alexandre Agustini   Pontifícia Universidade Católica do Rio Grande do Sul, Brazil
Sandra Aluisio   Universidade de São Paulo, Brazil
Amália Andrade   CLUL, Universidade de Lisboa, Portugal
Jorge Baptista   Universidade do Algarve, Portugal
Plínio Barbosa   Universidade Estadual de Campinas, Brazil
Dante Barone   Universidade Federal do Rio Grande do Sul, Brazil
Steven Bird   University of Melbourne, Australia
Antonio Bonafonte   Universitat Politècnia de Catalunya, Spain
António Branco   Universidade de Lisboa, Portugal
Luís Caldas de Oliveira   INESC-ID/IST, Portugal
Nick Campbell   NiCT/ATR, Japan
Diamantino Caseiro   INESC-ID, Portugal
Berthold Crysmann   Bonn University, Germany
Gaël Dias   Universidade da Beira Interior, Portugal
Bento Dias da Silva   Universidade Estadual Paulista, Brazil
Marcelo Finger   IME- USP, Brazil
Diamantino Freitas   Faculdade de Engenharia, Universidade do Porto, Portugal
Pablo Gamallo   Universidade de Santiago de Compostela, Spain

Caroline Hagège	Xerox Research Centre Europe, France
Julia Hirschberg	Columbia University, USA
Isabel Hub Faria	Universidade de Lisboa, Portugal
Tracy Holloway King	Palo Alto Research Center, USA
Eric Laporte	Université Paris-Est Marne-la-Vallée, France
Gabriel Lopes	Faculdade de Ciências e Tecnologia, Universidade Nova de Lisboa, Portugal
Saturnino Luz	Trinity College Dublin, Ireland
Lúcia Machado Rino	Dep. de Computação, Universidade Federal de São Carlos, Brazil
Sandra Madureira	Pontifícia Universidade Católica de São Paulo, Brazil
Belinda Maia	Faculdade de Letras, Universidade do Porto, Portugal
Ranniery Maia	ATR Spoken Language Communication Labs, Japan
Nuno Mamede	INESC-ID/IST, Portugal
Jean-Luc Minel	MoDyCo, CNRS, France
Climent Nadeu	Universitat Politècnica de Catalunya, Spain
João Neto	INESC-ID/IST, Portugal
Viviane Moreira Orengo	Universidade Federal do Rio Grande do Sul, Brazil
Manuel Palomar	Universidad de Alicante, Spain
Fernando Perdigão	Universidade de Coimbra, Portugal
Carlos Prolo	Pontifícia Universidade Católica do Rio Grande do Sul, Brazil
Paulo Quaresma	Universidade de Évora, Portugal
Violeta Quental	Pontifícia Universidade Católica do Rio de Janeiro, Brazil
Elisabete Ranchhod	Universidade de Lisboa, Portugal
Fernando Gil Resende Jr.	Universidade Federal do Rio de Janeiro, Brazil
António Ribeiro	IPSC, Italy
Irene Rodrigues	Departamento de Informática, Universidade de Évora, Portugal
Solange Rossato	University of Grenoble 3, France
Diana Santos	SINTEF, Norway
Luís Seabra Lopes	DETI/IEETA, Universidade de Aveiro, Portugal
António Serralheiro	INESC-ID and Academia Militar, Portugal
Vera Strube de Lima	Pontifícia Universidade Católica do Rio Grande do Sul, Brazil
António Teixeira	DETI/IEETA, Universidade de Aveiro, Portugal

Ana Maria	
Tramunt Ibaños	Pontifícia Universidade Católica do Rio Grande do Sul, Brazil
Isabel Trancoso	INESC-ID/IST, Portugal
João Veloso	Universidade do Porto, Portugal
Renata Vieira	UNISINOS, Brazil
Aline Villavicencio	Universidade Federal do Rio Grande do Sul, Brazil
Fábio Violaro	Universidade Estadual de Campinas, Brazil
Maria das	
Graças Volpe Nunes	Universidade de São Paulo, Brazil
Dina Wonsever	Universidad de la Republica, Uruguay
Nestor Yoma	Universidad de Chile, Chile

Additional Reviewers

Petra Wagner	Bonn University, Germany
Luísa Coheur	INESC-ID, Portugal
José Adrián	
Rodríguez Fonollosa	Universitat Politècnica de Catalunya, Spain
Thiago Pardo	Universidade de São Paulo, Brazil

Table of Contents

Machine Learning Applied to Natural Language Processing

Speech Recognition and Applications

Natural Language Processing Tools and Applications

POSTERS

Natural Language Processing

Speech and Language Processing

Event Detection by HMM, SVM and ANN:
A Comparative Study

Carla Lopes[1,2] and Fernando Perdigão[1,3]

[1] Instituto de Telecomunicações
[2] Instituto Politécnico de Leiria-ESTG
[3] Universidade de Coimbra - DEEC
Pólo II, P-3030-290 Coimbra, Portugal
{calopes,fp}@co.it.pt

Abstract. The goal of speech event detection (SED) is to reveal the presence of important elements in the speech signal for different sound classes. In a speech recognition system, events can be combined to detect phones, words or sentences, or to identify landmarks with which a decoder could be synchronized. In this paper, we introduce three popular classification techniques, HMM, SVM, ANN and Non-Negative Matrix Deconvolution (NMD) for SED. The main purpose of this paper is to compare the performance of (1) HMM, (2) hybrid SVM/NMD (3) hybrid SVM/HMM and (4) hybrid MLP /HMM approaches to SED and emphasize approaches to reaching lower Event Error Rates (EER). It was found that the hybrid SVM/HMM approach outperformed the HMM system. Regarding EER, an improvement of 6% was achieved. The hybrid MLP/HMM got the best EER rate. Improvements of 11% and 8% were found in comparison with the HMM and hybrid SVM/HMM event detector, respectively.

Keywords: Speech recognition, event detection, HMM, SVM, ANN.

1 Introduction

Despite the continuous nature of speech, standard automatic speech recognition systems describe it as a sequence of discrete units, usually phonemes. Since speech is a result of changes in both the excitation source and the vocal tract system, it may be described as a sequence of events. These events may be related to the signal acoustics, the signal production, the speaker, etc, because any significant change may itself be treated as an event. What is most interesting is the fact that these events are common to all languages, and so they can be studied in different contexts and languages. In the literature, event-based systems are described in several contexts, these being: the classification of the signal into broad classes according to the presence of some specific features in the acoustic structure of the signal, [4],[1]; the detection of landmarks where some specific changes like syllabic dips, glottal closures or vowel onset points occur [10]; the finding of structural events like sentence boundaries, filled pauses, discourse markers, and edit disfluencies, [15], etc. Notwithstanding this fuzzy concept of speech events, all event-based systems have the same goal: to detect both the occurrence of important elements and the time when they occur. Several authors have

A. Teixeira et al. (Eds.): PROPOR 2008, LNAI 5190, pp. 1–10, 2008.

already focused on the goal of detecting events using Hidden Markov Models (HMMs) [4],[15], Support Vector Machines (SVMs), [1], Artificial Neural Networks (ANNs) [4] and also hybrid architectures, [7],[8],[3]. Nevertheless, to the best of our knowledge, no comparative study of the performance of these techniques has yet been done. In speech recognition systems, events may be used as additional information that aims at correcting the errors made by an existing recognizer. They can be used as input features, they can be combined to detect phones, words or sentences, or to identify landmarks with which a decoder could be synchronized.

Hidden Markov Models are, without doubt, the leading technology for Automatic Speech Recognition (ASR). In HMMs the acoustic-level decisions are taken based on the likelihood maximization criterion: an HMM that best matches a current input pattern is selected. Thus, everything seems to point to the success of HMMs in event-based detection, too. On the other hand, event detection relies mainly on a classification problem, and this could perhaps more successfully be tackled by means of discriminative approach. Consequently, two other technologies were introduced to detect events in the speech signal: ANNs and SVMs.

ANNs stand for an important class of discriminative techniques, very well suited for classification problems. Their ability to be used as a detection mechanism which learns from observed data completely suits our goal. Also, its discriminative learning capability, where there is no need to make assumptions about the class statistical distributions, is a remarkable feature since statistical distributions may change with each event class.

SVMs are also an important discriminative technique with several outstanding properties. Their ability to learn from a relatively small amount of high dimensional data, while at the same time providing a solution with maximum margin, mark SVMs out for success. The purpose of this paper is to compare the performance of event detection systems using HMMs, SVMs and ANNs to emphasize approaches to reach lower event error rates.

2 Event-Based System Description

A front-end which performs utterance segmentation in terms of a sequence of events over time is proposed. For that purpose four attributes to be detected were defined: silence, frication, stops and sonorancy, in such a way that the output signal of the proposed front-end is a segmented signal in terms of four broad classes: silences, fricatives, stops and sonorants.

The experiments were carried out using the TIMIT database, [5]. The training set consisted of all si and sx sentences of the original training set (3698 utterances) and the test set consisted of all si and sx sentences from the complete 168-speaker test set (1344 utterances). The 61 TIMIT-labeled phones were divided into 4 broad classes (sonorant, fricative, stop and silence) according to the phoneme sets presented in Table 1. The performance is evaluated for Correctness (Corr), Accuracy (Acc), and Event Error Rate (EER). The expressions for these measures are: $Corr=(N_T-S-D)/N_T$, $Acc= (N_T - S - D - I)/N_T$, and $ERR=1 - Acc$, where N_T is the total number of labels in the reference utterance and S, D and I are the substitution, deletion and insertion errors, respectively. The results were computed using the HTK Hresult tool, [11].

TIMIT database were used not only because it provides manually segmented pho-neme boundaries but also because it is widely used and thus allow us to compare results with other works.

Table 1. 61 TIMIT-labeled phones division into 4 broad classes of events

Broad classes	TIMIT-labeled phones
Fricatives	z, zh, s, sh, jh, ch, th, f, dh ,v
Silences	h#, epi, pau, bcl, dcl, gcl, pcl, kcl, tcl, q
Stops	b, d, g, p, t, k
Sonorants	dx, hv, l, m, n, ng, nx, r, w, y, hh, aa, ae, ah, ao, aw, ax-h, axr, ay, ax, eh, el, em, en, eng, er,ey, ih, ix, iy, ow, oy, uh, uw, ux

3 Baseline HMM Classifier

Hidden Markov Models are extensively used in speech recognition. Their success relies on their ability to model both the acoustic and temporal features of the speech signal. HMMs express the speech signal statistically, and at the same time they model the temporal evolution of the speech signal. In order to develop an HMM-based event classifier acoustic models were built for each class: sonorant, stop, fricative and si-lence, using HTK3.4, [11]. Each class was modelled by a three-state left-to-right HMM and each state by a single Gaussian. The input features were 12 MFCCs plus energy, and their 1st and 2nd order time derivatives, computed at a rate of 5ms and within a window of 15ms. The maximum likelihood criterion was used for training. Only acoustic models were employed in event recognition: no language model was used. Better results are obtained when adjacent events of the same class are merged. Results for the HMM system are in Table 2. With one Gaussian mixture we got a Correctness rate of 88.47% and an Accuracy rate of 73.14%. Despite the differences in training and testing conditions, these results outperform the results of a 5-class classifier described in [1]. With 8 mixtures the results are significantly better (89.36%, 77.57% for Correctness and Accuracy, respectively), however the number of training parameters also increased.

Table 2. HMM Classifier' results

HMM Classifier	Correctness	Accuracy	Number of training parameters
1 Gaussian mixture	88.47%	73.14%	979
8 Gaussian mixtures	89.36%	77.57%	7615

Although the HMMs work very well in speech recognition applications, the training maximum likelihood criterion has some limitations. This criterion maximizes the prob-ability of a given model generating the observation sequence, but does not minimize the probability of other models generating the same sequence. Even if each HMM model has the correct distribution for the corresponding speech, it has no knowledge about the distribution of the competing speech classes. These limitations lead to the appearance of other training techniques (discriminative) were the training is based on comparisons of the likelihood scores estimated for the speech units: SVM and ANN.

4 SVM Classifier

Binary classification using Support Vector Machines is a well known technique, [14]. Due to various complexities, a direct solution of multiclass problems using a single SVM formulation is usually avoided. To solve our multi-event problem, a binary classifier was built for each of the classes in a "one-versus-all" strategy. Figure 1 a) shows the SVM classifier's modular structure and Table 3 shows the acoustic features used to train each SVM classifier. In contrast to the HMM approach, where a temporal modulation of the features is made, the SVM operates in static mode: only parameters describing the recognized frame are used as input features. All SVM classifiers use only four acoustic static features, except the stop classifier, which uses a further eight dynamic features (first and second order time derivatives of the static features), within a context of 9 frames. These small sets of features seem to characterize each class well and lead to a smaller number of support vectors, thus shortening classification time. The features were computed at a frame rate of 5ms using a Hamming window of 15ms.

Table 3. Static acoustic feature set for each class of events

	Fricatives	Silences	Stops	Sonorants
5 ms log-energy	×	×	×	
Max amplitude		×		
Spectral Flatness Measure	×	×	×	
Spectral Centroid	×			
Log energy ratio at high/low frequencies	×	×	×	
Median of energy in a 9th filter bank			×	
Energy <500Hz				×
500<Energy <1500Hz				×
Voice evidence				×
Peakiness				×

The SVM software package SVM^{light} [13] was used for training and classification. It is common to use SVM with non-linear kernels 0[6], but despite the good classification capabilities of non-linear kernels, training and (especially) classification, are extremely time-consuming when the number of support vectors is high. It was thus decided to use linear kernels. Details of the parameters used in the SVM training, as well as training and testing statistics, can be found in [8].

The proposed event-based system has a modular structure as shown in Figure 1. The first module consists of the described SVM classifier. The outputs of this module provide membership predictions for each event class, for each frame. Since SVMs do not naturally give out posterior probabilities, the predictions were normalized by a *softmax* function, ensuring that a number between 0 and 1 is allocated to each class.

To turn the SVM outputs, which are frame-based, into a signal segmented in terms of events, an event merger is required. Two methods of event merging were tested and compared: The first proposes to generate events using Non-Negative Matrix Deconvolution, [12]. The second is a hybrid HMM/SVM architecture, [7]. Both methods are described below. Figure 1 a) illustrates the SVM classifier and Figures 1 b) and c) illustrate the merger methods.

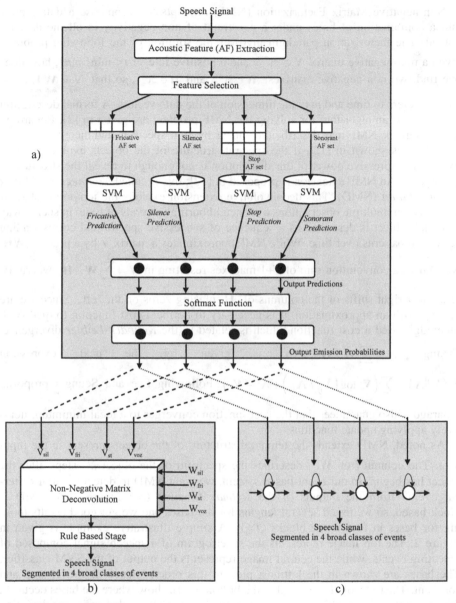

Fig. 1. Event-based system's modular structure. a) SVM classifier structure; b) Non-Negative Matrix Deconvolution Merger; c) Hybrid SVM/HMM system

4.1 Speech Event Detection by Non Negative Matrix Deconvolution

The speech signal was segmented into distinct classes of events by using Non-Negative Matrix Deconvolution, [12] which is an extension of Non-Negative Matrix Factorization, [2]. A brief description of both methods is given below and more details can be found in [8].

Non-negative Matrix Factorization (NMF) attempts to decompose a data matrix into a combination of bases under a constraint of non-negativity in all the data involved. The factorization paradigm involves the solution of the following problem: given a non-negative matrix $\mathbf{V} \in \mathfrak{R}^{m \times n}$ and a positive integer $r< \min\{m,n\}$, how does one find two non-negative matrixes, $\mathbf{W} \in \mathfrak{R}^{m \times r}$ and $\mathbf{H} \in \mathfrak{R}^{r \times n}$, so that $\underset{m \times n}{\mathbf{V}} \approx \underset{m \times r}{\mathbf{W}} \underset{r \times n}{\mathbf{H}}$. The index n refers to time and m to the dimension of the data vectors. A formal description of NMF and an algorithm for solving the NMF problem are given in [2]. Regarding speech signals, NMF may describe the objects by their spectrum and their energy over time, but, as shown in [12], if the spectral structure of the objects evolves distinctively, the expressive power of this description is not enough to reveal the structure of the objects. An NMF extension is proposed in [12]; this is called *Non-negative Matrix Deconvolution* (NMD). This deconvolutive extension exploits the temporal relationship between multiple observations over neighbouring intervals of time, in such a way that each object is described as a sequence of successive spectra and corresponding activation patterns over time. While NMF approximates a matrix $\mathbf{V}$ by a product $\mathbf{WH}$, NMD uses a convolution sum of $T+1$ matrixes resulting in $\mathbf{V} \approx \sum_{t=0}^{T} \mathbf{W}^{(t)} \overset{t\rightarrow}{\mathbf{H}}$, where $\overset{t\rightarrow}{\mathbf{H}}$ refers to t right shifts of the columns of $\mathbf{H}$, placing zeros on the left. Since we are dealing with an approximation, it is necessary to define a cost function to qualify it. Smaradgis used a cost function which is related to the *Kullback-Leibler* divergence. Taking $\Lambda = \sum_{t=0}^{T} \mathbf{W}^{(t)} \overset{t\rightarrow}{\mathbf{H}}$, it is given by the next expression

$$D(\mathbf{V}\|\Lambda) = \sum_{ij} \left(\mathbf{V}_{ij} \log\left(\mathbf{V}_{ij} / \Lambda_{ij} \right) + \Lambda_{ij} - \mathbf{V}_{ij} \right).$$ Following Lee and Seung's proposal,

Smaragdis also guarantees that the cost function converges to a local minimum, iteratively applying update functions.

As noted, NMD extends the temporal structure of the objects present in the input data. The column i of $\mathbf{W}^{(t)}$ describes the spectrum of the object i, t steps after the object has begun. In our event-based system, we want NMD to detect the four different events from the outputs of the *softmax* function (see Figure 1b)). NMD is block-based, so we tested several lengths for the bases and we got good results looking for bases in four-frame blocks ($T=3$). A simple illustrative example is given in Figure 2. The top image represents the spectrogram of a speech signal composed of repeating events, while the central image represents the output of the SVM classifier. The bases are shown in the leftmost plot, in this order: silence, fricative, stop and sonorant. The rows of $\mathbf{H}$ (depicted in the bottom plot), show where the bases occur. In this example all the objects were correctly detected. Nevertheless, some insertion errors do persist. These insertions were further reduced by applying simple rules inferred from the analysis of the event sequence and the duration of some events. Simple rules like deleting very short sonorant events or analyzing very improbable situations (ex: silence-fricative-stop or sonorant-stop-silence) improved the performance of the event-detector.

Using this procedure as our event merger we obtained 88.9% and 74% for correctness and accuracy, respectively. The results are present in the second row of Table 4. The achieved rates are significantly superior to Juneja's, [1] , and also outperformed the results from the HMM baseline. Event Error Rate (EER) improved about 3.24% compared with HMMs. Although such a setting makes only a small use of contextual information, we show that the performance achieved is better than for HMMs.

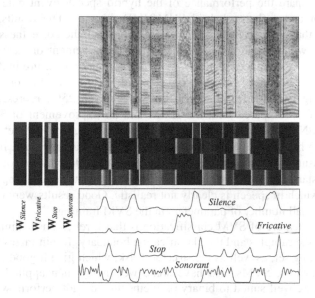

Fig. 2. Example of NMD event detection

4.2 Hybrid SVM /HMM Speech Event Detector

In order to combine the time warping abilities of HMMs with the discrimination capabilities of SVMs, we propose a hybrid system that combines an overall HMM structure with the class predictions given by SVM classifiers. The proposed system also performs utterance segmentation in terms of a sequence of the four broad classes. It uses a Markov process to temporally model the speech signal, but instead of using *a priori* state-dependent observation probabilities defined by a Gaussian mixture, it uses *a posteriori* probabilities estimated by SVMs, keeping the overall HMM topology unchanged. Figure 1c) shows the proposed system. In the hybrid system the normalized output predictions of the SVM are interpreted as the *a posteriori* event probabilities of ith class, $P(C_i \mid \mathbf{X})$, with C_i representing the ith class and $\mathbf{X}$ the feature observation vector. The likelihood ratio, $P(\mathbf{X} \mid C_i) / P(\mathbf{X})$, used in the HMM framework, are replaced with the posterior probabilities, using Bayes's rule,

$$\frac{P(\mathbf{X} \mid C_i)}{P(\mathbf{X})} = \frac{P(C_i \mid \mathbf{X})}{P(C_i)}. \tag{1}$$

The *a priori* class frequencies $P(C_i)$ are estimated off-line from the training data. Since the number of SVM outputs is equal to the number of classes, a one-state Markov model could be used for each class, but to allow some temporal modulation each class was modeled by a three-state HMM. We used HTK 3.4, [11] for testing, with some changes in order to replace the usual Gaussian mixture models by the normalized SVM outputs values.

We now compare the performance of the hybrid speech event detector with the HMM detector and with the hybrid SVM/NMD detector. The results, depicted in Table 4, show that with the hybrid SVM/HMM approach the correctness rate drops if we compare it with the HMM. Nevertheless the improvement on accuracy is quite noticeable in this case. Since Accuracy is the most common measure in ASR systems, and it is more precise than correctness, we consider it more useful for evaluating results. Regarding Event Error Rate (EER), we achieved 25%, representing an improvement of 6% compared with the HMM and an improvement of 3% compared with the SVM/NMD detector. The results show that speech event detection can be improved using both hybrid SVM/NMD and SVM/HMM architecture. In these architectures no assumptions are made about the statistical distribution of the acoustic space, unlike standard HMMs which assume that all the subsequent input frames are independent, which in speech is clearly not realistic. Good results were obtained even considering a small number of parameters in the SVM training.

An attractive feature of SVM classification is that it relies on maximizing the distance between the samples and the classification boundary. It minimizes the empirical risk in the training set, as well as the structural risk, which give it good generalization ability. Nevertheless, SVMs also have some weakness when applied to ASR. Although SVMs are well suited to binary problems they do not perform well enough in multi class problems (ASR). Also, the large memory requirement and computation time of SVM training algorithms means that it cannot use all the training data available in ASR databases. These SVM limitations led us to try another discriminative technique: Artificial Neural Networks.

5 ANN Classifier

Artificial Neural Networks are an important class of discriminative techniques. Because they learn according to a global discriminative criterion, they are appropriate for classification problems. ANN assigns discriminative weights for each of the input vectors, resulting in discrimination among the involved classes at the frame level.

Again unlike HMMs, ANN do not require any assumptions about the underlying statistical properties of the input data, but despite the good performance of ANN architectures in terms of frame error rate, they have also limitations. One significant drawback of ANN is that they are originally static classifiers. They do not inherently model the temporal evolution of data and in ASR they have to deal with the duration variability of speech units. A proposal for a hybrid system (ANN/HMM) emerged as a possible solution to this problem. The idea underlying the Hybrid ANN/HMM Speech Event Detector is as explained in Section 3.2: combine HMMs and ANN into a single system to profit from the best properties of the two approaches.

The proposed system is similar to the one described in section 3.2. It performs utterance segmentation in terms of a sequence of the four broad classes, but, instead of using the *posteriori* probabilities estimated by ANN, it uses the output values given by an ANN. The system used in the experiments consists of a Multi-Layer Perceptron

(MLP) network consisting of an input layer and an output layer, and it was trained for event frame classification. Since the main goal of this paper is to compare the performance of all the proposed systems we used the same features as in the SVM system. The first column of Table 3 specifies the 10 features used in ANN training. For a fair comparison with SVMs, no context window was considered for training. The targets derive from the phoneme boundaries provided by the TIMIT database. The *softmax* function is used as activation function of the output layer, so that the output values are interpreted as a posterior probability for each class. All the weights and bias of the network are adjusted using batch training with the resilient back-propagation (RP) algorithm [9] so as to minimize the minimum-cross-entropy error between the network output and the target values.

The experimental results are presented in Table 4. The hybrid ANN/HMM system outperformed all the other results. Correctness rose to 81.9% and Accuracy to 76.16%. Comparing it with the HMM event detector, improvements of 11.2% in EER were achieved. Comparing it with the hybrid SVM/HMM event detector the improvements were 8.3%, despite the fact that this last hybrid system has a very low number of training parameters. We trained an MLP with the same features used in the HMM event detector (12 MFCCs plus energy, and their 1st and 2nd order time derivatives), using a context window of 9 frames for training. The number of training parameters is similar to the HMM event detector with 8 Gaussian mixtures. The results were again very promising. Correctness rose to 87.12% and Accuracy to 80.17%, representing and improvement in EER of 11.6% compared with the HMM event detector with 8 Gaussian mixtures.

Table 4. Performance results for event detection

Performance	%Corr	%Acc	%EER	Number of training parameters	Improvements (%)		
HMM Event Detector (1mix)	88.47	73.14	26.86	979			
Hybrid SVM/NMD	88.86	74.01	25.99	<100	3.2		
Hybrid SVM/HMM	81.30	74.77	25.23	<100	6.1	2.9	
Hybrid ANN/HMM	81.93	76.16	23.84	983	11.2	8.3	5.5
HMM Event Detector (8 mix)	89.36	77.57	22.43	7615			
Hybrid ANN/HMM	87.12	80.17	19.83	7513	11.6		

6 Conclusions

In this paper, four systems were introduced for speech event detection. The first system is based on a traditional HMM. As a second approach we used SVM classifiers with a linear kernel in the SVM^{Light} implementation. To arrive at a segment-based detection, 2 methods were tested in combination with SVM: Non-Negative Matrix Deconvolution and a hybrid SVM/HMM. Both methods outperformed the traditional HMM results. Respective improvements of 3% and 6% were achieved in EER; another tested approach was based on a hybrid single layer MLP/HMM. This last approach achieves notable results. Improvements of 11% and 8% were found in comparison with the HMM and hybrid SVM/HMM event detector, respectively. The paper has compared the performance of the four classifiers, showing the strengths and weaknesses of each. Despite an English database were used for testing we believe that similar results can be achieved applying the methods to other languages, namely the Portuguese language, since the detected events are common to all languages.

Acknowledgments. Carla Lopes would like to thank the Portuguese foundation: Fundação para a Ciência e a Tecnologia for the PhD Grant (SFRH/BD/27966/2006).

References

[1] Juneja, A., Espy-Wilson, C.: Segmentation of continuous speech using acoustic-phonetic parameters and statistical learning. In: Proc. ICONIP, Singapore (2002)

[2] Lee, D.D., Seung, H.S.: Algorithms for non-negative matrix factorization. Advances in Neural Information Processing Systems 13 (2000)

[3] Bourlard, H., Morgan, N.: Hybrid HMM/ANN Systems for Speech Recognition: Overview and New. Research Directions. Springer, Heidelberg (1997)

[4] Li, J., Lee, C.H.: On Designing and Evaluating Speech Event Detectors. In: Interspeech 2005, Lisbon (2005)

[5] Garofolo, J.S., et al.: TIMIT Acoustic-Phonetic Continuous Speech Corpus. In: NIST (1990)

[6] Schutte, K., Glass, J.: Robust Detection of Sonorant Landmarks. In: Interspeech (2005)

[7] Lopes, C., Perdigão, F.: Hybrid HMM/SVM Speech Event Detector. In: 6th Conference on Telecommunications, Conftele 2007, Peniche, Portugal, vol. 1, pp. 601–604 (May 2007)

[8] Lopes, C., Perdigão, F.: Speech Event Detection By Non Negative Matrix Deconvolution. In: EUSIPCO-2007, Poznan, Poland, vol. 1, pp. 1280–1284 (September 2007)

[9] Riedmiller, M., Braun, H.: A direct adaptive method for faster backpropagation learning: The RPROP algorithm. In: Proc. of the IEEE ICNN, San Francisco (1993)

[10] Prasanna, S.: Event based analysis of speech, in Dept. of Computer Science and Engineering, Ph.D. Thesis: Indian Institute of Technology Madras, India (2004)

[11] Young, S., et al.: The HTK book. Revised for HTK version 3.4. Cambridge University Engineering Department, Cambridge (December 2006)

[12] Smaragdis: Discovering Auditory Objects through Non-Negativity Constraints. In: Statistical and Perceptual Audio Processing (SAPA 2004), Jeju, Korea (2004)

[13] Joachims, T.: Making large-Scale SVM Learning Practical. In: Schölkopf, B., Burges, C., Smola, A. (eds.) Advances in Kernel Methods - Support Vector Learning. MIT-Press, Cambridge (1999)

[14] Vapnik, V.: Statistical Learning Theory. Wiley Inter-science, Chichester (1998)

[15] Liu, Y.: Structural Event Detection for Rich Transcription of Speech, Ph.D. Thesis: Purdue University (2004)

Frication and Voicing Classification

Luis M.T. Jesus[1] and Philip J.B. Jackson[2]

[1] Escola Superior de Saúde da Universidade de Aveiro, and
Instituto de Engenharia Electrónica e Telemática de Aveiro
Universidade de Aveiro, 3810-193 Aveiro, Portugal
lmtj@ua.pt
http://www.ieeta.pt/~lmtj/
[2] Centre for Vision, Speech & Signal Processing
University of Surrey, Guildford GU2 7XH, UK
p.jackson@surrey.ac.uk
http://personal.ee.surrey.ac.uk/Personal/P.Jackson/

Abstract. Phonetic detail of voiced and unvoiced fricatives was examined using speech analysis tools. Outputs of eight f0 trackers were combined to give reliable voicing and f0 values. Log-energy and Mel frequency cepstral features were used to train a Gaussian classifier that objectively labeled speech frames for frication. Duration statistics were derived from the voicing and frication labels for distinguishing between unvoiced and voiced fricatives in British English and European Portuguese.

1 Introduction

1.1 Background

The long term objectives of the work presented in this book chapter are to deliver novel analysis methods for characterizing speech. Parameters for describing frication and voicing in fricatives are used to facilitate analysis of phonation and frication interaction effects observed. In particular, we aim to develop a concise model of the duration of voice and frication sources in fricative consonants in British English (BE) and European Portuguese (EP). The present work incorporates the following tasks: (i) development of speech analysis methods; (ii) development of new measures of voicing and frication to extend the phonetic description of Portuguese and English speech data; (iii) application of these parameters to the automatic classification of speech sounds; (iv) application of techniques across English and Portuguese using selected measures most apt for analysis, classification and modelling of mixed source speech signals.

This study deals with sounds produced by the simultaneous combination of two aeroacoustic sources, which have very different natures (one is quasi-periodic and the other noiselike). To measure properties of sounds like fricatives, stops and affricates, we evaluated the feasibility of conventional temporal and spectral measures, to yield useful descriptions of speech events. Pre-recorded EP and BE corpora of contextually-balanced acoustic data were used (Jesus and Shadle 2002; Pincas 2004).

A. Teixeira et al. (Eds.): PROPOR 2008, LNAI 5190, pp. 11–20, 2008.

The accurate determination of voicing onset/offset and the extraction of the fundamental frequency are important for the quantification of differences between normal and pathological voices, and for the robust encoding of normal voicing information in speech analysis/synthesis systems, as well as automatic labeling and segmentation. Francis et al. (2003) compared acoustic measures of voicing onset and found methods based on the waveform and low-frequency "voicing bar" to be more accurate and consistent than methods based on formants. Time-domain (McCree et al. 2002; Droppo and Acero 2007) and frequency-domain (Quatieri 2001; Pelle and Estienne 2007) methods for fundamental frequency analysis, used for low bit rate speech coding, have typically aimed at delivering a binary voiced/unvoiced decision and very few researchers (Childers et al. 1989) have tried to identify three different voicing states, i.e., voiced, partially voiced and unvoiced. Estimation of fundamental frequency typically relies on the signal periodicity (Hess 1992), and some researchers have explicitly disregarded irregular voice segments (Cheveigné and Kawahara 2002).

Previous work on fricatives with mixed sources includes the identification of the unvoiced fricative duration (UFD) as an essential feature for voicing categorization in English fricatives (Stevens et al. 1992; Pincas 2004). One important interaction effect, the modulation of frication during voicing, has been studied (Jackson and Shadle 2000; Pincas and Jackson 2005), as have the voicing characteristics of Portuguese fricatives (Jesus and Shadle 2003).

1.2   Motivation

Here, we combine our knowledge about observable (in the acoustic signal) differences in production strategy between unvoiced, devoiced and voiced fricatives for the same place of articulation. Interactions between voicing and frication sources are characterized by relative timings of onsets and offsets of voicing and frication, the fundamental frequency (f0), and relative levels of voicing and frication.

We believe that a processing approach inspired by speech production (data driven and knowledge based) can contribute to the performance of speech technology systems.

In vowel production, the vocal tract is relatively unconstricted and vocal folds tend to vibrate easily. In voiced obstruent consonants (fricatives or stops), a strong simultaneous noise source can only be produced at the expense of weakened voicing or devoicing.

In a study of devoicing of Portuguese voiced fricatives (Jesus and Shadle 2003), a criterion based on the ratio of variances in the electroglottograph (EGG) signal was used, during the VF transition and during the fricative, to derive a two-way classification (voiced/devoiced). The EGG variance, calculated at the beginning, middle and end of the fricatives, can be compared to the present classification scheme based on the f0 tracks of the speech signal.

Although f0 trackers seek periodicity in ways often similar to those used for manual annotation, they tend to be least reliable at voice onset/offset. We decided to test a range of freely accessible algorithms and combine their outputs to achieve a more reliable set of measurements.

The aim of the work is in using statistical tools in the fine phonetic analysis of fricatives. We have devised experiments that use an HMM to automatically classify both voicing and frication.

2 Speech Data

2.1 European Portuguese

A speech corpus, containing 1304 words that included fricatives /f, v, s, z, ʃ, ʒ/ from two male and two female adult native EP speakers, was recorded in a sound treated room using a Bruel & Kjaer 4165 $\frac{1}{2}$ inch microphone located 1 m in front of the subject's mouth, connected to a B & K 2639 pre-amplifier, then amplified and filtered by a B & K 2636 measurement amplifier (22 Hz-22 kHz). Acoustic and EGG signals were recorded with a Sony TCD-D7 DAT (16 bits, 48 kHz sampling frequency) and digitally transferred to PC. The simultaneous EGG signal was not used in the present study. Corpora were devised that included Portuguese words containing fricatives in frame sentences (Corpus 3), and the same set of words in sentences (Corpus 4). The EP corpus has manual annotations of the fricative start and end times that mark the transitions into and out of each fricative. Phonetic and phonological details of the corpus are described in Jesus and Shadle (2002).

2.2 British English

Fricatives from eight subjects, four male and four female, were recorded, all native speakers of BE. Speech-like tokens were obtained using nonsense /VFə/ words, F=/f, v, θ, ð, s, z, ʃ, ʒ/, embedded in the phrase "What does /VFə/ mean?" with vowel V=/ɑ, i, u/. Mono recordings were made in an acoustically-sheltered cubicle by Beyerdynamic M59 dynamic microphone linked directly to PC with a Creative Audigy soundcard (16 bits, 44.1 kHz sampling frequency). Nine repetitions of each possible VF combination by each speaker made 1728 sentences. The BE corpus was manually annotated separately for voicing and frication (Pincas 2004).

2.3 Dividing the Data

The data was divided into eight sets, having equivalent dimensions, and an even distribution of fricatives according to their place of articulation and phonological voicing classification, as shown in Table 1. Each data-set also has approximately the same number of samples from each speaker, gender, and for EP data the same number of samples from Corpus 3 and Corpus 4 (Jesus and Shadle 2002).

We needed to divide the data up for jack-knife experiments, maintaining separation of the training and the test data, meanwhile providing the most informative test results from the limited total data. Given the fact, that the BE data are all in vowel context, any files in the EP corpus that contained consonantal contexts were excluded. This resulted in the loss of 9% of the data (a fairly small proportion overall).

Table 1. Number of fricatives in the BE and EP data-sets

Set	British English									European Portuguese						
	[f]	[v]	[θ]	[ð]	[s]	[z]	[ʃ]	[ʒ]	Total	[f]	[v]	[s]	[z]	[ʃ]	[ʒ]	Total
set1	38	8	56	30	32	16	24	32	236	22	33	32	26	26	27	166
set2	24	7	31	21	40	40	24	30	217	22	33	31	25	26	27	164
set3	32	37	32	30	24	24	22	31	232	22	33	31	26	27	27	166
set4	24	59	32	30	24	23	32	8	232	22	34	32	27	27	29	171
set5	24	22	23	14	40	40	32	24	219	22	37	33	27	27	27	173
set6	22	20	16	38	16	39	24	45	220	22	38	34	28	26	26	174
set7	8	32	16	18	16	8	32	24	154	22	39	32	27	26	28	174
set8	40	16	8	8	24	24	24	16	160	20	39	32	27	23	28	169

3 Extraction of Reference f0

Wave files were processed to give a set of eight f0 tracks each, from which a reference f0 track was calculated. These were analysed together to evaluate voicing and f0 errors, which were treated as either *gross* (e.g., halving or doubling octave errors) or *fine*.

3.1 f0 Determination Algorithms

Only open-source software was employed, which enabled investigation (and correction) of the algorithms and represented widely-used speech research tools. Our selection included a number of standard f0 determination algorithms available in the Speech Filing System (SFS v. 4.6), the Auditory Perception Toolbox by MARCS Auditory Laboratories (MARCS v. 1.01) and Praat (v. 5.0.02):

1. `fxrapt -isp ...` – autocorrelation algorithm similar to
   Secrest and Doddington (1983) and used in `get_f0` Entropics' ESPS/Waves.
2. `fxcep -isp ...` – cepstral algorithm by Whittaker, Howard and Huckvale using Noll (1967)'s rules .
3. `fxanal -isp ...` – autocorrelation algorithm similar to
   Secrest and Doddington (1983) and implemented by Huckvale.
4. `fxac -isp ...` – autocorrelation algorithm by Huckvale.
5. `extractfundamental(...,...,0.01,'threshamp',0.02)` – Matlab implementation by Morris of Yehia's LPC-based algorithm.
6. `To Pitch (ac)... 0.0 75.0 15 off 0.03 0.45 0.01 0.35 0.14 600.0`
   – autocorrelation method implemented by Boersma (1993).
7. `To Pitch (cc)... 0.0 75.0 15 off 0.03 0.45 0.01 0.35 0.14 600.0`
   – forward cross-correlation method (Boersma).
8. `To Pitch (shs)... 0.01 50.0 15 1250.0 15 0.84 600.0 48`
   – subharmonic summation algorithm (Hermes 1988).

3.2   Combining f0 Tracks

The output from each f0 tracker was treated as the product of two simultaneous tracks, a binary voicing decision and the estimated fundamental frequency. Gaps in the f0 data (i.e., during unvoiced segments) were filled by linear interpolation. Both pieces of information, typically provided every 10 ms, were upsampled to every 1 ms. Hence, each f0 track yielded a voicing state and f0 estimate at 1 kHz frame rate. The median[1] gave the majority voicing state and a robust f0 value (see Figure 1).

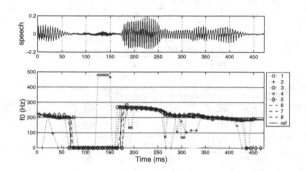

Fig. 1. Upper: acoustic signal of "a febra" [ɐ'fɛbɾɐ]. Lower: f0 tracks from 8 programs and the reference (ref).

The differences between the various f0 tracks and the reference track were analyzed to assess the consistency of the tracking methods, and hence an indication of the accuracy of the reference track. These differences fell into three broad categories: *voicing* errors, *gross* f0 errors and *fine* f0 errors. Voicing errors occurred when the voicing status of a given f0 track disagreed with that of the reference, and were classed as *false alarms* if the reference was unvoiced and as *false rejections* if it was voiced. With the same voicing status, a gross error indicated that the f0 track was closer (on a logarithmic scale) to either double or half of the current reference f0. The remaining voiced frames were considered *matched* and the fine errors were described for these by the RMS amplitude of the f0 difference (in Hz). A summary of the results of the error analysis is given in Table 2 for BE and EP data. The RAPT algorithm gave the best voicing decisions, while Boersma's methods provided most accurate f0.

4   Duration Analysis

In seeking an automatic and objective method for detecting and classifying the fine phonetic detail of fricatives, a series of hidden Markov models (HMMs) were built with Gaussian probability density functions. The MFCC and log-energy

---

[1] With eight values, the median was taken as mean of values ranked 4th and 5th; voicing status was rounded toward being voiced.

Table 2. f0 tracker (8 programs) error analysis (overall summary)

	1	2	3	4	5	6	7	8
Voicing error as proportion of entire corpus (%) – 69.8% voiced								
EP	4.7	30.0	6.7	9.5	12.0	6.0	6.2	14.0
BE	1.5	26.5	2.3	12.2	4.4	1.7	1.2	30.0
False alarm as proportion of unvoiced frames (%)								
EP	4.8	36.9	11.2	13.0	3.2	9.7	13.0	30.0
BE	1.3	24.3	1.9	13.5	0.7	0.5	0.5	36.9
False reject as proportion of voiced frames (%)								
EP	4.7	27.0	4.8	7.9	15.8	4.4	3.3	7.1
BE	2.3	34.7	3.5	7.2	18.5	6.2	4.0	4.3
Gross errors as proportion of voiced frames (%)								
EP	3.2	7.5	6.4	6.6	2.4	1.2	1.5	3.0
BE	3.1	8.5	9.6	11.2	2.8	1.4	3.4	3.9
Matched as proportion of voiced frames (%)								
EP	92.1	65.5	88.8	85.5	81.9	94.4	95.2	90.0
BE	94.7	56.8	86.9	81.5	78.7	92.4	92.6	91.9
RMS fine errors (Hz)								
EP	7.0	9.7	6.8	8.9	7.5	5.8	6.0	5.6
BE	7.2	10.1	5.9	10.5	9.3	6.3	6.2	7.0

features were obtained from the acoustic waveform (0.1 - 7.5 kHz) via HTK with 15 ms windows; only static features were used. The number of MFCCs was varied. The results of framewise classification accuracy against manual labels supported the use of 12 MFCCs plus log energy.

4.1   Method

Two experiments examined BE and EP respectively, using an HMM automatically to classify both voicing and frication. From the state alignment with respect to the acoustic features (i.e., the time spent in each state), we can derive an objective measure of devoicing, as well as other characteristics of the fricatives in our data sets.

Short audio clips containing one fricative plus 50 ms either side to give context and transitions, were extracted. Acoustic features (12 MFCCs and log energy) were computed with just 1 ms offset between frames, giving a 13 - D feature vector every 1 ms. Phonologically unvoiced fricatives typically start with a little or no overlap (<20 ms) between the voicing from the vowel to the onset of frication, then there is the main period of unvoiced frication until the onset of the following sound. For phonologically voiced fricatives, we expect there to be voicing throughout accompanied by the fricative source, although devoicing does sometimes occur. So, the state topology was defined to allow /V-uF-V/, /V-vF-uF-V/ or /V-vF-V/, where uF denotes unvoiced frication, vF denotes voiced frication, and V denotes the context of adjacent phonemes that were typically vowels (e.g., /ɑFə/ for BE). We have defined the topology to account for the state sequences that occur in our data set, so we do not allow /V-uF-vF-V/ because it does not normally occur, whereas there is often a short period of overlap between voicing and frication at the start of phonologically unvoiced fricatives. The timing of these transitions is critical to their categorical perception, because it carries important cues to whether the fricative should be considered voiced or unvoiced.

Models for the BE data provided two states for the preceeding vowel, two for the fricative (one voiced, one unvoiced), and two for the following schwa. In order to balance the amount of training data used for each of the model states, and to accommodate the increased variability of the contexts in the EP database, six separate 2 - state models were defined as follows: voiced frication (as with BE), unvoiced frication (as with BE), front, central and back vowels (and diphthongs starting with a front, central and back configuration), and silence. Nasalised and non - nasalised vowels were grouped together. This made a total of 12 states in the EP models, whereas the uniform context led to just 6 states in the BE models.

Initial state alignments were based on manual phone boundaries, dividing vowel segments, and using voicing decision from reference f0 for fricatives. One state was created for each of these with a 13- D Gaussian pdf. These initial definitions of state occupation were used to determine the mean and covariance for each state in Viterbi training. Training comprised of 10 further iterations in which the new state alignments were used to refine the models (allowing slight adjustments of the state boundaries for a better fit to the observed data).

The first set of multiple training iterations of jack - knife experiments, used *set2-8* for training and *set1* for testing. In the second set, we trained on *set1* and *set3-7* and tested on *set2*. The rest followed this pattern, i.e., the state alignment output from the HMMs were trained on 7/8 of the data and decoded on the remaining unseen files.

The final step consisted of using the trained models on the withheld test utterances to yield a completely automatic segmentation of the portion of the utterance around the fricative. This segmentation was then used to derive the duration statistics for final analysis of the data. The goal was a quantitative description of voiced and unvoiced periods during the phonological voiced and unvoiced fricatives.

4.2   Results

Manual annotations provided an initial alignment and the automatic ones were taken from the final alignment. These were used to extract the unvoiced frication duration (UFD) and the duration of frication with voicing, which we term the source overlap duration (SOD).

Figure 2 (top) shows the results of plotting SOD versus UFD for all eight English subjects, across all places of articulation. Voiced fricatives lie on the SOD axis, unvoiced lie on the UFD axis, and most of the data fall into the main area with some SOD and some UFD. The phonologically voiced and unvoiced fricatives tend to form two distinct clusters which are highlighted by the red and blue ellipses on those plots.

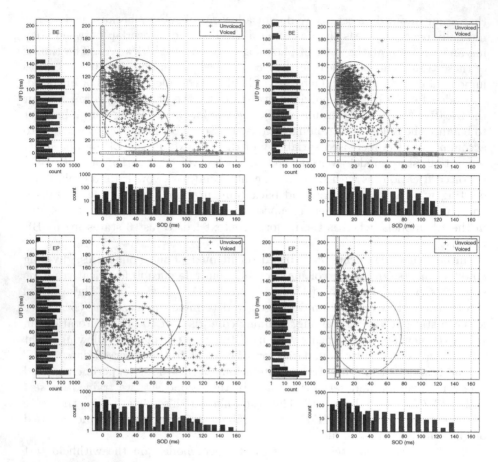

Fig. 2. Source overlap duration (SOD) and unvoiced frication duration (UFD) voicing classifications in BE (upper) and EP (lower) fricatives with manual (left) and HMM (right) alignments. Histograms show more clearly the distribution of data points.

Unvoiced fricatives cluster around (20, 100) ms, and a high classification accuracy of the phonological categories can be achieved by thresholding at UFD ≈ 60 ms (as reported previously by Pincas (2004)).

Considering the automatic voicing classification (Figure 2 top left), we see that the pattern is broadly consistent: SOD times have increased slightly at the expense of UFD. Figure 2 (top right) shows the output from the HMM annotation of states. The new clusters for unvoiced and voiced fricatives are centred at (10, 115) ms and (20, 50) ms respectively, suggesting a higher threshold UFD ≈ 70 ms.

Figure 2 (bottom) shows an analysis of Portuguese fricatives. As before, the left panel shows SOD versus UFD with manual frication annotation and voicing classification from the reference f0 track for the entire EP corpus. The distributions are similar to those from the BE corpus, however there is greater overlap including a large number of phonologically voiced fricatives that were devoiced. This variability may be attributed to differences in annotation procedure and the more natural context of the EP tokens.

5 Conclusions

In this book chapter we have proposed the development of an automatic method for phonetic analysis of the durational characteristics of voicing and frication features. Our experiments consider both British English and European Portuguese fricatives recorded as nonsense and real words respectively. By combining the outputs of eight publicly - available f0 determination algorithms, we obtained a more reliable reference f0 track for each utterance which was used to evaluate the accuracy of each technique, with an emphasis on fricative speech. Together with manual annotation of phone boundaries, we used the voicing state of the reference f0 track to define initial regions of voiced and unvoiced frication. Jack - knife experiments were then conducted, training HMMs to recognize these states in unseen test utterances. The final output was an objective annotation of voiced and unvoiced frication to 1 ms resolution, from which duration statistics were obtained.

We have shown that the technique can be applied across languages. It is relevant both to English and Portuguese, and enables objective investigation of the duration characteristics observed in various contexts. Further work is needed to extend the results of this pilot study to a wider range of speech data, and to encapsulate our knowledge of fricative duration characteristics. Such duration models could be made context - dependent and incorporated into model - based speech synthesis and articulatory - feature based speech recognition.

Acknowledgements

This work was partially supported by Fundação para a Ciência e a Tecnologia, Portugal, Conselho de Reitores das Universidades Portuguesas, Portugal, and British Council, UK (Treaty of Windsor Programme).

References

Boersma, P.: Accurate short - term analysis of the fundamental frequency and the harmonics - to - noise ratio of a sampled sound. In: Proc. Institute of Phonetic Sciences, U. Amsterdam, vol. 17, pp. 97–110 (1993)

Cheveigné, A., Kawahara, H.: YIN, a fundamental frequency estimator for speech and music. JASA 111(4), 1917–1930 (2002)

Childers, D., Hahn, M., Larar, J.: Silent and voiced/unvoiced/mixed excitation (four - way) classification of speech. IEEE Transactions on Acoustics, Speech and Signal Processing 31(11), 1771–1774 (1989)

Droppo, J., Acero, A.: A fine pitch model for speech. In: Proc. InterSpeech, pp. 2757–2760 (2007)

Francis, A., Ciocca, V., Yu, J.: Accuracy and variability of acoustic measures of voicing onset. JASA 113(2), 1025–1032 (2003)

Hermes, D.: Measurement of pitch by subharmonic summation. JASA 83(1), 257–264 (1988)

Hess, W.: Pitch and voicing determination. In: Furui, S., Sondhi, M. (eds.) Advances in Speech Signal Processing, pp. 3–48. Marcel Dekker, New York (1992)

Jackson, P., Shadle, C.: Frication noise modulated by voicing, as revealed by pitch-scaled decomposition. JASA 108(4), 1421–1434 (2000)

Jesus, L., Shadle, C.: A parametric study of the spectral characteristics of European Portuguese fricatives. J. Phon. 30(3), 437–464 (2002)

Jesus, L., Shadle, C.: Devoicing measures of European Portuguese fricatives. In: Mamede, N., Baptista, J., Trancoso, I., Nunes, M. (eds.) Comp. Processing of the Portuguese Language, pp. 1–8. Springer, Heidelberg (2003)

McCree, A., Stachurski, J., Unno, T., Ertan, E., Paksoy, E., Viswanathan, V., Heikkinen, A., Ramo, A., Himanen, S., Blocher, P., Dressler, O.: A 4kb/s hybrid MELP/CELP speech coding candidate for ITU standardization. In: Proc. ICASSP, pp. 629–632 (2002)

Noll, A.: Cepstrum pitch determination. JASA 41(2), 293–309 (1967)

Pelle, P., Estienne, C.: A pitch extraction system based on phase locked loops and consensus decision. In: Proc. InterSpeech, pp. 1637–1640 (2007)

Pincas, J.: The interaction of voicing and frication sources in speech: An acoustic study. M.Res. Thesis, University of Surrey, Guildford, UK (2004)

Pincas, J., Jackson, P.: Amplitude modulation of frication noise by voicing saturates. In: Proc. InterSpeech, pp. 349–352 (2005)

Quatieri, T.: Discrete-time Speech Signal Processing: Principles and Practice. Prentice Hall, Englewood Cliffs (2001)

Secrest, B., Doddington, G.: An integrated pitch tracking algorithm for speech systems. In: Proc. ICASSP, pp. 1352–1355 (1983)

Stevens, K., Blumstein, S., Glicksman, L., Burton, M., Kurowski, K.: Acoustic and perceptual characteristics of voicing in fricatives and fricative clusters. JASA 91(5), 2979–3000 (1992)

A Spoken Dialog System Speech Interface
Based on a Microphone Array

Gustavo Esteves Coelho[1], António Joaquim Serralheiro[1,3],
and João Paulo Neto[1,2]

[1] L2F – Spoken Language System Laboratory / INESC-ID
[2] IST – Instituto Superior Técnico / Technical University of Lisbon
[3] Academia Militar R. Alves Redol, 9
1000-029 LISBOA, Portugal
{gustavo.coelho,antonio.serralheiro,
joao.neto}@l2f.inesc-id.pt
www.l2f.inesc-id.pt

Abstract. In this paper we present a Spoken Dialog System (SDS) with a Microphone Array (MA). Our goal is to create a hands-free home automation system with a speech interface to control home devices. The MA interface enables to create ubiquitous speech acquisition for the SDS. The implemented system allows any user – in any position in a room – to establish a dialog with a virtual butler that is able to control a wide range of home appliances (room lights, air-conditioner, windows shades and hi-fi features). This virtual butler has a 3D animated face that is, while the dialog is engaged, able to steer to the user's position and respond to his/hers commands with synthesized speech. The presented results show that the MA, as distant talk interface, performs quite well and is a step towards a more realistic human-machine interaction.

Keywords: Home Automation, Microphone Arrays, Automatic Speech Recognition.

1 Introduction

Considering that speech is the most natural way of interaction between humans, it is reasonable to foresee that, in a near future, human-machine communication will comprise speech as well as the usual non-speech forms. To pursue this goal, adequately speech acquisition is imperative to provide the best recognition performances. Close-talking microphones (e.g. head-set, lapel) have the advantage of high Signal-to-Noise Ratio (SNR). However, they are intrusive and if the speaker needs to moves inside a large room, or to an adjacent one, other ways of communication with the computers are mandatory. Another approach is to use a single far-field microphone in a fixed place. However, preliminary tests show degradation on the recognition performances, whenever a user utters at increasing distances from that fixed microphone. For instance, in a quiet room, the Word Error Rate (WER) goes from circa 14% to 24% when the distance from the microphone is increased from 1 to 3.5 meters. If the acoustic environment now includes some noise sources (even at moderate levels,

A. Teixeira et al. (Eds.): PROPOR 2008, LNAI 5190, pp. 21–30, 2008.

typical in real acoustic environments) the WER increases to 95% at 1m distance. Briefly, a single far-field microphone is definitely not adequate for practical usage.

Seeking to create ubiquitous speech interfaces and to avoid the nuisance of wearing close-captioning microphones we used a suitably placed a Microphone Array (MA), as our speech acquisition front-end. MAs offer a principled approach to recovering a particular person's speech from a mixture of distant microphones signals. A MA is composed of a multiple omni-directional microphones arranged in purposeful geometries in a room. MAs filter the received signals according to the spatial configuration of speech sources and noise sources, enabling thus to focus on a sound originating from a particular location. Contrary to the single close-talk microphones, MAs are also capable of locating sound sources in reverberant enclosures, separation of the sources and enhancement of speech signals from desired sources.

One of the main problems with MA (in terms of speech recognition) is the robust acquisitions of the speech signal given the adverse conditions in most real acoustic environments. Real environments are often reverberant and they suffer from significant background noise. Close talking microphones alleviate many of these problems and give the highest accuracy from speech recognition system. However, MA processing techniques offers an increasingly viable alternative with overcomes many advantages of close-talk microphones. MA speech enhancement generally involves *Beamforming*, which consists of filtering and combining the individual microphone outputs in such way as to enhance signals coming from a specific location, while attenuating signals from other locations.

Projects like CHIL [1], AMI [2] and the recent DICIT [3], addressing the development of advanced technologies for speech/acoustic processing and interpretation based on MA devices, are examples of the wide spreading of this technology.

In this paper we evaluate the viability of a MA as the speech acquisition front-end of a Spoken Dialogue System (SDS) whose purpose is to control a set of home appliances. The SDS [4] comprises the following base technologies: Automatic Speech Recognition (ASR), Text-to-Speech (TTS) synthesis, Dialog Management (DM), Virtual Face Animation (FACE) and Microphone Array Processing. The main advantage of a SDS is the capability of interaction with the users to overcome recognition errors that can impair the execution of some uttered command.

This paper is organized as follows: in section 2 the description and implementation of the home automation system is presented; in section 3 experimental results with speech data are presented to evaluate our system and finally, in section 4, the conclusions are addressed.

2 The Virtual Butler System

The implemented SDS is currently tailored to work with Portuguese language[1], including both ASR and TTS systems. Our home automation demonstration system is based in a Virtual Butler (VB) that is always available to control the home devices (figure 1). Users can control a specific device with different speech commands - e.g. it is possible to turn on the ceiling light with either *"liga a luz"* (turn on the light), or

---

[1] The usage with other languages involves the modification of, at least, the acoustic models and the language models, not to mention the TTS.

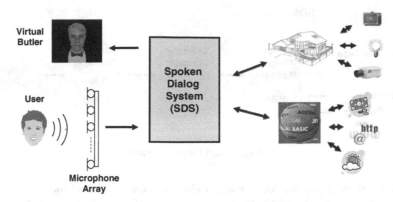

Fig. 1. Block diagram of the Home Automation system with a Virtual Butler

"*acende a luz*" (switch on the light), or "*ligar luz da sala* " (turn on the room light), or even "*liga-me a luz*" (switch me the light).

The user gets the VB "attention" by uttering its name "*Ambrósio*", followed by a command to control a specific device. The butler acknowledges the users request and, if more information is needed to disambiguate that order, automatically questions the user, engaging a dialogue. This ambiguity can arise, for instance, directly from the previous request example, since it is possible to control both table and ceiling lights in the room. Therefore, the VB needs to complete the command "*liga a luz*" (turn on the light) knowing which light will be switched on. So, the VB questions the user with the synthesized sentence "*qual a luz que pretende ligar?*" (which light do you want to switch on?); then, the user must answer "*da sala*" (room light) or "*da mesa*" (table light), to complete the command. Other cause of ambiguity can be erroneous recognition of uttered commands. The VB acknowledgements and/or questions are converted into speech by the TTS module and synchronized with a 3D animated butler face (including face expressions and movements of the lips).

The home automation system is divided in two main subsystems, the SDS and the MA processing unit, described is the following sub sections. The SDS provides the interface between the user's speech and the VB, briefly mentioned above. The MA front-end acquires the user's speech and performs the enhancement of the signal before delivering it to the SDS input. The MA processing unit also estimates the user's direction and signals the SDS with that information to steer the VB face towards the user.

2.1 Spoken Dialog System

The SDS module is divided in three main blocks, as depicted in figure 2.

The first block, the Input Output Manager (IOM), is where the interfaces of both the user and the butler are managed. The IOM comprises the following sub-blocks: the ASR (to recognize the user's speech commands), the TTS (to synthesize the speech of the butler) and the FACE to implement the 3D animated face of the VB. The second block of the SDS, the Dialog Manager (DM) module receives requests from the IOM in a XML format, determines the action(s) requested by the user,

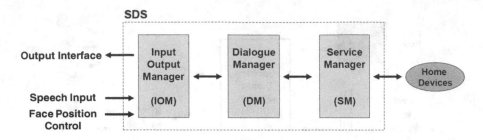

Fig. 2. SDS block diagram

and directs them to the Service Manager (SM) for the execution of that action(s). This last module provides the DM with the necessary interface with a set of heterogeneous home devices grouped by domains, which users can control or interact. This generic block approach enables our SDS to cope with different types of applications and, therefore, be fully tailored to other applications that require speech (or dialog) interaction. As an example, the SDS is currently applied to create a virtual personal assistant enabling automatic scheduling for meeting and other events, telephone answering and redirection, etc; and also a virtual home banking system, where users can access their banking information and services by telephone.

As mention earlier, one of the drawbacks of MA applied to ASR systems is the poor speech recognition results, namely when compared to close talk microphones, since speech data varies greatly with the acoustic environment, and therefore causes further degradation in the recognition performance. However, home automation systems are limited-domain ASR applications; we mitigate this drawback by tailoring the recognition vocabulary to the specific domain needs. Consequently, our speaker-independent (SI) home automation system with the MA interface is able to perform home automation tasks with no specific adaptation of the acoustic models. Nevertheless, it is possible to personalize the SDS system, tagging the butler commands with an activation word, namely the butler's name "*Ambrósio*". With this feature, the VB is able to respond only to the specific user's speech, while speech commands are processed in a SI basis.

To accomplish home automation tasks, a specific grammar is loaded into the SDS. This grammar was written according to SRGS specification format and contains a hierarchical structure defining all possible home automation commands rules. The SRGS specification format allows us to create flexible speech commands, enabling the user to order a specific command in many different ways. The vocabulary and lexicon of the SDS is automatically generated from the previous loaded SRGS grammar. The present vocabulary can be easily extended or modified and comprises 65 words, generating a total of 530 different sentences covering all current possible home automation speech commands.

The ASR is based on the Audimus [5], a hybrid speech recognizer that combines the temporal modeling capabilities of Hidden Markov Models (HMMs) with the pattern discriminative classification capabilities of multilayer perceptrons (MLPs). The ASR is used to recognize the enhanced speech processed by the MA.

The TTS module (DIXI +) [6] is a concatenative-based synthesizer, based on the Festival framework. This framework supports several voices and two different types of unit – fixed length units (such as diphones), and variable length units. This latter data-driven approach can be fine tuned to limited domain applications, by an adequate design of the corpus. The TTS is used to synthesize the VB speech output.

The FACE module [7] is a Java 3D implementation of a synthetic talking face with a set of visemes for the Portuguese phonemes and a set of emotions and head movements. The VB face representation is accomplished with this module.

This generic topology also allows the SDS to be independent from the input-output interface devices, and therefore the SDS can be accessed either locally or remotely from a wide range of devices, such as head-sets, PDAs, web browsers, mobile phones, just to mention a few.

2.2 Microphone Array Front-End

The MA, whose advantages were already mentioned [8-10], acquires the speech signal and outputs a multi-channel signal that is pre-processed in the Spatial Filtering Unit (SFU), for both Speech Enhancement and Direction of Arrival (DoA) estimation. Figure 3 depicts the block diagram of the SFU that interfaces the MA with the SDS. The main objective of the SFU is to virtually steer the directivity of the MA towards the sound source (the user) and, simultaneously, enhance the speech signal against environmental noise by means of spatial filtering (*Beamforming*). Furthermore, the estimation of the DoA, sent to the FACE unit, allows us to build a better visual interface, since the VB can "turn its face" into the direction of the speaker. This behavior, added to the automatic generation of synthetic speech, is a step towards a more realistic human-machine interaction.

A sixty four linear and uniformly spaced MA, based on the NIST MarkIII [11] MA, was built for both speech acquisition and DoA estimation [12]. The distance between microphones was set to 2cm to allow for a 16 kHz sampling frequency without spatial aliasing. The audio signal from all microphones is then 24-bit digitally converted with time-synchronized ADCs (simultaneous in-phase sampling). The MA module connects to a remote computer by an Ethernet interface. The communication and data transfer are based on the standard UDP protocol, which provides this MA a generic interface to any computer.

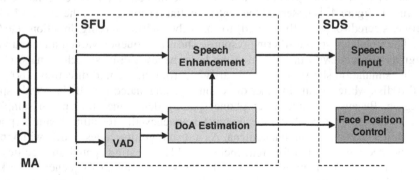

Fig. 3. SFU block diagram

Since the SDS input accepts a single channel input source, the multi-channel audio from the MA must be pre-processed. This task is done in real-time in the SFU. For speech enhancement, we apply the Delay-and-Sum Beamforming (DnSB) [13] algorithm that, when compared to the adaptive beamformers, has the advantage of providing less high-frequency spectral distortion to the desired speech signal and has a lower computational cost. The virtual steering process mention earlier is implemented by means of software, with the DnSB algorithm, maintaining the MA physically fixed in a pre determined location. The resulting enhanced signal from the DnSB output is then sent to the SDS input. For the DoA estimation, we apply the Generalized Cross Correlation with Phase Transform (GCC-PHAT) [14] algorithm. This estimation process is activated whenever the speech signal is above the Voice Activation Detector (VAD) threshold. The underlying idea of this procedure is to assure that the animated face of the VB only steers to the users when they speak, avoiding the VB to steer towards the noise sources, and to avoid *noise beam-steering* (aiming the MA virtual beam towards noise sources).

The MA works originally with a sampling frequency of 22.05 kHz, sending all 64 digital audio channels through an Ethernet connection to a remote SFU. The SFU is programmed in Java and splits the incoming audio channel to the DnSB, GCC-PHAT and VAD, respectively, since these algorithms concurrently process the incoming audio data. All audio data is windowed in 4096 samples ($\approx$190 ms) with no overlap. The GCC-PHAT implements the DoA estimation using only 2 of the 64 available microphones. This pair of microphones is chosen according to prior correlation and precision analysis, weighting two contradictory factors: microphones should simultaneously be close enough to assure that correlation coefficients are acceptable and, conversely, the pair must be separate enough to ensure precision in the DoA estimations. The VAD is implemented by calculating the energy over the windowed audio data from a single microphone in the MA, and sets a threshold to define the speech/non-speech decision. The estimated DoA is then sent from the SDS to the FACE unit also through an Ethernet connection.

The MA virtual beam steering direction is done according to the DoA estimations. The DnSB receives all 64 audio channels from the MA e returns a single audio channel with the enhanced speech data. The resulting single audio channel from the DnSB is down sampled to 16 kHz, since this the working sampling frequency of our ASR. This audio is sent also through Ethernet to the SDS, for ASR processing.

As an example of the spatial capabilities of the implemented speech enhancement algorithm, in figure 4, is observed the DnSB spatial response when the current MA is electronic steered (or virtually steered) towards the endfire steering direction (DoA = 180°). The simulated spatial filtering response show a frequency variant attenuation of the signals acquired with the MA, due the large bandwidth of speech signals. However, this simulation shows that signals arriving from the desired direction (180°) are passed (0dB), while signals in other directions are attenuated (<0dB) in a wide spectral region. Because the inter-microphone spacing determines the spatial sampling frequency, for frequency above the Nyquist frequency the resulting beam response will exhibit a spatial aliasing phenomena. As a result, grating lobes (0 dB) will appear out of the steering direction for frequencies > 8 kHz, creating spatial ambiguities, as depict in figure 4. As mentioned earlier, the working sampling frequency is 16 kHz and, therefore, the spatial aliasing does not constitute a problem to the overall system.

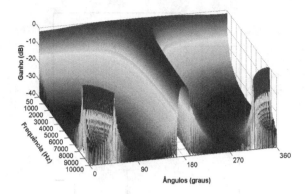

Fig. 4. Spatial filtering response of the implemented MA: DnSB aiming towards the endfire steering direction

3   Experimental Evaluation

In order to assess the recognition performance of our SDS with a MA interface we include, as a reference, results obtained with a close-talk (headset) microphone. Furthermore, we also present recognition results using one single microphone (#32 from the MA) in a far-talk setup. To begin with, all speech data was recorded in a clean acoustical environment using a headset. Our test corpora is composed of 73 different spoken Portuguese sentences (234 words), corresponding to the home automation task, e.g. "*diminuir a temperatura da sala*" (lower the room temperature). All experiments were obtained with off-line processing, using the previous described recordings. The recognition Word Error Rate (WER) for the close-talk microphone was 2.14%, and will be our base line for the ASR evaluation. Then, the recorded speech data was played with loudspeakers in 3 different locations, as depicted in figure 5. To assess the speech enhancement performance, the recorded speech audio was contaminated with a Gaussian white noise source, located in the same 3 positions. The objective of this experiment is to show that the DnSB is able to enhance the speech from a specific direction while attenuating the noise source in other directions. As a result, the DnSB should increase the WER, when compared with the clean speech recorded by the headset, and decrease when compared with the single far-talk microphone, validating thus the MA purpose for the far-field speech acquisition. The experimental results with a single microphone in far-field conditions were carried out in mild noise and reverberant conditions and the WER ranged from over 94% to 98%! These results do show how inappropriate a single far-field microphone is.

Table 1 depicts the WER results for both clean speech and noise source in different positions. It can be observed that position C achieves the lower WER, since it is the nearest to the MA. Conversely, the higher WER is achieved when the noise source is closest to the MA. The SNR gain, calculated from the #32 microphone signal and the DnSB output, is presented in column 4 of table 1. These results compare comfortably with the theoretical limit SNR gain for the DnSB of $10\log(N) \approx 18$dB, where N is the number of microphones of the MA. In practice, the DnSB is only able to attenuate spatial uncorrelated noise. Therefore, it was expected to observe a SNR gain less then 18dB.

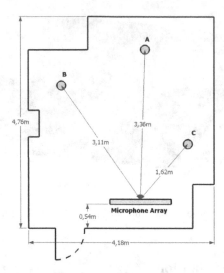

Fig. 5. Experimental setup with 3 different positions. The DoA is 92° for location A and 55° and 131° for B and C, respectively.

Table 1. DnSB experimental results

Speaker	Noise Source	DnSB DoA, °	SNR gain, dB	WER, %
A	B	92	10.6	12.8
B	A	55	11.0	18.0
B	C	55	12.6	24.8
C	B	131	12.9	6.4

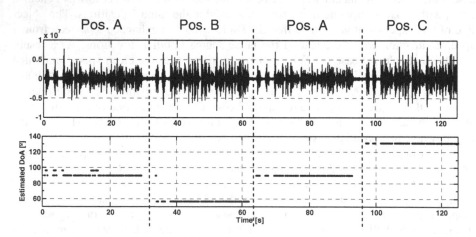

Fig. 6. DoA estimation results with GCC-PHAT: (above) audio from #29 microphone; (below) DoA results for the acquired speech in different positions

Finally, we present DoA estimation results (figure 6) using microphones #29 and #36. It can be observed that DoA estimation provides an accurate direction of the speech sources with a maximum error smaller than ±2.5 degrees. At 3.5m distance from the MA, this error corresponds to a 0.15 m location mismatch. Since the width of the loudspeaker (used to play the recorded speech data) is ≈0.2 m, the resulting error is within the physical size of the sound source. Considering that this error can occur due the user's face movements, this error is less than the normal length of the human face and, therefore, acceptable.

As mentioned, the VAD disables the GCC-PHAT estimation during silence periods, thus preventing erroneous beam-steering. As depicted in figure 6, the estimated DoA values are present only in speech intervals. During non-speech intervals, no estimation is done and the DnSB maintains beam steering to the previously estimated DoA.

4 Conclusions

In this paper we presented a Spoken Dialog System with a Microphone Array as the speech acquisition interface, being a step forward to a ubiquitous Home Automation system, where users can control some home devices establishing a dialog with the virtual butler. The presented home automation prototype has been deployed in our demonstration room and has been successfully tested with several users.

As expected, close-talk microphones achieve better results in terms if ASR performance but, obviously, they are not a practical solution. However, the presented results show that MAs, besides providing speech enhancement, achieve sufficiently small WER to enable home automation tasks.

Acknowledgments

This work was funded by PRIME National Project TECNOVOZ number 03/165.

References

[1] CHIL - Computers. In: the Human Interaction Loop, http://chil.server.de/
[2] AMI - Augmented Multi-party Interaction, http://www.amiproject.org/
[3] DICIT - Distant-talking Interfaces for Control of Interactive TV, http://dicit.fbk.eu/
[4] Neto, J.P., Cassaca, R., Viveiros, M., Mourão, M.: Design of a Multimodal Input Interface for a Dialog System. In: Vieira, R., Quaresma, P., Nunes, M.d.G.V., Mamede, N.J., Oliveira, C., Dias, M.C. (eds.) PROPOR 2006. LNCS (LNAI), vol. 3960, pp. 170–179. Springer, Heidelberg (2006)
[5] Meinedo, H., Caseiro, D., Neto, J., Trancoso, I.: AUDIMUS.media: a Broadcast News speech recognition system for the European Portuguese language. In: Mamede, N.J., Baptista, J., Trancoso, I., Nunes, M.d.G.V. (eds.) PROPOR 2003. LNCS, vol. 2721. Springer, Heidelberg (2003)
[6] Paulo, S., Oliveir, L.C.: Reducing the Corpus-based TTS Signal Degradation Due to Speaker's Word Pronunciations. In: Interspeech, ISCA, Portugal, pp. 1089–1092 (2005)

[7] Viveiros, M.: Cara Falante - Uma interface visual para um sistema de diálogo falado, Graduation thesis, Instituto Superior Técnico, Universidade Técnica de Lisboa (2004)
[8] Brandstein, M., Ward, D.: Microphone Arrays. Springer, Heidelberg (2001)
[9] Kellermann, W., Buchner, H., Herbordt, W., Aichner, R.: Multichannel Acoustic Signal Processing for Human/Machine Interfaces - Fundamental Problems and Recent Advances. In: ICA 2004. LNCS, vol. 3195, Springer, Heidelberg (2004)
[10] Buchner, H., Benesty, J., Kellermann, W.: Generalized Multichannel Frequency-Domain Adaptive Filtering: Efficient Realization and Application to Hands-Free Speech Communication. Signal Processing 85, 549–570 (2005)
[11] The Nist Mark-III Microphone Array,
     http://www.nist.gov/smartspace/cmaiii.html
[12] Coelho, G.E., Serralheiro, A.J., Neto, J.: Microphone Array front-end interface for Home Automation. In: Hands-free Speech Communication and Microphone Arrays (HSCMA), Trento, Italy, pp. 184–187 (2008)
[13] Johnson, D.H., Dudgeon, D.E.: Array Signal Processing: Concepts and Techniques. Prentice Hall, Englewood Cliffs (1993)
[14] Knapp, C., Carter, G.: The generalized correlation method for estimation of time delay. IEEE Trans. Acoust. Speech Signal Processing 24, 320–327 (1976)

PAPEL: A Dictionary-Based Lexical Ontology for Portuguese

Hugo Gonçalo Oliveira[1], Diana Santos[2], Paulo Gomes[1], and Nuno Seco[1]

[1] Linguateca, Coimbra node, DEI - FCTUC, CISUC, Portugal
[2] Linguateca, Oslo node, SINTEF ICT, Norway
hroliv@dei.uc.pt, Diana.Santos@sintef.no, pgomes@dei.uc.pt,
nseco@dei.uc.pt

Abstract. This paper describes a project aimed at creating a lexical ontology extracted (semi) automatically from a large Portuguese general dictionary. Although using machine readable dictionaries to extract semantic information is not new, we believe this is the first attempt for the Portuguese language. The paper describes a (to be) freely available resource, dubbed PAPEL, explaining the process used and the tools developed, and illustrating it with one specific relation: Causation.

1 Introduction

PAPEL (Palavras Associadas Porto Editora Linguateca) is a lexical resource for natural language processing (NLP) of Portuguese, based on the (semi) automatic extraction of relations between the words appearing in the definitions of a general language dictionary of Portuguese - the Dicionário da Língua Portuguesa (DLP) [1] developed and owned by the largest Portuguese dictionary publisher, Porto Editora. Similar lexical resources for English are Princeton WordNet [2] widely used by NLP researchers, and MindNet [3]. When it comes to Portuguese, despite projects with similar aims (WordNet.PT [4] and WordNet.BR [5]), there is no publicly available lexical ontology for our language (i.e., that one can download and use in its entirety).

Also, and differently from the two aforementioned projects, which are done from scratch resorting to extensive manual linguistic labour, we follow the approach of creating PAPEL from a machine readable dictionary (MRD).

This paper starts with a description of the most important works that, since the 1970's, have used MRDs as a source of information to solve the lexical bottleneck in NLP, pointing out the similarities of PAPEL compared to these earlier attempts. It then describes, in Section 3, the methodology employed in the creation of PAPEL, and the tools developed in the project. Section 4 explores in some detail the example of Causation, while Section 5 ends with a description of some further work.

2 Related Work

This section presents an overview on the most important works that used MRDs as a source of information for NLP, especially for the extraction of relations

A. Teixeira et al. (Eds.): PROPOR 2008, LNAI 5190, pp. 31–40, 2008.

between words or concepts and the creation of organized structures containing those relations. Similar resources are then described.

2.1 Relation Extraction from MRDs

The process of using MRDs in natural language processing (NLP) started back in the 1970's, with the work of Nicoletta Calzolari [6] [7]. Definitions were explored in order to organize the dictionary into a lexical database where morphological and semantic information about the defined words could be obtained directly. Similar work took place for English when the electronic versions of the Longman Dictionary of Contemporary English (LDOCE) and the Merriam-Webster Pocket Dictionary (MPD) were used as a source of information to build such a structure. The analysis of the structure of those MRDs showed that they made use of a very limited defining vocabulary [8] and that the text of the definitions often consisted of a *genus* an a *differentia* [9]. The *genus* identifies the superordinate concept of the defined word. The *differentia* presents the properties responsible for the distinction between this "instance" of the superordinate concept and other instances of the same concept. Amsler [10] suggested that the identification of the *genus* could lead to the construction of a taxonomy. Bearing in mind the definition structure, Chodorow [11] took advantage of its restricted vocabulary and developed semi-automatic recursive procedures to extract and organize semantic information into hierarchies. These heuristics didn't need to parse the whole definitions, due to their predictability. However, the human user played an important role when it came to disambiguation. Other approaches [12] [13] took advantage of the simple vocabulary of the definitions and used string patterns to extract semantic information from them.

Further approaches [14] [15] used patterns based on the structural level (i.e., syntactic phrases) of the analysed text, instead of string patterns. After some discussion about the advantages and the drawbacks of using structural patterns or string patterns to extract semantic information contained in the definitions, Montemagni and Vanderwende [16] concluded that although string patterns are very accurate for identifying the *genus*, they cannot capture the variations in the *differentia* as well as structural patterns, and they proposed the use of a broad-coverage grammar to parse the dictionary definitions in order to obtain rich semantic information. In spite of seeming an overkill to use a broad-coverage parser for definition text, the authors make the point that in some cases (relative clauses, parenthetical expressions, and coordination) its use is warranted. Although dictionaries have been explored for several purposes, such as parsing or word sense disambiguation, to our knowledge they have not been converted into an independent resource of its own before MindNet [3], which therefore can be said to be a sort of independent lexical ontology in a way that previous work was not.

2.2 Related Resources

Princeton WordNet [2] is probably the most important reference when it comes to lexical ontologies in English. It is freely available and it is widely used in NLP research. In the WordNet's lexicon, the words are clearly divided into nouns, verbs, adjectives, adverbs and functional words. The basic structure in WordNet is the *synset*, which is a set of synonym words that can be used to represent one concept. The *synsets* are organized as a network of semantic relations, such as Hyponymy and Meronymy (between nouns) and Troponymy and Entailment (between verbs).

WordNet.BR [5] is a Brazilian version of the "wordnet concept", started in 2002. Their database is structured around Synonymy and Antonymy manually extracted from a reference corpus where several dictionaries are included, and plans for adding more relations in the future have been reported in [5]. WordNet.PT [4] is another attempt of creating a Portuguese lexical resource from scratch, which started in 1999. The authors of WordNet.PT explicitly claim that the available resources for Portuguese NLP are not suitable for the automatic construction of such a resource. They use a set of 35 relations and are explicitly interested in cross-categorical relations such as those linking adjectives to nouns.

MindNet [17] is a knowledge representation resource that used a broad-coverage parser to build a semantic network, not only from MRDs but also from encyclopedias, and free text. MindNet contains a long set of relations, including Hypernymy, Causation, Meronymy, Manner, Location and many more. One interesting functionality offered by MindNet is the identification of "relation paths" between words[1]. For example, if one looks for paths between *car* and *wheel* a long list of relations will be returned. The returned paths include not only simple relations like *car is a modifier of wheel* but also more complex ones like *car is a hypernym of vehicle and wheel is a part of vehicle*.

Another kind of lexical resource is FrameNet [18], which constitutes a network of relations between semantic frames, extracted from corpora and from a systematic analysis of semantic patterns in corpora. Each frame corresponds to a concept and describes an object, a state or an event by means of syntactic and semantic relations of the lexical item that represents that concept. A frame can be conceived as the description of a situation with properties, participants and/or conceptual roles. A typical example of a semantic frame is *transportation*, within the domain *motion*, which provides the elements *mover(s), means of transportation* and *paths* and can be described in one sentence as: *mover(s) move along path by means*.

3 Building PAPEL

In this section, we describe the set of relations included in PAPEL, the parser used to analyse the definitions, some quantitative studies about the content of the definitions, the incremental nature of the work and the regression testing tools developed in this project.

3.1 Relations

The overview of the resources referred in Section 2.2, together with an exploration of the most common n-grams in the dictionary, led us to choose the first

---

[1] http://atom.research.microsoft.com/mnex/

set of relations that we want to have in PAPEL. Note that the decision of working on a relation means also the detection of its inverse.

Let us start by explaining that the most basic semantic relationship between words is, of course, identity of meaning (**Synonymy**, and in fact *synsets* in wordnets are simply a set of words having the same meaning), but we started by assuming that other semantic relations would be more interesting for general NLP applications and that their discovery would facilitate the identification of the set of final concepts. This is related to the often made remark that word sense disambiguation is an ill-defined task and is very dependent on the purpose [19]. Different lexicographers, or system developers, divide senses differently [20]. So we consider the task of ambiguating a dictionary [21] a task more germane to our interests than word sense disambiguation.

Table 1 shows some of the relations we are planning to include in PAPEL, their representation and some examples. These relations include the *is-a* relation (**HIPONIMO_DE**), the causation relation (**CAUSADOR_DE**) and the *part-of* relation (**PARTE_DE**).

Table 1. Relations we are planning to include in PAPEL

Relation	Inverse	Example
HIPERONIMO_DE(X,Y)	HIPONIMO_DE(Y,X)	**HIPERONIMO_DE**(animal, cão)
CAUSADOR_DE(X,Y)	RESULTADO_DE(Y,X)	**CAUSADOR_DE**(vírus, doença)
PARTE_DE(X,Y)	INCLUI(Y,X)	**PARTE_DE**(roda, carro)
MEIO_PARA(X,Y)	FINALIDADE_DE(Y,X)	**MEIO_PARA**(chave, abrir)
LOCAL_DE(X,Y)	OCORRE_EM(Y,X)	**LOCAL_DE**(restaurante, comer)

We are also planning to deal with other kinds of relations that should be easy to extract and that we thought would considerably increase the usefulness of the resource are words related to places (`lisboeta` related to `Lisboa`) and words describing affect (positive or negative connotation).

3.2   Parsing the Definitions

In order to parse the dictionary definitions, we used PEN, a chart parser freely available under a BSD license[2] which is a Java implementation of the well known Earley Algorithm [22]. PEN parses the text according to a grammar file it gets as input and it can yield several analysis for the same text. So far, we have used specific different grammars to identify different relations between the defined entities corresponding to words in the dictionary.

The relation extraction method starts with an empirical analysis of the patterns present in the definitions and which might suggest relations between the entry and other entities. Having a relation in mind, a selection of patterns (e.g. `tipo de X`) that can imply the relation is made.

An SQL table containing information about the n-grams in the definitions of the dictionary was created, providing us with the frequency of each n-gram

---

[2] http://linguateca.dei.uc.pt/index.php?sep=recursos

in the whole dictionary, its position inside the definition, and the grammatical category of the defined word. Guided by the frequency of the candidate patterns in the definitions, we look at a selection of entries where the patterns are actually used to make sure their selection makes sense and to possibly find more refined criteria as well.

After finding a set of patterns indicating a relation, we can start the construction of a specific grammar for the extraction of that relation in the dictionary.

To deal sensibly with multiple analyses of a same definition according to the same relation, we implemented the following heuristic in every grammar: the selected derivation is the one with less unknown tokens.

3.3   The Results

After having devising and debugging the grammars with the help of a set of hand-selected definitions (about 5000), we apply them to the whole dictionary, comprising 237,246 definitions.

We then analyse the results for the whole dictionary in order to classify the relations obtained into "correct" and "incorrect". This classification is made by a human user and can be very time consuming. That is why we have created a program to automate part of the process. We can feed the program with a set of correct and a set of incorrect relations from previous runs. The human user then only has to classify the relations which are not in any of the previous sets, making time spent to obtain the first division pay off in the following runs of new versions of the grammar(s) for the same relation.

In fact, the number of **new** kinds of problems drastically diminishes as more relations are classified, because since the dictionary definitions use simple and not very diverse vocabulary (though not as restricted as LDOCE), most of the problems detected are systematic (see Section 4.2 for examples of obtained errors). The number of "correct" and "incorrect" candidate relations extracted give us an idea of when to stop developing further the grammars.

3.4   Regression Testing

After analysing the relations considered correct and the incorrect ones, it is easier to find out the origin of the problems. This analysis helps us deciding what changes should be made in the grammar. The new version of the grammar is then tested, before processing the whole dictionary.

The results obtained with different versions of a grammar for the extraction of the same relation can be compared with a system we have developed especially for regression testing. This system identifies differences between two sets of results and can be used to obtain information about, and quantify:

- the relations in one set and not in the other;
- the relations that remained the same in both sets;
- the entries that have at least one relation in one set but any in the other;
- the changes to the relations obtained for each entry;

4 Detailed Example: Causation

We proceed by describing in some detail the process and results obtained for
CAUSADOR_DE relation, namely defined by us as a relation between an
agent (the causer) and a result (the caused). We have considered the inverse
relation, **RESULTADO_DE**, to be the same as effect/result, taking thus so far
a naive approach to this philosophical debate (see e.g. [23] or [24]).

As described above, we developed several grammars to parse the dictionary
definitions that included these relations, and went on testing them incrementally.
When it comes to this relation, we currently have a 96% success rate (precision)
in a total of 5,657 **CAUSADOR_DE** relations extracted and 91% in a total
of 1,693 **RESULTADO_DE** relations. These numbers were calculated after
manual analysis of the results. We are starting to look into corpus-based methods
to evaluate recall.

4.1 The Patterns

The grammars designed for the extraction of this relation are primarily based on
the verbs `causar`, `originar`, `provocar`, `produzir`, `motivar`, `gerar`, `suscitar`
and `resultar` and on the expressions `devido a` and `efeito de`.

The following patterns are used for the extraction of the **CAUSADOR_DE**
relation.

```
1c - causad{o|a|os|as} FREQ* PREP CAUSADOR
     originad{o|a|os|as} FREQ* PREP CAUSADOR
     provocad{o|a|os|as} FREQ* PREP CAUSADOR
     produzid{o|a|os|as} FREQ* PREP CAUSADOR
     gerad{o|a|os|as} FREQ* PREP CAUSADOR
     motivad{o|a|os|as} FREQ* PREP CAUSADOR
     suscitad{o|a|os|as} FREQ* PREP CAUSADOR
2c - devido {a|ao|à|aos|às} CAUSADOR
3c - efeito PREP CAUSADOR
```

CAUSADOR is a sub-pattern that denotes a **CAUSADOR_DE** relation between words it catches (which will be the cause) and the entry (which will be
the result): **CAUSADOR_DE**(cause, entry). The cause can be preceded by
specific words like determiners, pronouns, quantifiers, other modifiers or constructions like `acção de/do/dos/da/das`.

PREP denotes a preposition and FREQ a (optional) quantifier, such as `normal`
`mente` or `frequentemente`.

The following patterns are used for the extraction of the **RESULTADO_DE**
relation:

```
1r - que {causa|origina|provoca|produz|motiva|gera|suscita} RESULTADO
2r - {causar|originar|provocar|produzir|motivar|gerar|suscitar} RESULTADO
3r - resultar PREP_EM RESULTADO
```

The sub-pattern RESULTADO is similar to CAUSADOR, but catches the results in
the definition instead of catching the causes: **RESULTADO_DE**(result, entry).

Table 2. Examples of relations extracted by the grammars for the Causation grammars. 'ID' identifies the pattern matched by the definition.

ID	Entry	Grammar	Definition	Extracted relation
1c	quase-delito	s. m.	dano **causado** por negligência, sem intenção malévola	CAUSADOR_DE(negligência, quase-delito)
1c	concussão	s. f.	choque violento **originado** por uma explosão	CAUSADOR_DE(explosão, concussão)
1c	toxicose	s. f.	doença **provocada** pela presença de produtos tóxicos no organismo	CAUSADOR_DE(produtos ,toxicose)
1c	ecfonema	s. m.	elevação súbita da voz, **motivada** por surpresa ou comoção violenta	CAUSADOR_DE([surpresa, comoção], ecfonema)
1c	tisne	s. m.	cor **produzida** pelo fogo ou pelo fumo sobre a pele	CAUSADOR_DE([fogo, fumo], tisne)
2c	engasgo	s. m.	incapacidade de respirar devido a obstrução da garganta	CAUSADOR_DE(obstrução, engasgo)
3c	maximização^3	s. f.	**efeito de** maximizar	CAUSADOR_DE(maximizar, maximização)
1r	diplodoco	s. f.	bactéria que **causa** as meningites cerebrospinais	RESULTADO_DE(meningites, diplodoco)
1r	osteoporose	s. f	porosidade excessiva dos ossos, que **origina** a sua fragilidade	RESULTADO_DE(fragilidade, osteoporose)
1r	tentação	s. f.	coisa ou pessoa que **provoca** desejo	RESULTADO_DE(desejo, tentação)
2r	penalizar^3	v. tr.	**causar** pena, dor, aflição a	RESULTADO_DE([pena, dor, aflição], penalizar)
2r	inimizar^3	v. tr.	**provocar** inimizade entre	RESULTADO_DE(inimizade, inimizar)
3r	displasia	s. f.	desenvolvimento anormal de um órgão ou de um tecido, de que podem **resultar** deformidades graves	RESULTADO_DE(deformidades, displasia)

In pattern 3c `PREP_EM` denotes the preposition `em` contracted or not with a determiner.

Note that we also deal with enumeration of causes or effects/results separated by commas or conjunctions using a recursive rule that overcomes the "conjoined heads" problem, which is one of the limitations of using string patterns pointed by [16].

4.2 Results

Table 2 shows some examples of the relations extracted by the Causation grammars.

Manual inspection of the obtained relations yielded 4% erros in **CAUSADOR_DE** and 8% in **RESULTADO_DE** relations. Examples of the most common errors are:

---

[3] Note that these patterns discover relations between nouns, as well as between nouns and verbs, which may probably be better modelled by other relation names such as ACCAO_QUE_CAUSA and RESULTADO_DA_ACCAO.

1. definitions that mention the relation between two words of the definition, and not relative to the entry word: estetoscópio, s. m. - instrumento para auscultar a respiração, as batidas do coração e outros sons produzidos pelo corpo, **CAUSADOR_DE**(corpo, estetoscópio);
2. definitions where the pattern is preceded by a negative word, making the entity a "non-cause": respeitar, v. tr. - não causar dano, **RESUL-TADO_DE**(dano, respeitar);
3. definitions using brackets: inspirar, v. tr. - provocar (ideias, pensamentos, projectos), **RESULTADO_DE**((, inspirar);
4. definitions using commas: heterocarpo, adj. - que produz, espontaneamente ou por intervenção do homem, flores ou frutos diferentes, **RESULTADO_DE**(, , heterocarpo).

Items 1 and 3 are pointed out by [16] as limitations of using string patterns instead of structural patterns to extract relations from text.

5 Conclusions and Further Work

This project intends to create a computationally tractable ontology from mining a particular (general language) dictionary, and not provide THE ontology for Portuguese. In further (separate) projects we might investigate overlap with other sources for ontology (other dictionaries, reference works, corpora etc.) but this is outside the scope of PAPEL. So, corpus-based validation of PAPEL is simply a way of detecting further patterns in the dictionary to add rules for the particular relations, and not any general corpus-based ontology creation.

We are doing improvements to PEN in order to be able to decouple morphological and lexical information from the grammar. In this respect, we intend to try out a broad-coverage parser such as PALAVRAS [25].

We are also devising a system to help humans revising the residual examples that are not amenable to automatic parsing, so that they will be easily included in the final resource and possibly also feed the dictionary proper.

After the extraction of the relations, we will have a network of words linked by relations. We are considering the hypothesis of performing a process similar to the one described in [21] to identify groups of related definitions inside the same entry (word) and use them for the ultimate construction/detection of synonyms and *synsets*.

Acknowledgments

We would like to thank the group of R&D of Porto Editora for making their dictionary available for this research. The project PAPEL is supported by the Linguateca project, jointly funded by the Portuguese Government and the European Union (FEDER and FSE), under contract ref. POSC/339/1.3/C/NAC.

References

1. Dicionário PRO da Língua Portuguesa. Porto Editora, Porto (2005)
2. Miller, G.A., Beckwith, R., Fellbaum, C., Gross, D., Miller, K.J.: Introduction to WordNet: An on-line lexical database. International Journal of Lexicography 3(4), 235–244 (1990)
3. Richardson, S.D., Dolan, W.B., Vanderwende, L.: Mindnet: Acquiring and structuring semantic information from text. In: COLING-ACL, pp. 1098–1102 (1998)
4. Marrafa, P., Amaro, R., Chaves, R.P., Lourosa, S., Martins, C., Mendes, S.: Wordnet.pt new directions. In: Sojka, P., Choi, K.-S., Fellbaum, C., Vossen, P. (eds.) Proceedings of GWC 2006: 3rd International Wordnet Conference, Jeju Island, Korea, pp. 319–320 (2006)
5. Dias-da-Silva, B.C.: Wordnet.br: An exercise of human language technology research. In: Sojka, P., Key-Sun Choi, C.F., Vossen, P. (eds.) Proceedings of the Third International WordNet Conference — GWC 2006, South Jeju Island, Korea, January 22–26 (2006)
6. Calzolari, N.: An empirical approach to circularity in dictionary definitions. In: Cahiers de Lexicologie, pp. 118–128 (1977)
7. Calzolari, N.: Detecting patterns in a lexical data base. In: Proceedings of the 22nd annual meeting on Association for Computational Linguistics, pp. 170–173. Association for Computational Linguistics, Morristown (1984)
8. Michiels, A., Mullenders, J., Noël, J.: Exploiting a large data base by Longman. In: Proceedings of the 8th conference on Computational linguistics, pp. 374–382. Association for Computational Linguistics, Morristown (1980)
9. Amsler, R.A.: The structure of the Merriam-Webster Pocket dictionary. PhD thesis, The University of Texas at Austin (1980)
10. Amsler, R.A.: A taxonomy for english nouns and verbs. In: Proceedings of the 19th annual meeting on Association for Computational Linguistics, pp. 133–138. Association for Computational Linguistics, Morristown (1981)
11. Chodorow, M.S., Byrd, R.J., Heidorn, G.E.: Extracting semantic hierarchies from a large on-line dictionary. In: Proceedings of the 23rd annual meeting on Association for Computational Linguistics, pp. 299–304. Association for Computational Linguistics, Morristown (1985)
12. Markowitz, J., Ahlswede, T., Evens, M.: Semantically significant patterns in dictionary definitions. In: Proceedings of the 24th annual meeting on Association for Computational Linguistics, pp. 112–119. Association for Computational Linguistics, Morristown (1986)
13. Alshawi, H.: Analysing the dictionary definitions. Computational lexicography for natural language processing, 153–169 (1989)
14. Vanderwende, L.: Algorithm for automatic interpretation of noun sequences. In: Proceedings of the 15th conference on Computational linguistics, pp. 782–788. Association for Computational Linguistics, Morristown (1994)
15. Vanderwende, L.: Ambiguity in the acquisition of lexical information. In: Proceedings of the AAAI 1995 Spring Symposium Series, pp. 174–179 (1995); symposium on representation and acquisition of lexical knowledge
16. Montemagni, S., Vanderwende, L.: Structural patterns vs. string patterns for extracting semantic information from dictionaries. In: Proceedings of the 14th conference on Computational linguistics, pp. 546–552. Association for Computational Linguistics, Morristown (1992)

17. Vanderwende, L., Kacmarcik, G., Suzuki, H., Menezes, A.: Mindnet: An automatically-created lexical resource. In: HLT/EMNLP. The Association for Computational Linguistics (2005)
18. Baker, C.F., Fillmore, C.J., Lowe, J.B.: The berkeley framenet project. In: Proceedings of the 17th international conference on Computational linguistics, pp. 86–90. Association for Computational Linguistics (1998)
19. Wilks, Y.: Is word sense disambiguation just one more nlp task? Computers and the Humanities 34, 235–243 (2000)
20. Kilgarriff, A.: Word senses are not bona fide objects: implications for cognitive science, formal semantics, nlp. In: Proceedings of the 5th International Conference on the Cognitive Science of Natural Language Processing, Dublin, pp. 193–200 (1996)
21. Dolan, W.B.: Word sense ambiguation: clustering related senses. In: Proceedings of the 15th conference on Computational linguistics, pp. 712–716. Association for Computational Linguistics, Morristown (1994)
22. Earley, J.: An efficient context-free parsing algorithm. Communications of the ACM 6(8), 451–455 (1970)
23. Vendler, Z.: Causal relations. The Journal of Philosophy 64, 704–713 (1967)
24. Vendler, Z.: Effects, results and consequences. Linguistics in Philosophy 64, 147–171 (1967)
25. Bick, E.: The Parsing System PALAVRAS: Automatic Grammatical Analysis of Protuguese in a Constraint Grammar Framework. PhD thesis, Arhus University, Arhus (2000)

Comparing Window and Syntax Based Strategies for Semantic Extraction*

Pablo Gamallo Otero

Departamento de Língua Espanhola, Faculdade de Filologia
Universidade de Santiago de Compostela, Galiza, Spain
pablogam@usc.es

Abstract. In this paper, we describe and compare two different approaches for extracting similar words from large corpora. In particular, we compared a method based on syntactic contexts with two strategies relying on windows of tagged words, one using word order and the other bags of words. On a Portuguese corpus of 12 million words, syntactic contexts produce significantly better results for both frequent and not very frequent words.

1 Introduction

Finding semantically related words from large text corpora is one of the most popular tasks in Information Extraction. This is required to achieve more ambitious objectives, such as thesaurus construction, ontology design, question-answering enrichment, etc. The basic idea underlying the different techniques to find semantic similarity states that words are semantically related if they share a large number of contexts. There are basically two methods for defining word contexts. On the one hand, the context of a word is defined as the n words surrounding it (n-grams), where n stands for a window size. The methods using this type of word contexts are known as window-based approaches. On the other hand, the context of a word is determined by grammatical dependency relations. In this case, contexts are defined making use of syntax-based techniques.

It is broadly assumed that window-based approaches offer some advantages with regard to syntactic strategies: concerning *speed*, they are much less time consuming, while parsing large corpora is expected to be less computationally efficient. As far as *portability* is concerned, windowing techniques do not require contexts to be defined using specific grammars of particular natural languages; they are not language dependent. In addition, it has not been clearly demonstrated that syntactic contexts perform better that window contexts for discovering word similarity. On the contrary, it is assumed that the semantic relationships generated by approaches which use windowing techniques put words together according to associative relations, e.g., *doctor* and *hospital*. These relations are difficult to grasp by syntactic based methods, since related words such as *doctor* and *hospital* do not appear in the same syntactic contexts.

---

* This work has been supported by Ministerio de Educació y Ciencia of Spain, within the project ExtraLex, ref: PGIDIT07PXIB204015PR.

A. Teixeira et al. (Eds.): PROPOR 2008, LNAI 5190, pp. 41–50, 2008.

In this paper, we propose a syntax-based method which is provided with some of the advantages of windowing techniques: it is computationally efficient since the parsing strategy is robust and uses basic regular expressions. It is not language dependent since it relies on a multilingual parser whose grammar consists of very generic rules aimed to analyze texts in several languages. On the other hand, unlike window techniques, it is not aimed at discovering generic semantic associations between words, but only relationships between words belonging to the same class/kind of entities (i.e., *co-hyponyms*). We will demonstrate that methods using syntactic information have the tendency to find similarities between words that belong to the same semantic class, e.g., *doctor* and *nurse*, *teacher* and *pupil*. This specific semantic information is much more appropriate for many NLP applications, namely: ontology design by word clustering, word sense disambiguation, question-answering, pp-attachment, etc.

The main contribution of this paper is to define a protocol evaluation to compare the accuracy of our syntactic strategy against other window based techniques. Accuracy is defined with regard to the specific task of discovering word class relations. In order to perform this evaluation, cooccurrence data for different types of proper names (named entities), taking into account both syntactic dependencies and windowing relations, was collected from a Portuguese corpus of 12 million words. The corpus consists of articles of *O Público*, a general purpose Portuguese newspaper.

The paper is organized as follows: section 2 describes some related work. Then, section 3 briefly introduces two window-based methods, while section 4 describes more accurately our syntax based strategy, which relies on a very simple dependency parser. Finally, in Section 5, some experiments will be performed against a Portuguese corpus in order to evaluate the performance of the methods described in the previous sections.

2   Earlier Comparisons between Both Approaches

Despite the growing interest in semantic extraction, there exist still few previous works aimed to evaluate and compare the two strategies at stake. In [6], a syntax-based method is carefully compared to a windowing technique, with regard to the general task of word similarity extraction. The former is shown to perform better for high-frequency words, while the windowing method is the better performer for low-frequency words. This evaluation was focused on associative links between words, since both methods were compared against online thesauri which are provided with all kind of semantic relations. So, we cannot know which method is more reliable for discovering cohyponymy relations between words. Moreover, the experiments performed made use of very small text corpora, probably due to the low efficiency of the syntactic techniques available at that time.

In [9], the two techniques are compared with regard to the extraction of several semantic relations. In most tasks, the syntax approach was clearly better than the bag-of-word model. However, the latter was defined in a very restrictive way: only the 200 most frequent words were considered as dimensions of context vectors. In [10], it is described a similar comparative experiment against a Dutch corpus. The authors conclude again that a full syntactic context model outperforms all other approaches.

[12] proposes a comparative evaluation of the two techniques with regard to a different task: extraction of multiwords and collocations. In the conclusion, they state that syntax-based methods outperform windowing techniques thanks to a drastic reduction of noise. The main problem is that experiments were performed with a parser which is not robust and time consuming (130 word/second [14]).

3   Window-Based Contexts

Contexts can be defined using the immediately adjacent words, within a window of n words. Two different techniques can be applied: one defining contexts as bag of words and the other taking into account word order.

The technique based on bag of words builds context vectors considering simple words as dimensions, regardless of their positions within the window. For instance, let's suppose that the Portuguese adjective *pequeno* ("small") cooccurs twice with *homem* ("man"): once to the left and once directly following the noun: *homem pequeno* and *pequeno homem*. Table 1 shows the contexts vectors of *homem* and *pequeno*. The value of each dimension is the number of cooccurrences without taking word position into account.

Table 1. Bag of words

	pequeno	homem
pequeno		2
homem	2	

Table 2. Word order

	$(p, < 1)$	$(p, < 2)$	$(p, > 1)$	$(p, > 2)$	$(h, < 1)$	$(h, < 2)$	$(h, > 1)$	$(h, > 2)$
pequeno						1		1
homem	1			1				

On the other hand, Table 2 depicts the context vectors of the same two words when taking word order into account. Each dimension represents the position of the context word within a window of size 2. For instance, $(p, < 1)$ means that *pequeno* is 1 word ahead of the main word. $(p, > 2)$ represents two positions to the right. Using this technique, the vector size grows while frequency counting decreases. It results in a more sparse matrix. According to Rapp [11], word order is a statistical clue useful to simulate syntactic behavior. This window technique is, then, closer to the syntax-based approach.

4   Syntax-Based Contexts

The second technique to define word contexts relies on the identification of syntactic dependencies. So, context vectors will be provided with syntactic information.

Table 3. Dependency triplets and patterns of POS tags

Dependencies	Patterns of POS tags
$(\text{green}_5, mod_<, \text{jacket}_6)$ $(\text{big}_{10}, mod_<, d\text{dog}_{11})$	*R_1: $s/(\mathbf{A_i})(\mathbf{N_j})/\mathbf{N_j}/$
()	*R_2: $s/(\mathbf{N_i})(\mathbf{N})_j/\mathbf{N_i}/$
$(\text{man}_2, with_3, \text{jacket}_5)$	*R_3: $s/(\mathbf{N_i})(\mathbf{P_k})(\mathbf{N})_j/\mathbf{N_i}/$
$(\text{see}_6, obj_>, \text{dog}_{11})$	R_4: $s/(\mathbf{V_i})(? : D_k\|R_n) * (\mathbf{N})_j/\mathbf{V_i}/$
$(\text{see}_6, obj_<, \text{man}_2)$	R_5: $s/(? : D_k) * (\mathbf{N_i})(? : R_n) * (\mathbf{V})_j/\mathbf{V_j}/$
()	R_6: $s/(\mathbf{V_i})(? : R_n) * (\mathbf{P_k})(? : \|D_m\|R_r) * (\mathbf{N})_j/\mathbf{V_i}/$

4.1 Dependency Parsing with Generic Regular Expressions

Instead of searching for windows positions around words or lemmas, we make use of regular expressions to identify syntactic dependencies. Regular expressions represent basic patterns of POS tags which are supposed to stand for binary dependencies between two lemmas. Our parsing strategy consists of a sequence of syntactic rules, each rule being defined by a specific pattern of tags that stands for a binary dependency. This strategy is implemented as a finite-state cascade [1]. So far, our grammar is focused on dependencies with verbs, nouns, and adjectives, since it is assumed that these dependencies are useful for semantic extraction. Let's take an example. Suppose our corpus contains the following tagged sentence:

a_D_1 man_N_2 with_P_3 a_D_4 green_A_5 jacket_N_6 see_V_7 yesterday_R_8 a_D_9 big_A_{10} dog_N_{11}

The aim is to identify dependencies between lemmas using basic patterns of POS tags. Dependencies are noted as triplets: $(head, rel, dependent)$. The first column of Table 3 shows the 5 triplets generated from the sentence above using the patterns appearing in the second column. Patterns are organized in a sequence of substitution rules in such a way that the input of a rule R_n is the output of a rule R_m, where $m \leq n$. A rule substitutes the POS tag of the head word (right side) for the whole pattern of tags representing the head-dependent relation (left side). The first rule, R_1, takes as input a string containing the ordered list of all tags in the sentence:

$D_1 N_2 P_3 D_4 A_5 N_6 V_7 R_8 D_9 A_{10} N_{11}$

The left pattern in this rule identifies two specific adjective-noun dependencies, namely "$A_5 N_6$" and "$A_{10} N_{11}$". As a result, it removes the two adjective tags from the input list, and produces as output:

$D_1 N_2 P_3 D_4 N_6 V_7 R_8 D_9 N_{11}$

Then, rule R_3 is applied to the output of R_1. The left pattern of this rule matches "$N_2 P_3 D_4 A_5 N_6$" and rewrites the following ordered list of tags:

$D_1 N_2 V_7 R_8 D_9 N_{11}$

This list is the input of the following applicable rule, R_4, which produces:

$D_1N_2V_7$

Finally, rule R_5 is applied and gives as result only one tag, V_7, which is associated to the root head of the sentence: the verb "see". As this verb does not modify any word, no rule can be applied and the process stops. This is in accordance with the main assumption of dependency-based analysis, namely, a word in the sentence may have several modifiers, but each word may modify at most one word [8]. In sum, each application of a rule, not only rewrites a new version of the list of tags, but also generates the corresponding dependency triplet. So, even if we do not get the correct root head at the end of the analysis, the parser generates as many triplets as possible. This strategy can be seen as a particular case of partial and robust parsing [1], which is as faster as identifying contextual words with a window-based technique (over 7000 words/second).

The 5 triplets in Table 3 where generated from 4 substitution rules, each matching a type of dependency: adjective-noun, noun-prep-noun, verb-noun, and noun-verb. The sentence analyzed above does not contain triplets instantiating noun-noun and verb-prep-noun dependencies. Wildcards (? : $D|R$)∗ stand for optional determiners and adverbs, that is, they represent optional sequences of determiners or/and adverbs that are not considered for triplets. Rules with an asterisk can be applied several times before applying the next rule (e.g., when a noun is modified by several adjectives). Subscript numbers allow us to link tags in the patterns with their corresponding lemmas in the sentence. To represent triplets, we use 4 types of binary relations: prepositions, left modifiers (noted as $mod_<$), right objects ($obj_>$), and left objects ($obj_<$). Note that the patterns of tags in Table 3 work well with English texts, but they are so generic that they also can be used for many languages. To extract triplets from texts in Romance languages such as Portuguese, Spanish, French, or Galician, 2 tiny changes are required: to provide a new pattern with dependent adjectives at the right position of nouns ($mod_>$), and to take as the head of a noun-noun dependency the noun appearing at the left position. The experiments that will be described later were performed over a Portuguese corpus. To date, our parser can be applied on text previously tagged with either Treetagger[1] and Freeling [2].

4.2   Lexico-Syntactic Contexts

The second step of our syntax-based method consists in extracting lexico-syntactic contexts from the dependencies and counting the occurrences of lemmas in those contexts. This information is stored in a collocation database. The extracted triplets of our example allow us to easily build the collocation database depicted in Table 4. The first line of the table describes the entry "man". This noun occurs once in two lexico-syntactic contexts, namely that representing the left position ($obj_<$) of the verb "see", (see, $obj_<$, N), and that denoting the noun position being modified by the prepositional complement "with a jacket". The second line

---

[1] http://www.ims.uni-stuttgart.de/projekte/corplex/Tree-Tagger/
DecisionTreeTagger.html

Table 4. Collocation database of lemmas and lexico-syntactic contexts

Lemmas	Lexico-Syntactic Patterns and freqs.
man	$< (see, obj_<, N), 1 >$
	$< (N, with, \text{jacket}), 1 >$
see	$< (V, obj_<, \text{man}), 1 >$
	$< (V, obj_>, \text{dog}), 1 >$
big	$< (\text{dog}, mod_<, A), 1 >$
dog	$< (N, mod_<, \text{big}), 1 >$
	$< (see, obj_>, N), 1 >$
green	$< (\text{jacket}, mod_<, A), 1 >$
jacket	$< (N, mod_<, \text{green}), 1 >$
	$< (\text{man}, with, N), 1 >$

describes the entry "see", which also occurs once in two different lexico-syntactic contexts: $(V, obj_<, man)$ and $(V, obj_>, dog)$, i.e., it co-occurs with both a left object, "man", and a right object: "dog". The remaining lines describe the collocation information of the remaining nouns and adjectives appearing in the sentence above.

Notice we always extract 2 complementary lexico-syntactic contexts from a triplet. For instance, from $(man, with, \text{jacket})$, we extract:

$(N, with, \text{jacket})$ $(\text{man}, with, N)$

This is in accordance with the notion of co-requirement defined in [5]. In this work, two syntactically dependent words are no longer interpreted as a standard "predicate-argument" structure, where the predicate is the active function imposing syntactic and semantic conditions on a passive argument, which matches such conditions. On the contrary, each word in a binary dependency is perceived simultaneously as a predicate and an argument. In the example above, $(man, with, N)$ is seen as an unary predicate that requires nouns denoting parts of men (e.g. jackets), and simultaneously, $(N, with, \text{jacket})$ is another unary predicate requiring entities having jackets (e.g. men).

Finally, syntax-based context vectors are easily built from the collocation database. As in [7] and [5], we use several types of dependencies to define syntactic contexts, and not only objects and subjects.

5   Experiments

5.1   Corpus

Experiments have been carried out using a Portuguese corpus with 12 million tokens extracted from the general-purpose journal *O Público*. Before building the window and syntax based contexts, texts were lemmatized and *POS* tagged with TreeTagger.[2] In the case of window contexts (both bag of words and word order), function words were previously removed.

---

[2] For Portuguese, see in *http://gramatica.usc.es/~gamallo/tagger.htm*

5.2   Vector Similarity

The similarity coefficient used in our experiments to compare vector contexts is a particular version of Dice score: $dice\dagger$. Similarity between the context vectors of two lemmas, $lemma_1$ and $lemma_2$, is computed as follows:

$$dice\dagger(lemma_1, lemma_2) = \frac{2 * \sum_i min(f(lemma_1, cntx_i), f(lemma_2, cntx_i))}{F(lemma_1) + F(cntx_2)}$$

where $f(lemma_1, cntx_i)$ represents the number of times $lemma_1$ cooccurs with $cntx_i$. $F(lemma_i)$ stands for the absolute frequency of $lemma_1$. We use this coefficient because it produced the best results in related work [3,13,4].

5.3   Initial List of Seed Proper Nouns

Our objective is to design an evaluation protocol avoiding unclear and fuzzy judgments about word similarity. For this purpose, we only consider a reduced sample list of proper nouns. For each member of the list, we compute its $dice\dagger$ similarity with all proper nouns in the corpus, and produce a ranked list with its top 5 most similar nouns. The test list was built by hand and consists of 28 proper nouns divided in 7 semantic categories: countries, capitals of countries, Portuguese towns, politicians, organizations, press agencies, and football teams. As we selected 5 similar candidates for each test noun, the final list to be evaluated contains 140 proper nouns. The evaluator is just required to classify each new proper name as a member of the 7 categories enumerated above. For instance, if *Washington* is selected as a similar noun to *Bruxelas*, which belongs to the category of capitals within the test list, the evaluator only needs to decide if *Washington* is or not a capital.

Furthermore, the 28 test proper nouns were selected according to their position in the list of all nouns ranked by frequency. They were required to be distributed in 4 ranges: (1) very frequent words, ranked between 1 and 1000, with frequency > 479; (2) quite frequent words, ranked between $1,000$ and $3,000$, and whose frequency is $> 100 < 479$; (3) not very frequent words, ranked between $3,000$ and $5,000$, with frequency $> 50 < 100$; (4) quite rare words, ranked between $5,000$ and $10,000$, with frequency $> 20 < 50$.

5.4   Results

We measured the precision of three methods: two window based techniques, one relying on bag of words and the other on word order, and our syntax based method. For each method, we computed the total precision and that obtained for each one of the 4 frequency ranges considered.

Table 5 shows the list of candidates obtained from 3 test proper nouns: the first one is *Peru*, designing a country, and situated in range (3), that is, its frequency in the corpus is $> 50 < 100$. The second one is a capital, *Belgrado*, also situated in range (3), and finally, a Portuguese town, *Viseu*, situated in range (2). Correct candidates are in bold.

Table 5. Similar words according to the 3 methods

name	bag of words	word order	syntax
Peru	Alberto Fujimori, PRESIDENTE, Abimael, Vargas Llosa	Resultados, Política, Humanidade, Fernando Sousa, Argentina	**Tchetchénia**, Sul de Espanha, **Guiné-Bissau**, **Libéria**, **Guatemala**
Belgrado	SPS, EF, Solidariedade, Pedro Caldeira Rodrigues, Unidades	SPS, Líbia, **Krajina**, Sérbia, **Jacarta**	**Moscovo**, **Washington**, **Jacarta**, **Zagreb**, **Argel**
Viseu	Juventude, Pereira, **Montijo**, Teatro, **Guarda**	Intervenção, Olivieria de Azeméis, **Guarda**, Australia, **Castelo Branco**	**Bragança**, **Beja**, **Guarda**, **Santarém**, **Leiria**

Notice that the method based on bag of words does not select any country from *Peru*, but is able to retrieve two individuals associated to this country: both *Alberto Fujimori* and *Vargas Llosa*. This is in accordance with one of our initial assumptions: window-based techniques are not suitable to extract word class relations (co-hyponymy), but rather any kind of associative link between words.

Table 6 depicts results on precision for the 3 methods, taking into account the 4 frequency ranges of lemmas as well as all lemmas with frequency > 20. The results show that the syntax-based method performs much better than the windowing techniques, whereas the strategy based on word order is quite better than that relying on just bag of words (see Figure 2). So, structural information (dependencies and word order) helps identifying meaningful contexts. On the other hand, the former method is the only one that clearly improves when it is being applied on more and more frequent lemmas (see Figure 1). Hence, it follows

Table 6. Results of the 3 evaluated methods

Range / freq	Syntax 117, 266 cntxs Prec-%	Word order 190, 228 cntxs Prec-%	Bag of words 148, 422 cntxs Prec-%
1-1000 > 450	97	63	29
1000-3000 > 100 < 450	97	49	37
3000-5000 > 50 < 100	80	57	31
5000-10000 > 20 < 50	57	54	22
Total > 20	83	55	36

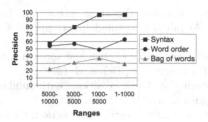
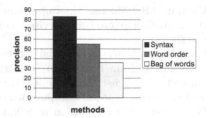

Fig. 1. Precision scores by ranges of frequency

Fig. 2. Precision scores for proper nouns with frequency > 20

that the syntactic strategy would perform better as corpus size grows, which is not true for the two windowing techniques. This is somehow in accordance with the experiments performed by Grefenstette [6], where the window-based method was the better performer for low-frequency words.

In Table 6, we also show the number of syntactic contexts used by each method. Let's note that the number of syntactic contexts $(117, 226))$ is much smaller than that of window based contexts. As the size of context vectors in the syntactic approach is not very large, the process of computing similarities turns out to be more efficient.

6 Conclusion

We consider that syntactic analysis of source corpora is more suitable for extraction of co-hyponymy semantic relationships, and that the syntactic structure of source text has to be taken into account in order to ensure the quality of results for both frequent and not frequent words. In addition, our syntax-based method is more computationally efficient than the windowing techniques since it defines and uses smaller context vectors. On the other hand, the syntactic strategy defined in this paper can be considered as knowledge-poor as the window-based approach, since the robust parsing described here relies on few generic regular expressions. Moreover, as the generic knowledge underlying the parser is used to identify basic dependencies for several natural languages, our multilingual strategy turns out to be almost as language-independent as any windowing technique. In sum, in order to extract co-hyponymy, it seems to us there are no strong arguments to use window techniques instead of syntactic contexts.

References

1. Abney, S.: Part-of-speech tagging and partial parsing. In: Church, K., Young, S., Bloothooft, G. (eds.) Corpus-Based Methods in Language and Speech. Kluwer Academic Publishers, Dordrecht (1996)
2. Carreras, X., Chao, I., Padró, L., Padró, M.: An open-source suite of language analyzers. In: LREC 2004, Lisbon, Portugal (2004)

3. Curran, J.R., Moens, M.: Improvements in automatic thesaurus extraction. In: ACL Workshop on Unsupervised Lexical Acquisition, Philadelphia, pp. 59–66 (2002)
4. Gamallo, P.: Learning bilingual lexicons from comparable english and spanish corpora. In: Machine Translation SUMMIT XI, Copenhagen, Denmark (2007)
5. Gamallo, P., Agustini, A., Lopes, G.: Clustering syntactic positions with similar semantic requirements. Computational Linguistics 31(1), 107–146 (2005)
6. Grefenstette, G.: Evaluation techniques for automatic semantic extraction: Comparing syntactic and window-based approaches. In: Workshop on Acquisition of Lexical Knowledge from Text SIGLEX/ACL, Columbus, OH (1993)
7. Lin, D.: Automatic retrieval and clustering of similar words. In: COLING-ACL 1998, Montreal (1998)
8. Lin, D.: Dependency-based evaluation of minipar. In: Workshop on Evaluation of Parsing Systems, Granada, Spain (1998)
9. Padó, S., Lapata, M.: Dependency-based construction of semantic space models. Computational Linguistics 33(2), 161–199 (2007)
10. Peirsman, Y., Heylen, K., Speelman, D.: Finding semantically related words in dutch. co-occurrences versus syntactic contexts. In: CoSMO Workshop, Roskilde, Denmark, pp. 9–16 (2007)
11. Rapp, R.: Automatic identification of word translations from unrelated english and german corpora. In: ACL 1999, pp. 519–526 (1999)
12. Seretan, V., Wehrli, E.: Accurate collocation extraction using a multilingual parser. In: COLING-ACL 2006, pp. 953–960 (2006)
13. van der Plas, L., Bouma, G.: Syntactic contexts for finding semantically related words. In: CLIN 2004 (2004)
14. Wehrli, E.: Fips, a deep linguistic multilingual parser. In: 5th Workshop on Important Unresolved Matters, pp. 120–127 (2005)

The Mitkov Algorithm for Anaphora
Resolution in Portuguese

Amanda Rocha Chaves and Lucia Helena Machado Rino

Universidade Federal de São Carlos, Brazil
amandarc@icmc.usp.br, lucia@dc.ufscar.br

Abstract. This paper reports on the use of the Mitkov´s algorithm for resolution of third person pronouns in texts written in Brazilian Portuguese. A system for anaphora resolution was built that embeds most of the Mitkov's features. Some of his resolution factors were directly incorporated into the system; others had to be slightly modified for language adequacy. The resulting approach was intrinsically evaluated on hand-annotated corpora. It was also compared to Lappin & Leass's algorithm, also customized to pronoun resolution in Portuguese. Success rate was the only evaluation measure used.

Keywords: Pronoun resolution, anaphora resolution.

1 Introduction

A major problem in Natural Language Processing (NLP) is to recognize or build text segments that convey coherent information. Amongst the linguistic devices for that, referential cohesion is one of the most significant for acknowledging, and guaranteeing, coherence. In this paper we address such a phenomenon aiming at identifying cohesive mechanisms that help automatically resolving referential links. Anaphoricity, i.e., pronoun resolution (PR), is the only linguistic construction under focus here. An anaphoric pronoun signals a relationship between two or more text components that share with each other their meanings. It comes after its antecedent referent in the text, which is usually a noun phrase (NP), as shows Example (1)[1]:

(1) *O parlamentar*ᵢ, porém, é alvo de acusação em outro escândalo. *Ele*ᵢ será investigado sobre as denúncias de corrupção (...).
   [*The member of the parliament*]ᵢ, however, is enrolled as guilty in other scandal. *He*ᵢ will be investigated on the corruption accusations (…).

Above, the pronoun 'Ele' (*He*) is the anaphor with an NP antecedent given by 'O parlamentar' (*The member of the parliament*). This conveys a full meaning, while the pronoun itself is issued its antecedent meaning. Differently from this, other types of anaphors also may happen, e.g., the generalization introduced by the NP *'the corruption accusations'* in (1). To resolve this anaphor, ontological and other referential links are usually needed (e.g., *'other scandal'* enables the plural *'accusations'*).

---

[1] In this paper, all the examples have been extracted from a BP corpus of authored texts. Their literal English translations are supplied for readability.

A. Teixeira et al. (Eds.): PROPOR 2008, LNAI 5190, pp. 51–60, 2008.
© Springer-Verlag Berlin Heidelberg 2008

Usually, ontological links do not entail the same but similar meanings that make them connect to each other. Contrarily, anaphoric pronouns convey no meaning, but their antecedent ones. So, to resolve the anaphoric 'Ele' above, retrieving the former NP is of utmost importance.

PR may be very complex when there are many antecedent candidates. To resolve that, varied linguistic features may be needed, such as morphological, syntactic, semantic, or pragmatic ones. Several computational approaches have been undertaken for English or Spanish in this line. Examples focusing on the former are [7], [8], [10] [11], and [1]; an example of the latter is [14]. There are few approaches to Portuguese, namely, modified versions of both Lappin and Leass' [3] and Hobbs' algorithms [17], and a model that uses heterogeneous knowledge [15]. Actually the only system currently available is the first one. The work based on Hobbs' model is still under progress; the other has not been implemented. Especially, Coelhos's system [3] operates on information produced by the Xtractor tool [5], which in turn modifies syntactic trees produced by the parser PALAVRAS [2]. It first calculates the so-called salient NPs conveyed by the syntactic structure of a sentence. Then it uses a dynamic model of attentions to pinpoint the anaphor antecedent from the NP candidates.

Our PR system uses intermediate results by Coelho as input (mostly resulting from preprocessing), to focus on the PR module itself. We adopted Mitkov's model [10] for the following reasons: (1) it has been explored for several languages, being thus language-independent and portable; (2) it is heuristics-based and does not depend upon deep knowledge. Instead, it applies surface or empirical information to determine candidate antecedents of an anaphor; (3) it also adopts usual parsing and morphological preprocessing tools, which are largely available for most languages, as they are for Brazilian Portuguese (BP). In our implementation, Mitkov's original algorithm has been modified to handle only 3rd person pronouns that convey NPs as antecedents.

Mitkov classifies the so-called anaphora resolution (AR) factors to signal antecedent candidates as *restrictive* and *preferential*. Restrictive factors signal mandatory properties of the antecedent candidates, in order for them to resolve the anaphor. The ones that do not convey such properties are thus discarded. *Preferential factors* do not discard candidates; they just classify them according to their likelihood of resolving the anaphor the best. Usually, both factors are applied altogether: classification takes place only after filtering those potential candidates that satisfy the restrictions. The highest the probability of a candidate, the more likely it is to be the antecedent of the anaphor under focus.

There are other, more recent approaches, which also address similar factors. For example, in [16] some preferential and restrictive factors coincide with Mitkov's ones (e.g., morphological, nearest NP, and syntactic parallelism), but addressing the French language. [19] also addresses Lappan & Leass' method, but for German. Further experiments after Mitkov's approach are carried out with MARS, Mitkov Anaphora Resolution System [9], which evaluate the impact of AR for NLP systems. Especially, automatic summarization, term extraction, and text categorization are explored.

In what follows, we first present Mitkov's proposal for AR (Section 2), then we introduce our approach for PR in BP (Section 3). In Section 4 we describe our PR assessment. Final remarks are presented in Section 5.

2   The Mitkov's Algorithm

Mitkov avoids complex syntactic and semantic constructions by adopting a set of heuristics – the antecedent indicators – that are capable of pinpointing potential antecedents of an anaphor based on surface indicators. The text under focus is first parsed and its NPs are extracted. For AR, (1) at most two previous sentences to the anaphor are examined, as the referential context for its antecedent NP. The result is a set of NPs. (2) to narrow the number of candidate NPs, gender and number concordance with the anaphor are verified, yielding a smaller candidate set, or the actual set of potential antecedents. (3) Each potential NP is thus scored for the likelihood of being the antecedent of the anaphor. In this step, each antecedent indicator is used to issue the NP an integer value ranging from -1 to 2 and all the values are summed up for the final NP score. The highest scored NP is finally chosen as the antecedent. In a tie case, the closest candidate to the anaphor is chosen instead.

3   Adapting Mitkov's Algorithm for PR in Brazilian Portuguese

Our system embedding Mitkov's indicators for BP is named RAPM, for *Resolução Anafórica do Português baseada no algoritmo de Mitkov* (Anaphora Resolution for Portuguese based on Mitkov's algorithm). RAPM differs from the original algorithm in that it aims at BP and, most importantly, its input texts are automatically annotated. In Mitkov's approach morphosyntactic annotations are manually corrected before going into AR itself instead. Moreover, to resolve morphological dependencies RAPM looks up an XML onomastic file with correct information on gender and number of proper nouns, and the antecedent search scope is of three sentences, instead of two. The XML file conveys proper nouns extract from a text corpus and aims at minimizing preprocessing problems. In the absence of such information, they would be assigned both genders and numbers. A last, but minor, distinction from the original is that currently RAPM does not incorporate modules for preprocessing, as we shall describe in Section 4.

RAPM processes in the following way: it identifies the NPs that appear previously to the pronouns using the three-sentence window, and then it produces the set of potential NP candidates. Since antecedent indicators may endorse or prevent an NP candidate of being the antecedent of a pronoun, the total sum of the scores may be correspondingly positive or negative. Only five out of eleven antecedent indicators by Mitkov were incorporated into RAPM, along with three others that we found interesting to add, as follows (the last three are the novel ones):

First NP (FNP). A +1 score is issued to the first NP of each sentence. This heuristic may be either justified on syntactic, or on communication terms, in that human beings usually express meanings through distinct language levels. According to Mitkov, for example, in declarative sentences the FNP occupies the subject position. In the absence of a parse tree, theories of both communication and discourse organization might help determining which should be such FNP, provided that they signaled the underlying communicative or discourse structure of the focused text. If Centering Theory [6] were considered, e.g., the FNP would be the actual center of the sentence.

Conversely, if [18] or [4] were used, the corresponding given-new or thematic-rhematic information should signal that. In either case, given-new or theme-rheme units could provide coreferential links as do a first NP and a pronoun, at the text surface. It is worthwhile to notice that, since declarative sentences convey a default discourse organization, the theme/given unit can be the very NP occupying the subject position, as pinpointed before.

Lexical Reiteration (LR). A +2 score is issued to NPs that occur twice or more within the search scope; a +1 score is issued to an NP otherwise. LR assumes that a greater score signals NPs that are more salient and, thus, more likely to be the anaphor antecedent than those that score less. In RAPM reiterated lexical items are identified through direct string matching.

Indefinite NP (INP). Indefinite NPs are assigned a -1 score because very often they are supposed to be less likely to be antecedents of pronominal anaphors than definite ones [12]. RAPM regards an NP as definite if its nucleus is modified by a definite article or by demonstrative or possessive pronouns.

Prepositional NP (PNP). A -1 score is issued to those NPs that occur in a prepositional phrase. Such a demoting score may be justified by the Centering Theory [6]: the sentence main constituents are classified according to their salience (e.g., subject, direct, and indirect objects in this decreasing salience order) and the most salient units provide the center of a text segment. Moreover, a sentence center is more likely to be a candidate antecedent of an anaphor than an NP occurring in a prepositional phrase sentences.

Referential Distance (RD). This antecedent indicator may promote or demote a candidate according to its distance from the anaphor: NPs in the immediate antecedent clause, in the very same sentence as the anaphor, are scored +2; NPs in the previous sentence are scored +1; NPs in a sentence that is two sentences apart from that of the anaphor are scored 0; NPs still farther than those are scored -1.

Syntactic Parallelism (SP). A +1 score is issued to an NP that conveys the same syntactic function as the corresponding anaphor.

Nearest NP (NNP). A positive +1 score is issued to the nearest NP to the anaphor. This indicator is used as a baseline by Mitkov, and so it is in RAPM, in which case it corresponds to the so-called 'Baseline_NP'[2] .

Proper Noun (PN). Proper nouns are scored +1 in RAPM because they occurred with relative frequency as anaphors antecedents in our corpus. The assumption behind such a score was that promoting PNs could improve PR performance.

Using the above antecedent indicators to resolve pronouns in BP aims only at predicting language behavior, and not at constraining automatic PR, hence the so-called *preference factors*: they are not intended to be definite. Although such indicators may punctuate anaphor antecedents incorrectly, usually PR is improved when they are used altogether, as we shall show when we apply them to BP. Adding the indicators SP, NNP, and PN to RAPM resulted from a corpus analysis that aimed at filtering out those Mitkov's antecedent factors that didn't apply to BP. They were chosen for the

---

[2] Baselines are used for assessment and will be described in Section 4.

following reasons: (a) since the input texts to RAPM are already morphosyntactically annotated, syntactic parallelism (SP) could be readily verified; (b) a nearest NP (NNP) to the anaphoric pronoun tended to be its antecedent; (c) proper nouns (PN) were highly frequent in the corpus as anaphor antecedents. So, using them should be advantageous for PR in RAPM. Excluding the remaining six Mitkov's indicators from RAPM was due to their inadequacy to the corpus under study. For example, the original indicator 'Section heading preference' did not apply to our corpus, because this conveys only non-structured or non-titled texts. Amongst RAPM eight indicators, two are impeding indicators (INP and PNP) and one (RD) may be either an impeding or a boosting indicator, yielding thus negative or positive values to a candidate. The remaining indicators can only boost the candidate.

Example (2) illustrates the use of RAPM processing:

(2) O flúor fortifica o esmalte, uma espécie de capa protetora dos dentes. Com a difusão de seu uso, outro problema surgiu: a fluorose, o excesso de flúor no organismo. Afinal, a *substância*$_i$ não se encontra apenas em cremes dentais: *ela*$_i$ também está presente em diversos alimentos (...).

The fluorine fortifies the enamel, a sort of protective cape of the teeth. With the diffusion of his use, another problem appeared: the fluorosis, the excess of fluor in the organism. After all, the *substance*$_i$ is not only in toothpastes: *it*$_i$ also is present in several foods (...).

In this case the algorithm generates the set of potential candidates that agree in gender and number[3] with the anaphor 'ela' (she){F,S}, resulting in the members NP_1: [uma espécie de capa protetora dos dentes]{F, S}, NP_2: [capa protetora dos dentes]{F,S}, NP_3: [a difusão de seu uso]{F,S}, NP_4: [a fluorose]{F, S} NP_5: [a substância]{F, S}. Finally, RAPM assigns the indicators scores to each candidate NP and sums them up (Table 1). In the example, the biggest sum allows RAPM to pinpoint NP_5 – 'a substância' – as the anaphor antecedent.

Table 1. Individual and total scores of each candidate NP

NP candidate	Antecedent indicators								
	FNP	LR	SP	NNP	PN	INP	PNP	RD	$\sum$
NP_5	0	0	1	0	0	0	0	1	2
NP_4	0	0	0	0	0	0	0	0	0
NP_3	0	0	0	0	0	0	-1	0	-1
NP_1	0	0	0	0	0	-1	0	-1	-2
NP_2	0	0	0	0	0	-1	-1	-1	-3

4   Assessing RAPM

We used the success rate measure [10] to assess RAPM. It measures the ratio between the total number of correctly resolved anaphors and the total number of anaphors that are present in the whole corpus of texts. A reference corpus of the same input texts previously annotated for anaphors by human experts was thus used.

---

[3] Signalled by the set {G,N}, being F=Feminine, M= Masculine; S=Singular, P=Plural.

According to Mitkov, the success rate should mirror exactly the performance of the anaphora resolver itself, with no interference of any problem resulting from preprocessing. He emphasizes that the real success rate of a system may only be achieved if the input data are correct. Because oft this, he hand-edits any possible wrong output of his system. However reasonable his arguments may be, we consider such pre-editing unrealistic. So, we did not apply any correction procedure to our input data, aiming at a more realistic black-box approach in the future[4]. More importantly, we already consider PR to be fully automated, as if we had just plain texts as input. A possible drawback of this is that miss-annotated data may contribute negatively to the PR performance, as we shall discuss below.

In assessing RAPM, we used the very same three corpora of distinct genres adopted in [3]: a law, a literary, and a newswire one. Although we aimed at BP, we chose to fully replicate Coelho's experiment by including the law corpus, which is composed of 16 Portuguese Attorney General documents (c.a. 110,610 words; 260 3rd person pronominal anaphors). Most texts convey long and complex sentences. The literary corpus consists of the whole book 'O alienista', by the Brazilian author Machado de Assis (c.a. 16,530 words; 573 3rd person pronominal anaphors). The newswire corpus is composed of 14 texts of the Veja magazine (c.a. 13,217 words; 222 3rd person anaphoric pronouns) and it conveys simpler sentences than the others.

We also fully used Coelho's setting in our assessment. The extra gain in doing so was that we could keep preprocessing apart from RAPM. We just used the data files formerly automatically annotated by Coelho as input to RAPM. Such input is produced in the following way: raw texts are parsed by PALAVRAS [2] and converted to XML by the Xtractor tool [5]. In adopting such a setting, we just reproduced Coelho's results on success rates to compare with RAPM.

The reference corpus is built by enriching the same XML files produced by PA-LAVRAS with hand annotations for co-reference. This task is supported by the MMAX tool [13]. Such annotation is illustrated in Fig. 1 for Example (1).

```
<markable id="markable_406" span="word_147..word_148" form="antecedent" />
<markable id="markable_229" span="word_160" form="pronome" pointer="markable_406" />

<pron chunk_id="chunk_229" gender="M" number="3S" reci="no" refl="no" span="word_160" />

<np chunk_id="chunk_212" gender="M" head_span="word_148" number="S"
    span="word_147..word_148" />
```

Fig. 1. Snapshot of a reference text hand-annotated for co-reference

RAPM assessment consisted of comparing the XML file of each text produced automatically (a sample given in Fig. 2) with the corresponding reference XML file. RAPM output file contains, for each anaphoric pronoun, its marked antecedent. In Fig. 2, 'Ele' corresponds to the former and 'O parlamentar', to the latter. This is identified as the antecedent of the pronoun by the tag *IdAntecedente* in Fig. 2. The link between such information in both files is given by the identifier *chunk_212*. Notice that, in the reference file (Fig. 1), such a component is conveyed with all its

---

[4] Reminding that RAPM does not present an integrated preprocessing module yet.

morphosyntactic and location information in the text. The assessment is entirely automatic in two situations, allowing for pinpointing a correct anaphor: (1) exact match between the automatic solution and the reference one (*chunk_212* in both files is an example); (2) coincidence between the nucleus of the antecedent noun phrase with the nucleus of the reference solution, or with a term that is embedded in that reference solution (which is also a noun phrase in turn). If none of these apply, manual assessment is carried out to find other correct PRs. The remaining cases are considered unresolved.

```
<RAPM>
  <IdAnafora>chunk_229</IdAnafora>  <Anafora>Ele</Anafora>
  <IdAntecedente>chunk_212</IdAntecedente>  <Antecedente>O parlamentar</Antecedente>
</RAPM>
```

Fig. 2. Snapshot of the file conveying Example (1) automatically annotated by RAPM

Aiming at a broad assessment, we derived several versions of RAPM by combining the antecedent indicators. Each version is identified by "RAPM_n", n signaling the amount of antecedent indicators considered. Overall, eight distinct versions were provided, as follows (IS stands for the Indicators Set considered):

- RAPM_2: IS = {INP, RD}
- RAPM_3: IS = {INP, PNP, RD}
- RAPM_4: IS = {INP, PNP, RD, NNP}
- RAPM_5: IS = {FNP, LR, INP, PNP, RD}
- RAPM_6_SP: IS = {FNP, LR, INP, PNP, RD, SP}
- RAPM_6_NNP: IS = {FNP, LR, INP, PNP, RD, NNP}
- RAPM_6_PN: IS = {FNP, LR, INP, PNP, RD, PN}
- RAPM_8: IS = {FNP, LR, INP, PNP, RD, SP, NNP, PN}

As shown, the antecedent indicators combinations vary in size. They were configured through corpus analysis by choosing those that seemed most promising for PR. The analysis consisted of comparing RAPM performance for isolated antecedent indicators with reference annotations. Success rate helped discriminating potential indicators – those with the highest scores yielded the first 3 versions of RAPM[5]. Differently from these and RAPM_8, which conveys all the indicators, RAPM_5 considers only those also managed by Mitkov. Each of the 3 RAPM_6 versions was built adding to the RAPM_5 set each new indicator we introduced for BP (SP, NNP, and PN), one at a time.

RAPM assessment was undertaken in three different ways: firstly, we measured the average success rate of each system depicted above, when running on the newswire corpus (Table 2). The strategy with the best success rate (RAPM_8) was then used in two other experiments: we compared its performance with the results by Coelho [3] (Table 3) and finally we used again RAPM_8 results on the newswire corpus, but to compare it with two distinct baselines, namely, 'Baseline-NP' and 'Baseline_Subj'

---

[5] Success rates respectively classified INP>NNP>RD>PNP (X>Y indicating that the rate of the antecedent indicator X is greater than the rate of antecedent indicator Y).

(Table 4). In this case, we used the same baselines as did Mitkov in [10]. Baseline-NP pinpoints as the antecedent the closest NP to the pronoun, provided that the NP agrees in gender and number with the pronoun. Baseline_Subj adds to the Baseline-NP a third constraint: the antecedent NP must occupy the subject position in the sentence it occurs. The results of each assessment follow, in a decreasing success rate order.

Table 2. Overall assessment

RAPM version	Success rate (%)
RAPM_8	67.01
RAPM_3	66.02
RAPM_6_NNP	64.94
RAPM_6_PN	63.40
RAPM_2	62.50
RAPM_5	61.45
RAPM_4	61.21
RAPM_6_SP	60.26

Table 3. Comparison between RAPM_8 and Coelho's av. success rates

Corpus	RAPM_8	Coelho [3]
Newswire	67.01	43.56
Literary	38.00	31.32
Law	54.00	35.15

Table 4. Comparison between RAPM_8 and baseline strategies

PR systems	Success rate (%)
RAPM_8	67.01
Baseline-NP	55.49
Baseline_Subj	42.27

Although RAPM_8 performed better in the overall assessment, its use may be questionable because the system that was classified the second, RAPM_3, presented a close success rate (66%) using much fewer antecedent indicators. This result suggests that using impeditive indicators, i.e., INP and PNP, may well help resolving pronouns in BP, when newswire texts are considered, and is less costly. Even RAPM_6_NNP, which reached the 3rd best av. success rate, also performs closely to RAPM_8, and demands less indicators. Comparing the three RAPM_6 versions, adding NNP to the original Mitkov's indicators seems to be the only one that may slightly improve the success rate. Still, it does improve on RAPM_5 in c.a. 4 percentual points. Comparing now the success rates of our eight systems with those by Coelho, ours were consistently superior regarding the three corpora. Besides, our worst case for the newswire corpus – RAPM_6_SP (60.26% sucess rate) – performed much better than Coelho's (43.56% sucess rate). If we consider only RAPM_8, the outperforming is even more expressive: for the same corpus, RAPM_8 scored an average of 67.01% success rate. The third comparison (Table 4) also confirms RAPM_8 improvements on PR. Such variations are quite sutile and must be further analysed in the future.

5 Final Remarks

The assessment described before needs further exploration in several ways: (a) a proper error analysis should be carried out focusing not only in PR itself but also in the correctedness of the input data, as previously stated. (b) Statistical significance must be considered, for any other assessment to be relevant. The main errors generated at the preprocessing stage included wrong morphological annotations of both NPs and pronouns. During pronoun resolution itself, an expressive problem was that the system signaled pronoun antecedents that were not NPs. RAPM was not tailored to deal with such cases, even when such antecedents were correct. So, it is very likely that overcoming such obstacles will improve RAPM performance.

Concerning the original approach by Mitkov, RAPM_8 adds novel indicators that should be better analyzed aiming at their actual contribution for BP. For example, when considered separately in the three systems RAPM_6, they do not suggest considerable improvements on RAPM_5, but when put together, they lead to the best success rate. So, we must evaluate their isolated contribution to RAPM_8. Regarding Mitkov's av. success rate (89.7%), we could crudely say that RAPM_8 still has a significant room for improvement. It is important to notice, though, that we should mirror his approach of feeding the system with correct input data. Also, verifying the adequacy of the scores assigned by each antecedent indicator should be pursued. However, this is not straightforward to accomplish, for it involves scaling up the linguistic analysis we carried out, which was entirely dependent upon human expertise. It also involves considering other corpus-based means to verify the indicators adequacy or other assessment tasks.

Having RAPM_8 as the best system for PR in BP does not entitle us to say that it will work well when other data are used. There are many other ways of exploring further the current results, including verifying which combination of the indicator-based heuristics would be more profitable. Aiming at this seems quite reasonable, since our approach is entirely empirical. However, it is not less complex: we could have too many combinations of features to investigate. So, considering other statistical methods to pinpoint a more reliable feature combination should be also applicable. Overall, RAPM may be useful for several NLP applications, including Automatic Summarization and Information Retrieval, which are the ones focused more closely in our research.

Acknowledgments

The authors are grateful to the Brazilian agencies CNPq, CAPES for their support.

References

1. Bergsma, S., Lin, D.: Bootstrapping Path-based Pronoun Resolution. In: COLING-ACL, pp. 33–40 (2006)
2. Bick, E.: The Parsing System PALAVRAS: Automatic Grammatical Analysis of Portuguese in a Constraint Grammar Framework. Ph.D. Thesis. Århus University, Århus (2000)

3. Coelho, T.T.: Resolução de Anáfora Pronominal em Português Utilizando o Algoritmo de Lappin e Leass. Master's Thesis. University of Campinas (2005)
4. Firbas, J.: Functional Sentence Perspective in Written and Spoken Communication. Cambridge University Press, Cambridge (1992)
5. Gasperin, C.V., Vieira, R., Goulart, R.R.V., Quaresma, P.: Extracting XML Chunks from Portuguese Corpora. In: Proceedings of the Workshop on Traitement Automaticque dês Langues Minoritaries, Batz-sur-Mer (2003)
6. Grosz, B.J., Joshi, A.K., Weinstein, S.: Centering: a Framework for Modeling the Local Coherence of Discourse. Computational Linguistics 21, 203–225 (1995)
7. Hobbs, J.R.: Resolving Pronoun References. Lingua 44, 311–338 (1978)
8. Lappin, S., Leass, H.J.: An Algorithm for Pronominal Anaphora Resolution. Computational Linguistics 20, 535–561 (1994)
9. Mitkov, R., Evans, R., Orasan, C., Ha, A.H.: Anaphora Resolution: to What Extend Does It Help NLP Applications? In: Branco, A. (ed.) DAARC 2007. LNCS (LNAI), vol. 4410, pp. 179–190. Springer, Heidelberg (2007)
10. Mitkov, R.: Anaphora Resolution. Longman (2002)
11. Mitkov, R.: Robust Pronoun Resolution with Limited Knowledge. In: COLING-ACL, pp. 869–875 (1998)
12. Mitkov, R.: Factors in Anaphora Resolution: They Are Not the Only Things that Matter. In: A Case Study Based on Two Different Approaches. ACL-EACL Workshop on Operational Factors in Practical, Robust Anaphora Resolution, pp. 14–21 (1997)
13. Müller, C., Strube, M.: MMAX: A Tool for the Annotation of Multi-modal Corpora. In: The 2nd IJCAI Workshop on Knowledge and Reasoning in Practical Dialogue Systems, pp. 45–50 (2001)
14. Palomar, M., Moreno, L., Peral, J., Muñoz, R., Fernández, A., Martinez-Barco, P., Saiz-Noeda, M.: An Algorithm for Anaphora Resolution in Spanish Texts. Computational Linguistics 27, 545–567 (2001)
15. Paraboni, I.: Uma Arquitetura para a Resolução de Referências Pronominais Possessivas no Processamento de Textos em Língua Portuguesa. Master's Thesis, PUC, Rio Grande do Sul (1997)
16. Refoufi, A.: A Modular Architecture for Anaphora Resolution. Journal of Computer Science 3(4), 199–203 (2007)
17. Santos, D.N.A., Carvalho, A.M.B.R.: Hobbs' Algorithm for Pronoun Resolution in Portuguese. In: Gelbukh, A., Kuri Morales, Á.F. (eds.) MICAI 2007. LNCS (LNAI), vol. 4827, pp. 966–974. Springer, Heidelberg (2007)
18. Ventura, C.S.M., Lima-Lopes, R.E.: O Tema: Caracterização e Realização em Português. DIRECT Papers 47, 1–18 (2002)
19. Wunsch, H.: Anaphora Resolution – What Helps in German (Extended abstract). In: Pre-Proceedings of the International Conference on Linguistic Evidence, Tübingen, Germany, pp. 2–4 (2006)

Semantic Similarity, Ontologies and the Portuguese Language: A Close Look at the subject[*]

Juliano Baldez de Freitas, Vera Lúcia Strube de Lima,
and Josiane Fontoura dos Anjos Brandolt[†]

Programa de Pós-Graduação em Ciência da Computação - PPGCC
Avenida Ipiranga, 6681 - Prédio 32 - Partenon
CEP 90619-900 Porto Alegre - RS - Brasil
juliano.bfreitas@gmail.com,
{vera.strube,josiane.brandolt}@pucrs.br

Abstract. Semantic similarity and mapping between ontologies are a crucial subject, which is just starting to be researched for ontologies written in Portuguese. Our study begins with SiSe (*Similaridade Semântica*) measure, an extension for the Taxonomic Overlap proposed by Maedche and Staab [1], which compares the similarity between terms of distinct ontologies through the analysis of the hierarchies where they are placed. SiSe development and evaluation, even bringing some interesting conclusions, point to continuing efforts, what is discussed here in the context of more recent proposals presented in the ontology mapping domain.

Keywords: Ontologies, semantic similarity, mapping, natural language processing.

1 Introduction

Ontologies[1] play an important role for applications that involve engineering and representation of knowledge, as well as for applications that deal with precisely defined terms. However, with the generalized use of ontologies, some practical problems arise, related mainly to interoperability. For example, users and engineers of ontologies frequently have a main ontology they use for navigating or consulting data, though they need to extend, adapt or compare their ontologies with the vast set of other existing ones [1]. The research for similarity between ontologies refers to the comparing of whole ontologies or their subelements. To avoid semantic inconsistencies obtained while integrating and reusing information between ontologies built individually, the analysis of similarity between these ontologies and their elements

---

[*] This work was supported in part by CAPES under the FAROL Project, #0035050, and in part by CNPq under the projects PLN-BR #550388/2005-2 and PONTO # 490752/2006-3.
[†] This author is financially supported by CAPES.
[1] The term 'ontology' is used here in a broader sense standing for ontological structures like vocabularies, thesauri, lexical databases and ontologies themselves. The terms 'class', 'subclass' (more specific) and 'superclass' (more general) should also be contextualized in the sense of a hierarchy of items.

A. Teixeira et al. (Eds.): PROPOR 2008, LNAI 5190, pp. 61–70, 2008.
© Springer-Verlag Berlin Heidelberg 2008

uses measures that compare elements or groups of elements among these structures and, identify similarities between the same, avoiding inconsistencies.

The current work reports a study directed to the Portuguese language, in which we adapt methods of analysis of similarity between ontologies and their elements in a semi-automatic and automatic manner, adjusting a measure of semantic similarity between ontologies existing in the literature [1], adding a simple Natural Language Processing (NLP) resource and a heuristic to it, for lowering semantic inconsistencies. In this article, we analyze the obtained results and discuss the continuity of research in this area, with the objective of providing a measure that involves advances in terms of intrinsic and extrinsic similarity between ontologies.

This article is organized in five sections. Section 2 presents the concepts regarding similarity between ontologies and a few considerations about the challenges of the area. Section 3 describes the proposal of the semantic similarity (SiSe) measure, also presenting an example of the SiSe measure, the method used for its evaluation, and the obtained results. In Section 4, we present some new approaches concerning mapping between ontologies. Section 5 presents considerations on SiSe evolution directed to a measure that articulately subsumes the concepts of intrinsic and extrinsic similarity between ontologies.

2 Problems on Mapping between Ontologies

Many authors have their own definitions regarding the types of correspondence related to similarity between ontologies, knowingly: alignment, mapping and matching, among others. Ehrig, in [2], while considering the subject, alerts for those still disjointed positions.

These three processes – alignment, mapping and matching – have something in common: finding matches between elements of different ontologies. Alignment and mapping are considered synonyms (if we compare the Tous and Delgado [3] with the Kalfoglou and Schorlemer [4] descriptions). These questions on similarity are also reported by Ehrig in [2] and [5]. Regarding the matching process, we believe that the most adequate way of using it is as a tool for finding alignment/mapping, as defined by Euzenat and Shvaiko [6] and Castano and co-authors [8].

On mapping between individually built ontologies, terms with the same meaning might not be considered similar by the most usual similarity measures. That is due to two important factors: (*i*) Natural Language – as it has already been observed by Maedche and Staab, known real ontologies do not specify their conceptualizations only by logical structures, but through a natural language founded reference of terms; (*ii*) Taxonomical structure: ontologies may also contain bad matches at the structural model level as, for instance, on distinct taxonomies [8], as shown in Table 1. The taxonomies in Table 1 are different, even if they have many terms in common. Also, although the terms *partido político* and *partidos políticos* are represented by different strings of characters, they refer to the same entity (information on plural being discarded).

According to Noy [8], to allow automatic or semi-automatic mapping the tools must analyze the following characteristics on the definition of the ontologies: names of concepts and descriptions in natural language; class hierarchy (superclass/subclass relations); properties definitions (domain, coverage, restrictions); class instance; class descriptions. The construction of tools and techniques that allow automatic or semi-automatic mapping between ontologies is a research area currently receiving attention. Concerning the Portuguese language, the efforts are still very few [13, 9].

Approaches are found in the literature to measure similarity between parts or elements of ontologies, eminently focused on English and German. Until recently, we classified these approaches within two groups: lexical similarity and semantic similarity. The work of Andreas Hess [10] brought another look to the subject, classifying such measures as intrinsic and extrinsic, as it will be shown latter on.

Lexical similarity may be characterized as the approach that measures similarity of elements, trough the words that constitute them (their strings of characters). In this approach, solutions that measure similarity between chains of characters are normally used, some of them with heuristics. The measure is normally given through interval coefficients [0,1], which calculate the proximity of elements, structurally and lexically [9]. The semantic (or semantic-structural) similarity compares elements of ontologies through their meanings, searching for synonymy and other semantic relations between these elements. This measure compares elements according to the position of the same in the hierarchical structure, searching for the existing semantic relations [9]. Other than the hierarchical relations, properties and instances may also be analyzed in search of semantic relations and, thus, improve the performance of similarity measure.

Recent works can be found in the literature that approach lexical and semantic similarity in a deeper way, focusing the so called intrinsic and extrinsic similarities between ontologies. These terms congregate the same principles adopted in the lexical and the semantic processing approaches, however, they make use of more elaborated and deepened similarity techniques. Intrinsic similarity refers to inherent characteristics of ontology elements (such as textual presentation), and extrinsic similarity refers to relations existing between different elements of a same or of different ontologies (such as types of relationships among terms, constituents etc   [10]). These characteristics allow the isolated treatment of elements (going beyond the lexical scope and using features expressed through annotation patterns with RDF descriptions), passing normalization and editing distance to arrive at the relationship between elements through similarity calculation.

In order to measure the semantic similarity between ontologies in Portuguese we proposed the approach named SiSe (*Similaridade Semântica*), described in details in [9]. SiSe is an adaptation of the Taxonomic Overlap (TO) presented by Maedche e Staab [1]. SiSe adds Portuguese stemming to TO in the SC and CSC collected terms. This showed to be rewarding when results were compared with those obtained by the original TO measure, especially in the cases of inflections. The SiSe measure is described in the following section.

3   SiSe Measure

The most natural approach to be adapted to measure semantic similarity between ontologies in Portuguese was TO, which does not use additional resources of language processing for the similarity calculation. To define the SiSe measure we did some adaptations on the TO measure on what concerns the concepts of Semantic

Cotopy (SC) [1] and Common Semantic Cotopy (CSC) [11]. These adaptations use a stemming algorithm, and have been named SC' and CSC'. The use of stemming gives priority to the lexical level of knowledge with the intention of finding lexically similar terms to follow the semantic-structural comparison. The use of stemming helps finding similarity when words present lexical variations, such as *eleições* (elections) and *eleição* (election), but the same stem. However, the use of stemming is not appropriate to compare words when they present different roots or stems (example: for the synonyms *voto* (vote) and *sufrágio* (suffrage) stemming will not bring similar stems).

Table 1. Excerpts of Law domain hierarchies extracted from two distinct ontologies (terms and their respective stems)

Ontology 1 (O$_1$)	Ontology 2 (O$_2$)

1 direito constitucional (direitConstituc)
  - *constitutional Law*
1.1 direito eleitoral (direitEleitor)
  - *election Law*
1.1.1 campanha eleitoral (campanhEleitor)
  - *election campaign*
1.1.2 eleição (ele)
  - *election*
1.1.3 partido político (partPolitic)
  - *political party*
1.1.4 sistema eleitoral (sistemEleitor)
  - *election system*
1.1.5 voto (vot)
  - *vote*
	1 direito (direit)
	  − *Law*
	1.1 direito eleitoral (direitEleitor)
	  - *election law*
	1.1.1 crime eleitoral (crimeEleitor)
	  - *election crime*
	1.1.2 domicílio eleitoral (domiciliEleitor)
	  - *election precinct*
	1.1.3 eleições (ele)
	  - *elections*
	1.1.4 justiça eleitoral (justicEleitor)
	  - *electoral justice*
	1.1.5 partidos políticos (partPolitic)
	  - *political party*
	1.1.6 sistema distrital (sistemDistrit)
	  - *district system*
	1.1.7 voto (vot)
	  - *vote*

The TO measure compares ontology hierarchy, and the same happens with SiSe. Ontologies created by different specialists may differ on the hierarchical representation for a single concept, each specialist having a different vision over a certain domain, and those differences are visible through the construction of distinct hierarchies. This fact makes the TO measure, which is based on superconcepts and subconcepts of terms, able to hide or reinforce some similarities between ontology terms. Terms that are semantically similar may be arranged in hierarchy in such a way that their superconcepts and subconcepts are different, making the TO measure return a low similarity coefficient. This hierarchical similarity may reinforce, for instance, the existence of terms with different lexical representations, but that, however, possess similar superconcepts and subconcepts. This might indicate that these terms are semantically similar.

Following we present the SiSe measure and the CSC' (adapted from CSC) for hierarchies comparison between two terms of distinct ontologies. The SC' approach will not be exemplified in this article: its details can be found in [9], and its results are analyzed on Figure 2.

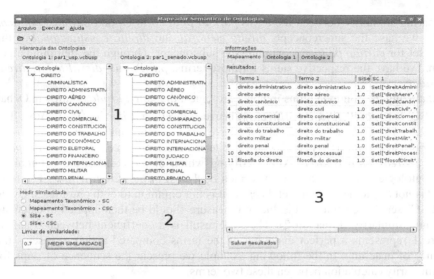

Fig. 1. SiSe user interface for similarity calculation and result collecting

Figure 1 presents the SiSe graphic user interface, which was developed for similarity calculation between ontologies and for collecting the results of this process. SiSe prototype was implemented in Python (www.python.org) and it offers the user two different semantic similarity measures: TO and SiSe (number 2 indicated in Figure 1), being able to compare terms in two different ontologies (number 1 indicated in Figure 1). The results produced by each approach as well as the sets of terms being compared are also presented in the interface (number 3 indicated in Figure 1).

We modified the original definition of CSC [11], what gave origin to CSC' used in SiSe, presented accordingly to Equations 1 and 2.

$$CSC'(c_i, O_1, O_2) = \{\Delta c_j \in \Delta c_1 \cap \Delta c_2 \mid c_i \leq_{c_1} c_j \text{ ou } c_j \leq_{c_1} c_i\} \tag{1}$$

$$SiSe(c_1, O_1, c_2, O_2) = \frac{CSC'(c_1, O_1, O_2) \cap CSC'(c_2, O_1, O_2)}{CSC'(c_1, O_1, O_2) \cup CSC'(c_2, O_1, O_2)} \in [1, 0] \tag{2}$$

On Equation 1, the symbol Δ represents the stem of the c_j term in question. The CSC' set associated to a term is formed based on the subconcepts and superconcepts of that term that are common to both ontologies. These common terms are represented through their stems Δc_1 and Δc_2 , forming a $\Delta c_1 \cap \Delta c_2$ set.

This way, the stem of a subconcept or superconcept c_i will be part of the CSC' set if the same appears in both hierarchies. The sets of terms of each ontology in CSC' are compared through Jaccard measure, according to Equation 2. As an example of similarity calculation we use the ontological structures represented in the two hierarchies (O_1 and O_2) in Table 1. Terms in each column in Table 1 are numbered and represented by their complete character string followed by their stem.

The set of common terms (represented by their stems) for these ontological structures, given by $\Delta c_1 \cap \Delta c_2$, is {*partPolitic, vot, direitEleitor, ele*}. For example, when comparing similarity between the terms *partido político* in O_1 and *partidos políticos* in O_2, we have the superconcepts *direito eleitoral* and *direito constitucional* as the CSC' set for *partido político*, containing no subconcept. Through the stemming algorithms we have the respective stems for those terms: *partPolitic, direitEleitor, direit-Constituc*.

Analyzing the common terms in both ontologies, the term *partido politico* in O_1 has the CSC' set {*partPolitic, direitEleitor*}, because *direitConstituc* $\notin \Delta_{C_1} \cap \Delta_{C_2}$. The term *partidos políticos* in O_2 also do not have subconcepts, and has the terms *direito eleitoral* and *direito* as superconcepts. The terms are represented by the stems *partPolitic, direitEleitor* e *direit*. After verifying which of these stems are common to both ontologies we come to the set formed by {*partPolitic, direitEleitor*}. The term *direit* is not an element of this set, because it is not common to both ontologies. Given the CSC' sets for each term, the Jaccard measure should be then applied to calculate semantic similarity (Equation 2). The final result is a coefficient between 0 and 1, where 1 represents a perfect matching of the terms compared and 0 represents the absence of matching. In the following example we present the sequence of steps for the similarity calculation between these two terms.

$$SiSe\ (partido\ político, O_1,\ partidos\ polítɪcos,\ O_2) =$$

$$\frac{CSC'\ (partido\ político, O_1, O_2)\ \cap CSC'\ (partidos\ políticos, O_1, O_2)}{CSC'\ (partido\ político, O_1, O_2)\ \cup CSC'\ (partidos\ políticos, O_1, O_2)}$$

$$= \frac{|\{direitEleitor, partPolitic\}|}{|\{direitEleitor, partPolitic\}|} = \frac{2}{2} \in [0,1]$$

The SiSe proposal has been evaluated according to the following methodology. Evaluation was done with the use of the SiSe user interface for pairs of terms present in two excerpts of ontological structures. Five pairs of extracts were selected from this excerpt, showing hierarchies of terms (details and complete hierarchies of terms in [9]). For each pair of extracts, TO (SC), TO (CSC), SiSe (SC') and SiSe (CSC') were calculated by the interface.

The evaluation of the similarity measures obtained was done against a Golden Mapping (GM), which was previously constructed with help of 3 human evaluators, in a four step methodology. First, each of the human evaluators received a document containing the hierarchical structure of the extracts and a mapping table for each pair of ontology hierarchies (excerpts). The evaluators were chosen from different areas of expertise: Law, Literature and Computing Sciences. Second, each human evaluator indicated, on the mapping table, the terms he considered similar, so filling the mapping table for each pair of ontology excerpts. Third, after this individual evaluation was concluded, we carried on an analysis in order to reach a consensus for the mappings. This final consensus was achieved according to the following rules: *the GM considers the mappings between pairs of terms that have been identified by at least two out of three humans. The mappings signaled by the Law graduate are always considered, independently of the other analysis, due to his specific knowledge of the domain that might allow him find mappings that the automatic measures and the other human evaluators might not detect.* By the end of this process, fourth step, we created a mapping reference between the extracts. The mapping generated by similarity measures was than confronted to this reference.

We got then to the results presented on Figure 2. Comparing the results obtained with the TO and SiSe measures, to the GM column, we observe that the measure that generally found most mappings for each pair of extracts was the SiSe measure using CSC'. For example, this measure detected 100% of the GM for the Pair 4 with SiSe (CSC').

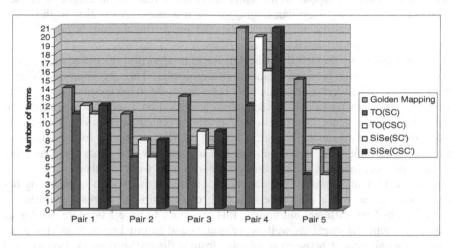

Fig. 2. Comparative graphic of mappings between the terms of two distinct hierarchies using SiSe and TO

The number of mappings didn't vary much between the measures, as was the case for Pairs 1, 2 and 3. We also noticed that, when the hierarchies had terms with minor lexical differences (for example, *eleições* (elections) and *eleição* (election) in Pair 4), the approaches of the SiSe measure that used stemming (SC' and CSC') obtained the best results. CSC and CSC' find a bigger number of mappings than SC and SC' when the hierarchies have different levels. However, they also find an elevated number of false positives (terms that are not semantically similar according to the GM). For example, 10 false positives were found on Pair 4 using CSC, and 26 false positives using CSC'. This analysis should take into account that the statistical relevance for these results has not yet been tested.

4   Other Approaches on Mapping between Ontologies

In this section we describe some approaches that deepen and combine processes in search of similarity between ontologies. The Tous and Delgado measure, described in [3], uses the Vector Space Model (VSM) for alignment between ontologies. This measure proposes that, in order to calculate the alignment between two ontologies, it is necessary to adapt the mapping based on an algorithm proposed by Blondel[2] and his co-authors [12]. VSM is an algebraic model proposed by Salton that allows to

---

[2] The algorithm introduces the similarity concept between directed graphs, defining a similarity matrix.

describe and compare objects represented by vectors in n-dimensions, where each dimension corresponds to an orthogonal characteristic of the object. Tous and Delgado show the importance of some semantic similarity measure and report that most alignment algorithms between ontologies are focused only on finding close elements. The measure proposed by them tries to find evidence to deduce that two different data items correspond to the same information (data items might be classes and properties of the ontology, but might also be instances). According to those authors, the generated results are satisfactory, and prove the measure's adequacy in situations where structure-based similarity exists.

Andreas Hess study in [10] describes an iterative algorithm to calculate intrinsic and extrinsic similarity between elements of ontologies. This algorithm establishes metrics of distance between strings that have been discussed in the literature and applied in structural similarity measures. These metrics are based on a vector representation of relations between elements, that may be used directly to calculate a similarity value. On Hess's proposal a third ontology is used, serving as training data to increase the performance of mapping. That author believes that a great potential exists for combining his ideas with other methods, setting different algorithms to work together.

Euzenat and Shvaiko studies (presented on [6]) refer to the process of matching between ontologies as being the process to find relationships or correspondences between elements from different ontologies. It is necessary to have 2 or more ontologies and a pre-existent alignment, as well as parameters and external resources. This process returns an alignment between ontologies from 3 dimensions: *entry*, *process* and *exit*. *Entry* depends on data and conceptual models in which the ontologies are expressed (for example: relational, object oriented, entity-relationship, XML and RDF models, etc). *Process* dimension is organized in 3 classes: syntactic, semantic and external. *Exit* refers to the way of producing the system's results. From these dimensions we can classify the matching techniques by their characteristics: (a) entry interpretation and granularity – based on matching granularity, for example: at the element or structure level; (b) Type of entry – this classification is based on the technique used on elementary matching. This approach seems to be didactical and practical, but it does not seem to bring new solutions to the problem.

To Castano and his co-authors in [7], strategies adopted for combining different similarity techniques can be summarized in two main categories: the first based on combining different measures and the second based on iteration of the matching process. In the case of combination of different similarity measures, each measure can be calculated independently, while the results are analyzed together. Other approaches are possible for measure combination, especially with the use of machine learning techniques. An example is APFEL (*Alignment Process Feature Estimation and Learning*), a machine learning approach that explores the validation of initial alignments, done by users, to optimize alignment methods. Combination of the similarity techniques by means of the iteration process is based on the idea that each similarity measure is founded on some basic type of similarity between the elements to be compared. As an example we can consider two graphs where similarity is calculated, taking into account the context of the nodes. A different approach is based on the idea of calculating similarities associated with different levels of semantic complexity in many stages. After each stage, the results are stored and registered for similarity evaluation in the subsequent task. This way, the process acts refining the results, step-by-step, with the advantage of being scalable for the level of semantic complexity required. After the measure composing, we have a set of results that contains pairs of concepts with a similarity value. From there on, a threshold must be established to classify the similarities found.

The studies in [3, 10, 6, 7] conduct to some general considerations. The more simple similarity measures like those based on the lexical characteristics of terms are being combined with measures based on the relationships among terms (revealed by the hierarchies, but also seen on the relationship graphs and specialized descriptions of concepts). In order to capture this wider evidence of similarity, the antecedents and the consequents of an item, together with the lexical presentation of this item, are barely sufficient. We should be able to deal with external resources (such as extra vocabulary sources) that could confirm, or suggest, connections between terms. And we should enrich these with relevance information provided by other techniques such as weighting, for example.

5 Conclusions

We believe that SiSe measure was a first effort toward the semantic similarity between ontological structures in Portuguese. Considering the studies presented on the previous section, the next step would be to incorporate other strategies to SiSe, adding to the prototype 2 groups of techniques: access to an external vocabulary and, association of weights to the different types of relationships among terms. As an external source of vocabulary, the synsets to be provided by the Wordnet.BR effort seem to be very promising. Other alternative already being studied is the access to an external search engine in order to create such sets. The association of weights with the different relations is independent from the external data techniques, and is already being considered. For instance, generalization and specialization, two relations that are taken into account to form the sets used by SiSe, could have different weights, leading to the insertion of a weight factor to adjust the SiSe similarity calculation.

Tous and Delgado [3] used VSM techniques, as well as ontologies composed of classes, properties and instances for semantic similarity calculation. Adapting this approach to Portuguese reminds us of the lack of ontologies built and represented on the formats selected by those authors. The use of a greater number of relations between the terms of ontologies, as well as properties and instances, add a semantic richness to similarity measures. However, ontologies available in Portuguese usually still have only simple hierarchy relations (superconcepts and subconcepts, so "is-one" or "part-of" relations), inhibiting the complete usage of this approach.

Andreas Hess [10] uses known ontologies and employs them in a specific methodology, as training data, before the similarity calculation. The ontologies used by Hess, in their turn, are described in RDF format. Again, it becomes impracticable at this time to apply this proposal to Portuguese, due to the lack of a repository of ontologies described on Semantic Web formats (RDF and OWL), making training data unavailable. The approaches by Euzenat and Shvaiko [6] and Castano and his co-authors [7] also use ontologies with complex relations for similarity calculation bringing to the same problem of the lack of data.

We believe that the limitations caused by this lack of more throughout and complex ontologies can be diminished, in a first moment and as already proposed by Castano, with an iterative and semi-automatic process, fed with the results of human intervention. Yet other steps can be taken, with the refining of similarity equations applied to Portuguese. However, as the ontology alignment or mapping techniques evolve, we face the urgency for the construction, or acquisition by methods of translation, of formally described ontologies, rich in semantic relations and throughout descriptions for the Portuguese language, which will be soon need for the Semantic Web applications.

References

1. Maedche, A., Staab, S.: Measuring Similarity between Ontologies. In: Gómez-Pérez, A., Benjamins, V.R. (eds.) EKAW 2002. LNCS (LNAI), vol. 2473, pp. 251–263. Springer, Heidelberg (2002)
2. Ehrig, M.: Ontology Alignment: Bridging the Semantic Gap, p. 247. Springer, New York (2007)
3. Tous, R., Delgado, J.: A Vector Space Model for Semantic Similarity Calculation and OWL Ontology Alignment. In: Bressan, S., Küng, J., Wagner, R. (eds.) DEXA 2006. LNCS, vol. 4080, pp. 307–315. Springer, Heidelberg (2006)
4. Kalfoglou, Y., Schorlemmer, M.: Ontology mapping: the state of the art. Cambridge Journals: The Knowledge Engineering Review 18, 1–31
5. Ehrig, M., Staab, M.: QOM - Quick Ontology Mapping. In: McIlraith, S.A., Plexousakis, D., van Harmelen, F. (eds.) ISWC 2004. LNCS, vol. 3298, pp. 683–697. Springer, Heidelberg (2004)
6. Euzenat, J., Shvaiko, P.: Ontology Matching, p. 333. Springer, Berlin (2007)
7. Castano, S., et al.: State of the Art on Ontology Coordination and Matching. BOEMIE: Bootstrapping Ontology Evolution with Multimedia Information Extraction 1, 1–5 (2007)
8. Noy, N.F.: Semantic integration: a survey of ontology-based approaches. SIGMOD Record 33(4), 65–70 (2004)
9. Freitas, J.B.: SiSe: Medida de Similaridade semântica entre ontologias em português. Dissertação de Mestrado. Programa de Pós-Graduação em Ciência da Computação, PUCRS (2007)
10. Hess, A.: An Iterative Algorithm for Ontology Mapping Capable of Using Training Data. In: Sure, Y., Domingue, J. (eds.) ESWC 2006. LNCS, vol. 4011. Springer, Heidelberg (2006)
11. Cimiano, P., et al.: Learning concept hierarchies from text corpora using formal concept analysis. Journal of Artificial Intelligence Research - JAIR 24, 263–303 (2005)
12. Blondel, V.D., et al.: A measure of similarity between graph vertices. Applications to synonym extraction and web searching. SIAM Rev. 45(4), 647–666 (2004)
13. Chaves, M.S.: Mapeamento e comparação de similaridade entre estruturas ontológicas. Dissertação de Mestrado, Programa de Pós-Graduação em Ciência da Computação, PUCRS (2003)

Boundary Refining Aiming at Speech Synthesis Applications*

Monique V. Nicodem, Sandra G. Kafka, Rui Seara Jr., and Rui Seara

LINSE – Circuits and Signal Processing Laboratory
Department of Electrical Engineering
Federal University of Santa Catarina, Brazil
{monique,kafka,ruijr,seara}@linse.ufsc.br
http://www.linse.ufsc.br

Abstract. In concatenative synthesis, speech is produced by joining segments automatically selected among units contained in a previously segmented database. The synthetic speech resulting from such a technique is often improved when accurate segmentation tools are considered. The performance of these tools is often enhanced by a hybrid approach resulting from the association of an HMM modeling with a boundary refining process. Such a refining has been carried out sucessfully by using techniques based on neural networks. This paper presents a set of networks that outperform other topologies discussed in the literature. These networks are trained by performing a clusterization of the training set taking into consideration phonetic transitions with similarities to each other.

Keywords: Concatenative speech, boundary refining, neural networks.

1 Introduction and Problem Statement

Most of the state-of-the-art speech synthesis systems are capable of converting any written text into speech. Such a conversion has been carried out by considering several techniques presented in the open literature. Among these techniques, the concatenative one has been highlighted by its ability to produce synthetic speech very close to the human speech [1].

The first stage towards generating concatenative speech consists in designing a reference speech database (an offline stage). In this case, a considerable amount of predefined sentences is firstly recorded by a professional speaker. Next, resulting utterances are subject to a segmentation process responsible for locating smaller acoustic units (usually phonemes) existing within the speech corpus. During the synthesis itself, speech is produced by joining units automatically selected from the previously designed database [2,3]. Anyway, even for the concatenative

---

* This work was partially supported by the Brazilian National Council for Scientific and Technological Development (CNPq), Studies and Projects Funding Body (FINEP), and Dígitro Tecnologia Ltda.

A. Teixeira et al. (Eds.): PROPOR 2008, LNAI 5190, pp. 71–80, 2008.

approach, the naturalness of the ultimate synthetic speech has been constrained by factors such as variability of the corpus chosen for recording, speaker's voice characteristics, and accuracy of segmentation methods. The latter factor may harm synthetic speech quality since an inaccurate segmentation process (i.e., providing imprecise boundary locations) may lead to deletions, insertions, or even substitutions of certain phonemes into the synthetic speech.

As a result, much research has been focused on methods capable of improving segmentation accuracy [4]. It is important to mention that manual methods are impractical to partition large speech databases since a considerable time would be required to manually segment them. Another problem would be a potential loss of consistency in manual segmentation, especially if the time required for the whole process were extremely large.

Nowadays, an automatic procedure often adopted for segmentation purposes consists in aligning the phonetic transcription of a given sentence with the corresponding speech signal (forced alignment). In this case, such an alignment is usually performed with the aid of speech recognition techniques. Therefore, a set of parameters extracted from the speech signal are taken as a reference to represent each acoustic unit by a hidden Markov model (HMM) [5]. In the following, the Viterbi algorithm [6] is employed to align and define segmental boundaries.

HMM-based systems may have even better segmentation results when they are associated with boundary refining processes. Some approaches to perform this refinement have been presented in the open literature [4, 7]. Among them, those based on artificial neural networks (ANNs) have led to satisfying results. For such, parameters extracted from manually segmented corpora are taken as a reference to train a set of neural networks. After training, these networks must point out (by considering a single neuron in the output layer) the probability of existing phonetic boundaries within analyzed frames.

In [7], a single ANN is trained to determine the required probability, being this ANN responsible for modeling all phonetic transition patterns (diphones). Another network configuration outperforming the latter is also proposed by [7]. In this case, four networks indicate the boundary existence probability, and each ANN models one of the following diphone classes: voiced/unvoiced, voiced/voiced, unvoiced/voiced, and unvoiced/unvoiced. Results achieved by the latter technique demonstrate that performance may be improved (for each trained network) by properly clustering diphones before training a set of ANNs. In [4], similar diphones are automatically separated into clusters. In this case, firstly, four ANNs are trained by randomly distributing the training set. For each diphone, it is verified which trained ANN leads to the lowest error in boundary probability estimation. After that, analyzed diphones are reclassified to the ANN providing lowest estimation error. Next, those initial four networks are retrained. Such a procedure is iteratively repeated until the ultimate ANNs reach an error variation smaller than a predefined threshold. One disadvantage of the technique proposed by [4] consists in defining how diphones should be partitioned into the four ANNs during the first algorithm iteration. Such an initialization aspect is not discussed in [4]. In addition, experimental results confirm that the final performance depends on initialization conditions.

In this paper, aiming to obtain both a better performance than [7] as well as an autonomy between performance and initialization conditions, the following procedure is proposed: the partition of diphones into clusters is carried out through a visual inspection of spectrograms whose phonetic transitions are similar to each other. In this case, the training set is separated into 36 subsets and 36 ANNs are trained. Results obtained by this ANN configuration outperform others presented in the open literature (here implemented as well).

It is important to consider that segmentation accuracy is often evaluated by computing the rate of phonetic boundaries whose segmentation errors are lower than 20 ms. This threshold value is considered here since segmentation errors lower than 20 ms are perceptually inaudible [4]. Even though the proposed approach outperforms others presented in the open literature, our focus is on determining an appropriate size of the training set, aspect that up to our knowledge is not discussed in previous papers considering boundary refining based on ANNs. Such a training size definition avoids that an excessive and unnecessary amount of sentences is manually segmented. Thereby, the time required for manual segmentation (a long and exhaustive process) is considerably decreased. Our experiments confirm that the segmentation performance may be improved by increasing the training set. However, after a certain amount of sentences is reached, there occurs a saturation in the obtained performance.

2 Segmentation Based on Hidden Markov Models

The first stage required for segmentation consists in obtaining the phonetic transcriptions of previously recorded sentences. Since our focus is on segmentation, canonical transcriptions are firstly obtained. After that, these transcriptions (of phonemes and pauses) are manually corrected by a phonetician. Phonetic transcriptions are considered both to build HMM models and carry out forced alignments with the corresponding speech signals. Depending on the existence or not of a manually segmented corpus, two distinct methodologies may be considered to effect model training. When no information is provided about manual segmentation, models are trained by using the Baulm-Welch [8] reestimation algorithm provided by HTK [9] (i.e., the model is obtained by a simple likelihood maximization). Otherwise, an HMM model that properly represents manual segmentation is obtained through a Viterbi algorithm (by using the HInit function from [9]). In the latter case, one has verified a higher segmentation performance [4].

3 Boundary Refining

After HMM-based segmentation, boundaries are refined by ANNs. Thus, a set of speech parameters are taken as ANN inputs aiming at network training. In this paper, one considers the following 56 parameters: 13 mel frequency cepstrum coefficients (MFCCs) extracted from four consecutive frames, zero crossing rates

(ZCRs) from the first and fourth frames, one spectral feature transition rate (SFTR) [10], and the symmetrical Kullback-Leibler distance (SKLD) [11].

For SFTR computation, a vector $\mathbf{x}(n) = [x_1(n) \quad x_2(n) \quad \ldots \quad x_l(n)]^{\mathrm{T}}$ of MFCCs is extracted from the nth frame, where $x_i(n)$ represents the ith MFCC, and L the total number of coefficients per frame. The SFTR measure is given by

$$s(n) = \sum_{i=1}^{L} \left[\frac{\displaystyle\sum_{m=-M}^{M} m x_i(n+m)}{\displaystyle\sum_{m=-M}^{M} m^2} \right]^2 . \tag{1}$$

Here one uses an M-value equal to 2.

SKLD distance is expressed as

$$D_{\mathrm{SKL}}(n) = \int_{0}^{\pi} [P_n(\omega) - P_{n+1}(\omega)] \log \left[\frac{P_n(\omega)}{P_{n+1}(\omega)} \right] d\omega \tag{2}$$

where $P_n(\omega)$ and $P_{n+1}(\omega)$ represent the spectral envelopes (Fourier transforms) of two consecutive central frames (n and $n+1$, respectively) taken from a set of four frames. These input parameters are similar to those presented in [4].

One output value is associated with every set of 56 input parameters. Allowable output values are 0, 0.5, and 1. A null value indicates the inexistence of phonetic boundaries within the four analysis frames. A 0.5-value indicates the existence of a phonetic boundary between the third and the last frames or between the first and second frames. Finally, a 1-value points out the existence of a transition between the second and third frames [4]. This procedure adopted to define the ANN output value is illustrated in Fig. 1.

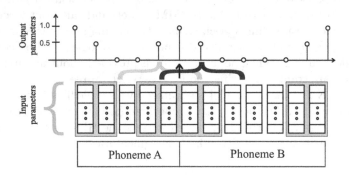

Fig. 1. Procedure adopted to define the ANN output value [4]

After ANN training, any set of 56 parameters extracted from four consecutive speech frames may be taken as input for the previously trained network. Thus, one corresponding output (score) is obtained for each set. The location providing the highest output score within an analyzed search interval is considered as the refined phonetic boundary. In this case, the search is performed in an interval located between the middle of two succeeding phonemes.

4   Context-Dependent Boundary Refining

It is well known that acoustic parameters of phonetic segments are influenced by their neighbor contexts. Because of that, more accurate boundary estimations may be achieved when phonetic information is provided during the refining process. Thus, a context-dependent boundary refining (i.e., a specialized network is trained for each diphone pattern) may be adopted to improve the performance in the segmentation process. Nevertheless, there exist discussions about which transitions should be considered in each specialized network. In [7] and [4], some strategies are presented to classify diphone transitions into a set of similar patterns. However, experimental results (shown in this paper) indicate that these methods present some performance deficiencies.

Thus, acoustic space partition is here carried out by considering a visual inspection of speech spectrograms. In this way, those phonetic transitions detected as similar to each other are grouped into a single network. For such, one assumes that the human visual system recognizes these patterns better than other automatic classifiers, such as [4]. Table 1 presents all phoneme classes (and labels) considered to define the transition patterns used in this paper [12]. Our focus is on Brazilian Portuguese (BP) phonemes.

Table 1. Phoneme classes

Labels	Phonemes	Labels	Phonemes
AFR	Africates	LTA	Laterals (alveolar)
AFV	Africates (voiced)	LTP	Laterals (palatal)
AFU	Africates (unvoiced)	PLO	Plosives
CN	Consonants (nasal)	PLV	Plosives (voiced)
FRI	Fricatives	PLU	Plosives (unvoiced)
FRA	Fricatives (alveolar/palatal)	ROT	Rothics
FRL	Fricatives (labiodental)	VS	Vowels/semivowels
FRC	Fricatives (unvoiced coda)	VSN	Vowels/semivowels (nasals)
FRV	Fricatives (velar)	VSO	Vowels/semivowels (orals)
LAT	Laterals		

Table 2 shows the transition classes considered here. The first one, for example, represents the transition between oral vowels/semivowels and voiced plosive consonants. This rule indicates that speech frames located in the interval between the middle of oral vowels/semivowels and voiced plosives are adopted to train a network specific for this transition pattern. Thereby, a total of 36 networks are trained. It is important to mention that the open literature already presents a

Table 2. Phonetic transition classes

1st phone	2nd phone	1st phone	2nd phone	1st phone	2nd phone
VSO	PLV	VSN	FRI	FRC	AFR
VSO	PLU	VSN	CN	PLV	VS
VSO	FRL	VSN	ROT	PLU	VS
VSO	FRA	VSN	FRV	FRL	VS
VSO	CN	VSN	AFR	FRA	VS
VSO	LAT	ROT	PLO	CN	VS
VSO	ROT	ROT	FRI	LTA	VS
VSO	FRV	ROT	CN	LTP	VS
VSO	AFU	ROT	AFR	FRV	VS
VSO	AFV	FRC	PLO	ROT	VS
VSN	PLV	FRC	FRI	AFU	VS
VSN	PLU	FRC	CN	AFV	VS

context-based boundary refining technique. However, contexts adopted are different from the ones proposed here and the referred technique has not been based on neural networks [13].

After ANN training, 36 networks are used to refine segmentation boundaries. Thus, for any search interval under analysis, its segmentation boundary is moved towards that position whose corresponding parameter set assumes the highest ANN output value. Boundary refining is carried out only when this maximum value exceeds a previously defined threshold. In our case, such a threshold is 0.75.

It is important to emphasize that only phonetic transitions shown in Table 2 are considered for boundary refining purposes. One pattern ignored in this table is the transition between vocalic phonemes. In this paper, vocalic phoneme boundaries are disregarded for refining purposes since their strong coarticulation hinders (even for an experienced listener) an accurate identification of where a phoneme ends and another starts. Thus, low consistency is obtained when manual segmentation is performed over these phonetic transitions.

5   Experimental Results

Aiming to assess the procedure proposed here for boundary refining, two speech corpora are considered. The first one is composed of a total of 450 sentences (total duration of 4871 seconds) recorded by a female professional speaker (in BP language) in an acoustically isolated room. These sentences are phonetically transcribed and manually segmented by an experienced linguist. The amount of distinct diphones and the total number of diphones contained in this corpus (for distinct subsets) are presented in Table 3.

The second corpus (recorded by the same professional speaker) is composed of approximately 11500 sentences. This corpus is taken as a reference to train an HMM model for each phoneme. In this case, manual segmentation is ignored

Table 3. Amount of diphones contained in the training set

Sentences	Distinct diphones	Total amount
50	985	5655
100	1222	12114
150	1318	17507
200	1370	21039
250	1468	28467
300	1504	33630
350	1568	41334
400	1596	48212

for modeling purposes. For such, modeling is carried out by using the HTK software [9]. Thus, sentences are segmented into frames by using a 25 ms-long Hamming window with an overlap of 15 ms. In the following, each speech frame is represented by a set of parameters by considering 13 MFCC coefficients [14] with the corresponding 13 delta and 13 acceleration components, totaling 39 coefficients. A preemphasis factor equal to 0.97 is adopted. After parameter extraction, a five-state model (being three emitting states) is considered to represent each BP phoneme. The number of phonemes modeled is 53. Additional models for short pauses and silence segments are also considered. Models present a left-to-right topology, being allowed a direct transition from state 2 (first emitting state) to state 4 (third emitting state). In this case, diagonal covariance matrices are used to obtain each model. The number of Gaussian mixtures adopted to model the probability density function of each emitting state is varied from 2 to 4. The rate of phonetic transitions whose segmentation errors are lower than 20 ms for 2, 3, and 4 Gaussians (considering 11500 training sentences and 50 test sentences) are 79.24%, 80.49%, and 80.57%, respectively.

The segmentation results achieved with four mixtures are refined through the following techniques: (a) Technique 1 – based on a single neural network that represents all transition patterns [7]; (b) Technique 2 – four networks with transitions defined by voicing status [7]; (c) Technique 3 – four networks with automatically obtained transitions [4]; (d) Technique 4 – the proposed approach. For these refining procedures, a set of parameters is firstly extracted from the training database. Mel-cepstral parameters are obtained by using the following configurations: 20 ms-long Hanning window, overlap of 15 ms, and preemphasis factor of 0.95. The SFTR and SKLD measures are computed by using the parameters extracted from the two central consecutive frames. Zero crossing rates of the first and fourth frames are also determined. At the end of this process, the 56 resulting coefficients are normalized by their estimated means and variances. Since these parameters are taken as input for network training, a number of 56 input neurons is required for each network. In this case, perceptrons composed of three neuron layers (input, hidden, and output) are trained by using the error retropropagation algorithm [15]. For such, an exhaustive search procedure is adopted to select the algorithm learning rate and the amount of neurons in the hidden layer (0.6 and 18, respectively). The computational tool considered in such a training is the Fast Artificial Neural Network (FANN) library written in C language [16]. Among the four previously mentioned techniques, only the third one possesses a result dependent on initialization conditions. So, for this specific technique, initial networks are partitioned by taking as a reference the voicing status of neighbor phonemes (unvoiced/unvoiced, unvoiced/voiced, voiced/unvoiced, and voiced/voiced).

Table 4 presents the rate of phonetic transitions whose segmentation errors are lower than 20 ms, considering emitting states modeled by a mixture of four Gaussians (the best previously analyzed condition). In addition, the number of training sentences is varied from 50 to 400 in steps of 50. By analyzing the results shown in Table 4, one can verify that the proposed technique provides a better performance than other techniques presented here.

Table 4. Segmentation results provided by a Baum-Welch-based HMM segmentation and a boundary refining process

Number	Technique 1 [7]	Technique 2 [7]	Technique 3 [4]	Proposed
50	77.72%	80.23%	81.34%	81.63%
100	82.42%	84.57%	81.23%	84.66%
150	82.31%	84.97%	81.76%	85.78%
200	83.49%	85.80%	81.92%	86.15%
250	83.23%	85.36%	82.05%	86.66%
300	83.69%	86.63%	81.98%	86.87%
350	83.24%	86.48%	81.96%	**87.16%**
400	80.58%	86.96%	78.40%	86.96%

Those 400 previously mentioned sentences are taken as a reference to build a phoneme-based HMM model. In this case, phonetic modeling is attained by taking into account information provided by manual segmentation. Table 5 shows the rates of phonetic transitions with segmentation errors smaller than 20 ms. These rates are obtained for a segmentation based on HMM with 1, 2, 3, and 4 Gaussians (to model each emitting state). The number of sentences is also varied from 50 to 400 in steps of 50. For each sentence set (composed of a distinct amount of sentences), the best performance configuration is highlighted (in bold letter) in Table 5.

It is important to emphasize that when reestimations are performed (after the model based on manual information is obtained) segmentation accuracy is degraded. For example, considering 400 training sentences and one Gaussian, the segmentation performance is decreased from 90.05% (as shown in Table 5) to 89.07%.

Finally, sentences segmented by considering information provided by manual segmentation are then submitted to a boundary refining process. Table 6 presents

Table 5. Results of HMM estimation for a manually segmented database

	1 mixture	2 mixtures	3 mistures	4 mixtures
50	88.98%	90.00%	**90.15%**	89.88%
100	89.71%	90.35%	**91.43%**	90.69%
150	89.91%	90.62%	**91.26%**	90.95%
200	89.71%	90.84%	90.67%	**91.11%**
250	89.68%	91.14%	91.20%	**91.26%**
300	89.79%	90.81%	91.20%	**91.56%**
350	90.04%	91.14%	91.24%	**91.58%**
400	89.92%	91.15%	**91.61%**	91.35%

Table 6. Results considering both an HMM estimation based on a manually segmented database and a boundary refining process

Number	Technique 1 [7]	Technique 2 [7]	Technique 3 [4]	Proposed
50	84.96%	82.20%	89.81%	86.25%
100	85.76%	87.57%	91.43%	89.84%
150	85.87%	87.92%	91.09%	90.13%
200	86.28%	88.58%	90.83%	90.26%
250	86.62%	88.28%	91.20%	90.88%
300	87.25%	89.56%	91.58%	91.50%
350	86.85%	89.56%	91.54%	92.01%
400	91.61%	90.88%	88.54%	**92.03%**

phonetic transition rates with segmentation errors (existing after boundary refining) lower than 20 ms.

It is verified by inspecting Table 6 that in some situations the performance is reduced after boundary refining. Such a reduction means that boundary refining is efficient only when a considerable amount of sentences is used for training. Until a certain number, the resulting refining may even harm segmentation accuracy. For a sentence number equal or higher than 350, one verifies that the proposed technique results in a segmentation improvement. Other assessed techniques often lead to a reduction in segmentation accuracy. The proposed approach, comparatively with that discussed in [4], is advantageous for being independent of initialization conditions.

6   Concluding Remarks

This paper presents a technique to refine segmental boundaries aiming at concatenative synthesis systems. The proposed approach outperforms previously presented techniques, i.e. increases the rate of phonetic transitions whose segmentation errors are lower than 20 ms. For future work, we intend to verify the performance of a hybrid approach in which the technique proposed by Lee [4] is initialized by the network configuration proposed here.

References

1. Chou, F.-C., Tseng, C.-Y., Lee, L.-S.: An Evaluation of Cost Functions Sensitively Capturing Local Degradation of Naturalness for Segment Selection in Concatenative Speech Synthesis. Speech Communication, 48 (1), 45–56 (2006)
2. Hunt, A. J., Black, A. W.: Unit Selection in a Concatenative Speech Synthesis System Using a Large Speech Database. In: ICASSP, pp. 373–376, IEEE Press, Atlanta (1996)
3. Kawai, H., Toda, H., Ni, J.: Ximera: A New TTS from ATR Based on Corpus-Based Technologies. In: SSW, pp. 179–184, ISCA Press, Pittsburg (2004)
4. Lee, K.-S.: MLP-Based Phone Boundary Refining for a TTS Database. IEEE Trans. Audio, Speech, Language Processing 14 (3), 981–989 (2006)
5. Rabiner, L. R.: A Tutorial on Hidden Markov Models and Selected Applications in Speech Recognition. Proceedings of the IEEE 77 (2), 257–286 (1989)
6. Huang, X., Acero, A., Hon, H.: Spoken Language Processing: A Guide to Theory, Algorithm and System Development. Prentice Hall, Upper Saddle River (2001)
7. Toledano, D. T.: Neural Network Boundary Refining for Automatic Speech Segmentation. In: ICASSP, pp. 3438–3441, IEEE Press, Istanbul (2000)
8. Deller Jr., J. R., Hansen, J. H. L., Proakis, J. G.: Discrete-Time Processing of Speech Signals. IEEE Press, New York (2000)
9. Young, S., Evermann, G., Kershaw, D., Moore, G., Odell, J., Ollason, D., Valtchev, V., Woodland, P.: The HTK Book (for HTK Version 3.1). Cambridge University (2001)
10. Athaudage, C. R. N., Lech, M.: On Optimal Modeling of Speech Spectral Transitions. In: ICICS, pp. 1330–1334, IEEE Press, Singapore (2003)
11. Klabbers, E., Veldhuis, R.: Reducing Audible Spectral Discontinuities. IEEE Trans. Speech Audio Processing 9 (1), 39–51 (2001)
12. Silva, T. C.: Phonetic and Phonology of the Portuguese Language: Study Script and Exercise Guide. Contexto, Sao Paulo (in Portuguese) (1999)
13. Wang, L., Zhao, Y., Chu, M., Soong, F. K., Zhou, J., Cao, Z.: Context Dependent Boundary Model for Refining Boundaries Segmentation of TTS Units. IEICE Trans. Information and Systems E89-D (3), 1082–1091 (2006)
14. Molau, S., Pitz, M., Schluter, R., Ney, H.: Computing Mel-Frequency Cepstral Coefficients on the Power Spectrum. In: ICASSP, pp. 73–76, IEEE Press, Salt Lake City (2001)
15. Haykin, S.: Neural Networks: A Comprehensive Foundation. Prentice-Hall (1998)
16. Nissen, S., Spilca, A., Zabot. A.: Fast Artificial Neural Networks (FANN), http://leenissen.dk/fann/

Evolutionary-Based Design of a Brazilian Portuguese Recording Script for a Concatenative Synthesis System*

Monique Vitório Nicodem, Izabel Christine Seara, Daiana dos Anjos,
Rui Seara Jr., and Rui Seara

LINSE – Circuits and Signal Processing Laboratory
Department of Electrical Engineering
Federal University of Santa Catarina, Brazil
{monique,izabels,daiana,ruijr,seara}@linse.ufsc.br
http://www.linse.ufsc.br

Abstract. Modifications of prosodic parameters in concatenative synthesis systems may lead to a degradation in speech quality, especially when significant pitch changes are accomplished. Aiming to avoid large changes in the speech signal parameters, the speech corpus should present segments with phonetic and prosodic features close to the predicted ones. This condition is more often fulfilled by a speech corpus specially designed to be both phonetic and prosodically rich. The design of this corpus is strongly dependent on the script chosen for recording. For such, a procedure to select the recording script of a TTS system is proposed for the Brazilian Portuguese language. Selected sentences include declarative, exclamatory, and interrogative ones. Phonetic and prosodic information are firstly represented as a set of feature vectors. Next, the amount of distinct feature vectors is used as a fitness value for a genetic-based sentence selection. Experimental results point out a considerable improvement in script variability for speech synthesis applications.

Keywords: TTS systems, recording script design, genetic algorithms.

1 Introduction

In state-of-the-art concatenative systems, speech recordings are often carried out in a soundproof room by using the voice of a professional speaker [1, 2, 3, 4, 5, 6]. In addition, these systems perform a procedure of automatic selection in which non-uniform units are collected from a corpus previously recorded. Systems based on this class of units have produced a synthetic speech very close to that of human speakers. However, the naturalness of the ultimate synthetic speech has been strongly constrained by factors such as the quality of the speech

---

* This work was partially supported by the Brazilian National Council for Scientific and Technological Development (CNPq), Studies and Projects Funding Body (FINEP), and Dígitro Tecnologia Ltda.

A. Teixeira et al. (Eds.): PROPOR 2008, LNAI 5190, pp. 81–90, 2008.

recordings, speaker's voice characteristics, intonation modeling algorithms [7], and text chosen for recording [8, 9].

While most of the current recording scripts are designed based on phonetically rich or balanced principles [10, 11], this work focuses on attaining a phonetic and prosodically rich script [12]. For the Brazilian Portuguese language (BP), phonetically balanced speech corpora have been discussed in [10] and [11]. In [10], a set of 200 phonetically balanced sentences (obtained by manual selection) are given. An automatic search procedure (based on a genetic algorithm) is proposed by [11], providing 1000 phonetically balanced declarative sentences. Both approaches disregard any aspects of prosodic representativeness.

In the approach proposed here, four main stages are considered, namely, grapheme-to-phoneme (G2P) conversion, prediction of prosodic patterns, feature vector representation and automatic selection per se. Since this latter stage is language independent, this paper aims to provide more details about it.

Grapheme-to-phoneme conversion is performed aiming to provide the information required to evaluate the phonetic variablity of those sentences which are candidates for selection. On the other hand, the prediction of prosodic patterns (text-to-prosody) is necessary to supply prosodic labels for each syllable of the sentence, being these labels needed to assess sentence prosodic variability.

Phonemes and prosodic labels resulting from the two previous stages are stored in feature vectors. In this way, each candidate sentence is associated with a set of feature vectors responsible for providing clues about variability. In the open literature, this selection stage has been considered the key point of the overall script design process. Such a stage has often been carried out by using the additional greedy algorithm. However, greedy techniques can prematurely select certain sentences which prevent them from either finding the best overall solution or getting closer to this solution [13]. These premature selections are avoided when genetic tools are considered [14]. Therefore, one proposes a selection based on a genetic algorithm whose fitness value is the number of distinct feature vectors. Such a genetic-based approach leads to a corpus with a higher variability. As a result, a set of 4000 phonetic and prosodically rich BP sentences (including declarative sentences, wh-questions, yes/no questions, alternative questions, and exclamatory sentences) have been selected to compose the required corpus.

2   Grapheme-to-Phoneme Conversion

Grapheme-to-phoneme conversion is required to evaluate the phonetic coverage of the designed corpus. For such a task, a phonetic transcription of each word existing in the corpus under analysis is carried out. Such a transcription is achieved by considering a lexicon containing canonical pronunciations and an ad hoc set of transcription rules for BP, being some of them described in [15] and in [16].

3   Prediction of Prosodic Patterns

Aiming to improve prosodic coverage, this stage intends to predict the prosodic pattern (in terms of prominent levels of pitch height) of the speaker under analysis. In this case, annotation rules are responsible for attributing prosodic labels to the phonemes resulting from the grapheme-to-phoneme conversion. The labels adopted are based on the symbols from intonational phonology. Thus, the tonal events "high (H)" and "low (L)" in distinct levels of pitch height (H+, H−, H, L, and L−) are attributed to some syllables of the sentence based on their following syntactic characteristics: lexical, phrasal, and sentence classification. A similar text-to-prosody approach has been presented in [17] for French.

The procedure proposed for prosody prediction considers the following main stages: lexical classification of each word existing in a given sentence, segmentation into phrases, sentence classification, and prosodic annotation.

3.1 Lexical Classification

In this stage, lexical classification of each word existing in a given sentence is determined. Such a determination is ambiguous in several languages. For an automatic solution of ambiguities [18], strategies used for part-of-speech tagging based on rules, statistics, artificial neural networks (ANNs), support vector machines (SVMs), or hybrid approaches [19, 20] can be considered.

3.2 Segmentation into Phrases

Since there exists a relation between syntactic phrasing and prosodic patterns, the procedure of prosody prediction proposed here takes as a reference the sentence division into syntactic phrases [21, 22, 23][1]. We consider the categories of determiner, verbal, adverbial, and prepositional phrases (represented by DP, VP, AP, and PP, respectively) which have as their "head", a noun, a verb, an adverb, and a preposition, respectively. For example, the utterance "The girl helped my friend" is composed of three syntactic phrases, being segmented as "The girl / helped / my friend" and composed, respectively, of the phrasal sequence DP, VP, and DP. Phrasal classification could be carried out by using similar approaches considered for lexical classification [19, 20].

3.3 Sentence Classification

Provided that a distinct pattern of intonation is observed for each sentence class, the input text must be processed in order to determine whether a given sentence is declarative, exclamatory, or interrogative. We also consider the three following categories of interrogative sentences: wh-question, yes/no question, and alternative question[2].

---

[1] A syntactic phrase is a group of words working as a single unit and fulfilling a hierarchy of grammatical constituents. Each phrase has a word (called "head") which determines the category of a phrase based on its lexical classification [23, 24].

[2] Wh-questions are interrogative sentences in which the expected answer is determined by interrogative pronouns (or wh-locutions) such as "what", "who", "how much", "why", among others. These interrogative pronouns may be located at the start, middle, or end of the sentence, which are, respectively, classified as initial, medial, or final wh-questions. In a yes/no question, the expected answer is a yes or no. On the other hand, in an alternative question, such an answer is one alternative.

3.4 Prosodic Annotation

In this phase, sentences are labeled in such a way that the language intonation patterns of each sentence class are properly represented. Peaks or valleys in the sentence pitch contour are associated with syntactic characteristics which are common among the recorded sentences, such as phrasal classification, sentence phrasal position, stress. For such, a recorded database is firstly analyzed to determine the relation between intonation and syntactic characteristics. Such an analyis could be automatic and/or manual. After that, prosodic labels (H+, H−, H, L, and L−) are attributed to the phonemes resulting from G2P conversion.

3.5 Application to the Brazilian Portuguese Language

Prosody prediction is based on rules specially designed for BP. Rules adopted for lexical classification, segmentation into phrases, and prosodic annotation have been described by [14, 25, 26]. The latter ones (specific for BP) are briefly described in Appendix A. The syllables which do not fit in any of the rules are labeled with the symbol N (neutral), which means that their phonemes may present any pitch contour (falling, rising, or neutral).

The whole prosody prediction stage (lexical classification, segmentation into phrases, sentence classification, and prosodic annotation) is exemplified here by the initial wh-question *"Qual é o seu nome?"* (What is your name?) presented in Table 1. In this case, the sentence is segmented into phrases based on the grammatical categories of the words within it. These words are divided into syllables and, after that, prosodic labels are associated with each syllable.

Table 1. Example of prosodic annotation

Sentence	What is your name?				
BP sentence	*Qual*	*é*	*o*	*seu*	*nome?*
Lexical category	INT	V	DT	DT	N
Phrasal segmentation	DP	VP		DP	
Syllabic division	Qual	é	o	seu	no me
Phonetic transcription	[kw'aw]	['ɛ]	[ʊ]	'sew	'no mI]
Prosodic label	H+	N	N	L	H L−

4   Feature Vector Representation

After prosody prediction, prosodic and phonetic information obtained in previous stages are adopted to represent each phone by a feature vector. Each vector contains the following four elements: previous phone, current phone, next phone, and prosodic annotation of the current phone. For the sentence of Table 1, one verifies the following vectors: [silence k w H+], [k w 'a H+], and so on.

5   Automatic Selection Based on Genetic Algorithms

After representing the phonemes and their prosodic annotation as a set of feature vectors, the automatic procedure proposed for sentence selection is carried out. Such a selection performs a search based on genetic algorithms of the sentence set (population) which has the highest amount of feature vectors excluding those vectors containing the prosodic label "N". Our algorithm only considers the feature vectors containing the labels H+, H, H−, L, and L− since they are related to those syllables which are key points (points in which falling and/or rising pitch contours usually occur) to determine the expressive style (declarative neutral, wh-question, yes/no question) of the sentence. In this case, a higher quantity of pitch movements would be covered by the speech database and a higher prosodic variability would be achieved.

Genetic algorithms are optimization tools based on natural selection and genetic inheritance. They are recommended when the search space for the optimal solution is considered large enough to make an exhaustive search prohibitive [11].

Firstly, in a genetic algorithm, the fitness value is computed for each chromosome of the initial population. Two chromosomes (parents) are selected among the existing population. They combine themselves through a crossover genetic operation generating children. A mutation may also occur instead of crossover. The fitness function is also evaluated for children. If the children present a higher fitness value than their parents, they replace their parents. Otherwise, children are discarded and their parents survive for the next generation. Such a cycle repeats itself until some stopping criterion is reached [27].

6   Experimental Results

The current research work addresses a possible solution to the problem of finding in a large text corpus (with approximately 1500000 sentences extracted from CETENFolha[3]), 4000 sentences phonetic and prosodically rich, being 1000 declarative, 1000 wh-questions, 1000 yes/no questions, 500 alternative questions, and 500 exclamatory. A smaller amount is considered for exclamatory sentences and alternative questions since they occur at a lower rate both in CETENFolha database and BP language.

In our experiments, the CETENFolha corpus has been divided into declarative, exclamatory, and interrogative sentences. Such a corpus has 1390000 declarative, 909 exclamatory, and 36166 interrogative sentences (being 23500 wh-questions, 11151 yes/no questions, 1415 alternative questions). The algorithm has been run separately for each of the following sentence classes: declarative, exclamatory, wh-questions, yes/no questions, and alternative questions.

The set of declarative sentences are firstly divided into 40 groups of 35000 sentences. In this case, a genetic algorithm which selects the 1000 sentences with

---

[3] The CETENFolha is a corpus obtained by compiling the texts of the Brazilian newspaper "Folha de São Paulo". Such a compilation has been made by the "Núcleo Interinstitucional de Lingüística Computacional (NILC)" located in São Carlos, Brazil.

the highest prosodic and phonetic variability is run for each group. In this way, 40000 sentences are obtained. These 40000 (40 groups of 1000) sentences are taken as a reference for another genetic algorithm responsible for obtaining the 1000 declarative sentences with the highest number of feature vectors.

In the case of exclamatory sentences, 500 sentences are selected by only taking into account the mutation of a single chromosome.

For interrogative sentences, 1000 sentences are selected among the 23600 wh-questions (23 groups of 1000 sentences plus 600 sentences taken for mutation). Other 1000 sentences are obtained by allowing for the 11151 yes/no questions (11 groups of 1000 sentences plus 151 for mutation). Finally, 500 alternative questions are selected among a total of 1415 (being two groups of 500 considered for crossover plus 415 for mutation).

In the considered algorithms, each group of sentences corresponds to a chromosome. A number of distinct feature vectors (excluding neutral ones) is determined for each chromosome. In a given generation, a genetic operation is carried out. This process iteratively continues until the best (with a higher fitness value) chromosome of a population remains unchanged for at least 1000 generations.

A mutation rate of 10% has been considered for declarative, wh-questions, and yes/no questions. For alternative questions and exclamatory sentences, a mutation rate of 50% and 100% has been adopted, respectively. It is important to notice that the mutation rate of alternative questions and exclamatory questions is higher than that of other sentences since their low occurrence in CETENFolha corpus leads to a small amount of chromosomes. In addition, one can notice that mutation performs better than crossover for a reduced chromosome amount. In an extreme case, we have the exclamatory sentences with only one chromosome, which makes the crossover operation impracticable without sentence repetition.

Parent selection is based here on the roulette wheel method [28]. In this method, those most adapted chromosomes (with a higher number of distinct feature vectors) have a higher probability of being selected as parents. The crossover point is obtained in a random way. After crossover operation, two children are obtained. If the best child has a higher number of distinct feature vectors than the best parent, children replace their parents. The mutation point and the amount of mutant genes are randomly obtained.

Algorithms are carried out by using the Python Programming Language [29]. The improvement obtained by using the proposed approach is summarized in Table 2. For example, the initial group of 1000 yes/no questions with the highest number of feature vectors has 5720 vectors. At the end of the procedure, the best resulting sentence set (chromosome with 1000 yes/no questions) provided 6191 vectors, indicating an improvement of approximately 8.2%.

Another interesting advantage of this automatic selection is the possibility of reducing database size while maintaining the same number of feature vectors. To determine the amount of database pruning which could be achieved, we have carried out another experiment. Sentences have been collected in random order from the CETENFolha database until the number of feature vectors obtained after the selection procedure (8413, 6469, 6191, 4145, and 1150 feature vectors

Table 2. Improvement achieved in prosodic and phonetic variability

	Declarative	Wh	Yes/no	Alternative	Exclamatory
Before	7268	6124	5720	3963	1069
After	8413	6469	6191	4145	1150
Improvement	15.7%	5.63%	8.23%	4.59%	7.04%

for declarative, wh-questions, yes/no questions, alternative questions, and exclamatory, respectively) have been reached for each sentence class. In this case, a number of 1249 declarative sentences, 1168 wh-questions, 1299 yes/no questions, 605 alternative questions, and 560 exclamatory sentences are selected, indicating that a pruning of, respectively, 19.94%, 14.38%, 23.02%, 17.36%, and 10.72% could be achieved by using the proposed selection procedure while maintaining the same number of feature vectors for each sentence class.

7 Conclusions and Future Work

As a result of the approach proposed here, a set of declarative, exclamatory, and interrogative sentences have been selected among a large speech database by using an approach based on genetic algorithms. Such an approach has shown to be useful to improve the phonetic and prosodic variability of text corpora. For future work, we intend to perform such a selection for "the big six" emotions leading to an emotional speech synthesis system with both a higher prosodic and phonetic variability.

References

1. Deller Jr., J. R., Hansen, J. H. L., Proakis, J. G.: Discrete-Time Processing of Speech Signals, IEEE Press, New York (2000)
2. Huang, X., Acero, A., Hon, H.-W.: Spoken Language Processing: A Guide to Theory, Algorithm and System Development, Prentice Hall PTR, Upper Saddle River, New Jersey (2001)
3. Hunt, A. J., Black, A. W.: Unit Selection in a Concatenative Speech Synthesis System Using a Large Speech Database. in: Proceedings of ICASSP, vol. 1, Atlanta, USA, pp. 373–376 (1996)
4. Schroeter, J.: Text-to-Speech Synthesis. in Circuits, Signals, and Speech and Image Processing, R. C. Dorf (Ed.), 3rd. ed., Taylor & Francis Group (2006)
5. Sak, H., Güngör, T., Safkan, Y.: A Corpus-Based Concatenative Speech Synthesis System for Turkish. Turkish Journal of Electrical Engineering, and Computer Sciences, 14 (2), 209–223 (2006)
6. Zhu, W., Zhang, W., Shi, Q., et al.: Corpus Building for Data-Driven TTS Systems. in: Proceedings of TTS, Santa Monica, USA, 199–202 (2002)
7. Pitrelli, J. F., Bakis, R., Eide, E. M., et al.: The IBM Expressive Text-to-Speech Synthesis System for American English. IEEE Transactions on Speech and Audio Processing. 14 (4), 1099–1108 (2006)

8. Nicodem, M. V., Seara, R., Pacheco, F. S.: Reducing the Natural Click Effect within Database for High Quality Corpus-Based Speech Synthesis. In: ISSPA, Sydney, Australia, pp. 607–610 (2005)

9. Nicodem, M. V., Seara, R.: Natural Click Processing Through Wavelet Analysis and Extrapolation for Speech Enhancement. In: ITS, Fortaleza, Brazil, pp. 600-605 (2006)

10. Seara, I. C.: Statistical Study of the Phonemes Spoken in the Capital of Santa Catarina for the Elaboration of Phonetically Balanced Sentences. Master's thesis, Federal University of Santa Catarina, Florianópolis, Brazil (in Portuguese) (1994)

11. Cirigliano, R., Monteiro, C., Barbosa, F., et al.: A Set of 1000 Brazilian Portuguese Phonetically Balanced Sentences Obtained Using the Genetic Algorithm Approach. In: SBrT, Campinas, Brazil, pp. 544–549 (in Portuguese) (2005)

12. Chou, F.-C., Tseng, C.-Y.: The Design of Prosodically Oriented Mandarin Speech Database. In: ICPhs, San Francisco, USA, pp. 2375–2377 (1999)

13. Li, Z., Harman, M., Hierons, R. M.: Search Algorithms for Regression Test Case Prioritization. IEEE Transactions on Software Engineering. 33 (4), 225–237 (2007)

14. Nicodem, M. V., Seara, I. C., Seara, R., dos Anjos, D.: Recording Script Design for a Brazilian Portuguese TTS System Aiming at a Higher Phonetic and Prosodic Variability. in: Proceedings of ISSPA, Sharjah, United Arab Emirates, pp. 1–4 (2007)

15. Seara, I. C., Pacheco, F. S., Seara Jr., R., et al.: Automatic Generation of Brazilian Portuguese Variants Aiming at Speech Recognition Systems. in: Proceedings of SBrT, Rio de Janeiro, Brazil, 1–6 (in Portuguese) (2003)

16. Silva, D. C., Lima, A. A. de, Maia, R., et al.: A Rule-Based Grapheme-Phone Converter and Stress Determination for Brazilian Portuguese Natural Language Processing. in: Proceeding of ITS, Fortaleza, Brazil, 992–996 (2006)

17. Malfrére, F., Dutoit, T., Hertens, P.: Automatic Prosody Generation Using Suprasegmental Unit Selection. In: SSW, Jenolan Caves, Australia, pp. 323–328 (1998)

18. Seara, I., Kafka, S., Klein, S., Seara, R.: Vowel Sound Alternation of Verbs and Nouns of the Portuguese Spoken in Brazil for Application in TTS Synthesis. Journal of the Brazilian Telecommunications Society. 17 (1), 79–85 (in Portuguese) (2002)

19. Hasan,M. M. and Lua, K.-T.: Neural Networks in Chinese Lexical Classification. In: PACLIC, Seoul, South Korea, pp. 119–128 (1996)

20. Ciaramita, M., Hofmann, T., and Johnson, M.: Hierarchical Semantic Classification: Word Sense Disambiguation with World Knowledge. In: IJCAI, Acapulco, Mexico, pp. 817–822 (2003)

21. Cagliari, L. C.: Phonological Analysis: Introduction to Theory and Practice with Special Emphasis to the Phonemic Model, Mercado Letras, Campinas, Brazil (2002)

22. Sândalo, M. F. S.: Prosodic Phonology and Optimality Theory: Reflexions about the Interface Syntax-Phonology in the Generation of Phonological Phrases. Revista de Estudos da Linguagem. 12 (2), 319–344 (2004)

23. Truckenbrodt, H.: On the Relation between Syntactic Phrases and Phonological Phrases. Linguistic Inquiry, 30 (2), 219–255 (1999)

24. Yoon, K.: A Prosodic Phrasing Model for a Korean Text-to-Speech Synthesis System. Computer, Speech, and Language, 20 (1), 69–79 (2006)

25. Nicodem, M. V., Seara I. C., Seara, R., dos Anjos, D., Seara Jr., R.: Automatic Selection of Text Corpus for Speech Synthesis Systems. In: SBrT, Recife, Brazil, pp. 1–6 (in Portuguese) (2007)

26. Seara, I. C., Nicodem M. V., Seara, R., Seara Jr, R.: Phrasal Classification Focusing Speech Synthesis: Rules for Brazilian Portuguese. In: SBrT, Recife, Brazil, 1-6 (in Portuguese) (2007)
27. Tang, K. S., Man, K. F., Kwong, S., et al.: Genetic Algorithms and their Applications. IEEE Signal Processing Magazine, 13 (6), 22–37 (1996)
28. Johnson, J. M., Rahmat-Samii, V.: Genetic Algorithms in Engineering Electromagnetics. IEEE Antennas and Propagation Magazine. 39 (4), 7–21 (1997)
29. Hetland, M. L.: Beginning Python: From Novice to Professional. Apress (2005)

Appendix A: Prosody Prediction in Brazilian Portuguese

For lexical classification, we have considered the following categories existing in BP language: adjective (AJ), adverb (AV), adverbial locution (AL), article (AR), conjunction (CO), conjunctive locution (CL), demonstrative pronoun (DE), indefinite pronoun (ID), interrogative pronoun (IT), nominative pronoun (NM), noun (NO), number (NU), objective pronoun (OB), possessive pronoun (PO), preposition (PR), prepositional locution (PL), relative pronoun (RE), verb (VE), and verbal (VER).

By taking into account the defined lexical categories, we have built a parser capable of determining phrasal classification (syntactic). Such a parser makes use of 42 rules shown in Table 3.

Table 3. Rules to classify Brazilian Portuguese sentences in syntactic phrases

Rules for phrasal classification	
DP -> PO|NU|DE|AR|ID + NO|AJ|VER + DP	VP -> VE + VER
DP -> NO|AJ|VER + NO|AJ|VER	VP -> VP + AV
DP -> NO|AJ|VER + DP	VP -> VE + AV|AL + VER
DP -> AR+ PO|NU|ID + DP+ DP	VP -> VE + VER + VER
DP -> PO|NU|DE|AR|ID +DP	VP -> AV|AL + VP
DP -> AR+ PO|NU|ID + VER	VP -> OB + VE|VER
DP -> IDS|DES|DE|NU|NM	VP -> VE|VER +OB
DP -> AR + PO|NU|ID + DP	VP -> AV + VP
DP -> ID + AR+ NO	VP -> CO|CL + VP
DP -> ID + ID + DP	AP -> AV|AL + DP
DP -> AR + ID + DP	AP -> AL + DP
DP -> NU + DP	AP -> AV|AL +PL|PR + DP
DP -> ID + DP	AP -> AV|AL + AP
DP -> AR + ID	AP -> AV|AL + AJ
DP -> AR + NU	AP -> AV
DP -> NO|AJ	AP -> CO|CL + AP
DP -> AR+ DP	PP -> PL|PR + VER|NO
DP -> NU	PP -> PL|PR + DP
DP -> ID	PP -> PP + DP
DP -> CO|CL + DP	PP -> PP + VER + DP
VP -> VE|VER	PP -> CO|CL + PP

Rules specially developed for BP are responsible for attributing prosodic labels to the phonemes resulting from the grapheme-to-phoneme conversion. These labels are presented in Table 4 in decreasing order of fundamental frequency (H+ > H > H− > L > L−).

Table 4. Rules for prosodic annotation of sentences − * Intermediate phrases (IP) are labelled whenever the sentence has more than five phrases

Declarative sentences	
H+	First syllable from the last word of the initial phrase of the sentence
H	Sentence next to last syllable if the preceding phrase is not AP and first syllable from AP last word
H−	Next to last syllable of the sentence if the preceding phrase is AP
L	Syllable which precedes H or H− from the final phrase of the sentence
L−	Last syllable from the final phrase of the sentence
Exclamatory sentences	
H	First syllable of the sentence
L	Last syllable from the final phrase of the sentence
Initial and medial wh-questions	
H+	Stressed syllable from the last word of a wh-locution or from the interrogative pronoun
H	Stressed syllable from the last word of the sentence
H−	Stressed syllable from the AP last word and from the last word of each IP*
L	Syllable which precedes the final H and H+ from the sentence wh-locution
L−	Syllable following the final H
Final wh-questions	
H+	First syllable from a wh-locution or interrogative pronoun
H	Stressed syllable from the last word of the initial phrase of the sentence
H−	Stressed syllable from the AP last word and from the last word of each IP*
L	Syllable which precedes H+ from the wh-locution and H from the initial phrase
L−	Syllable following the final H+ from the final wh-locution
Yes/no questions	
H+	Stressed syllable from the last word of the sentence
H	Stressed syllable from the last word of the initial phrase of the sentence
H−	Stressed syllable from the AP last word and from the last word of each IP*
L	Syllable which precedes the final H+ or syllable which precedes the H from the initial phrase
L−	Syllable following the final H+
Alternative questions	
H	Stressed syllable which precedes the conjunction *ou* (or)
H−	Conjunction *ou* (or)
L	Syllable which precedes H and syllable following the H if a pause occurs before such a conjunction
L−	Last stressed syllable of the sentence

DIXI – A Generic Text-to-Speech System for European Portuguese

Sérgio Paulo, Luís C. Oliveira, Carlos Mendes, Luís Figueira, Renato Cassaca,
Céu Viana[1] and Helena Moniz[1,2]

L^2F INESC-ID/IST, [1]CLUL/FLUL, [2]L2F INESC-ID
Lisbon, Portugal
{spaulo,lco,cmdm,luisf,rmfc,mcv,helenam}@l2f.inesc-id.pt

Abstract. This paper describes a new generic text-to-speech synthesis
system, developed in the scope of the Tecnovoz Project. Although it
was primarily targeted at speech synthesis in European Portuguese, its
modular architecture and flexible components allows its use for different
languages. We also provide a survey on the development of the language
resources needed by the TTS.

1 Introduction

This paper describes a new generic text-to-speech (TTS) synthesis system, developed in the scope of the Tecnovoz Project. Although it was primarily targeted
at speech synthesis in European Portuguese (EP), its modular architecture and
flexible components allows its use for different languages. Moreover, the same
synthesis framework can be used either for limited domain or generic speech
synthesis applications. The system's operation mode is defined by the currently
selected voice, enabling the user to switch from a limited domain to a general purpose voice, and vice-versa, with a single engine. Dixi currently runs on *Windows*
and *Linux*. The synthesis engine can be accessed, in both operating systems, by
means of an *API* provided by a set of *Dynamic Linked Libraries* and *Shared
Objects*, respectively.

Given the success enjoyed by the Festival Speech Synthesis System [3] and the
flexibility of its internal representation formalism, the heterogeneous relation
graphs [14], the Dixi's internal utterance representation follows approximately the
same scheme. However, the Festival system implementation has a large number of
drawbacks that led us to the implementation of a new system architecture. One
of the limitations of the Festival system is its inability to use multi-threading,
and thus incapable to profit from the multi-processing capabilities of nowadays
machines. Being multi-thread safe is a key feature of the new system. The system architecture is based on a pipeline of components, interconnected by means of
intermediate buffers, as depicted in Fig 1. Every component runs independently
from all others, loads the to-be-processed utterances from its input buffer and,
subsequently, dumps them into its output buffer. Buffers, as the name suggests,
are used to store the utterances already processed by the previous component
while the following one is still processing earlier submitted data.

A. Teixeira et al. (Eds.): PROPOR 2008, LNAI 5190, pp. 91–100, 2008.
© Springer-Verlag Berlin Heidelberg 2008

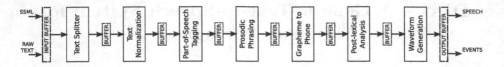

Fig. 1. Overview of the system architecture, where *SSML* stands for Speech Synthesis Markup Language

The capability of the system to use multi-processing and split large utterances into smaller ones, as will be explained later on in this paper, allows the streaming synthesis problem to be addressed more efficiently than in other well known synthesis systems [3,1]. Dixi comprises six components: text pre-processing, part-of-speech tagging, prosodic phrasing, grapheme-to-phone conversion, phonological analysis and waveform generation, as depicted in Fig 1.

1.1 Tecnovoz Project

The Tecnovoz project is a join effort to disseminate the use of spoken language technologies in different domains of application. The project consortium includes 4 research centers and 9 companies specialized in a wide range of areas like banking, health systems, fleet management, access control, media, alternative and augmentative communication, computer desktop applications, etc. To meet the goals of the project a set of 13 demonstrators are being developed based on 9 technology modules. Two of these modules are related with speech output: one module for limited domain speech synthesis and another for synthesis with unrestricted input. The first module will be used, for example, in banking applications were almost natural quality can be achieved by a proper design of the output sentences. An example of an application with unrestricted vocabulary is the oral feedback for a dictation machine.

We decided to adopt a single system to handle both requirements. The domain adaptation is performed at the level of the speech inventory used for each application. The inventory, usually called the system "voice", can have a wide or narrow coverage of the language. By using an inventory with very large number of carefully selected samples of a restricted domain, a very high quality can be achieved for sentences in that domain. A more general purpose system can use an inventory with a wider coverage but with fewer examples for each domain.

1.2 System Flexibility

The TTS users can adapt the system operation to their own needs by themselves. Accordingly, the users can create an addenda to the pronunciation lexicon, in order that words are rendered as desired. Moreover, specific normalization parameters[1] can be specified by the user, so that some particular text tokens are normalized according to the user-specific needs.

---

[1] Such as language, regional settings and text domain.

The speech signal can be produced by two distinct approaches. Parametric synthesis, using Hidden Markov Models based synthesis [15], or concatenative synthesis, using variable-length units [2].

1.3 Data-Driven Approaches

In order to accelerate the system's adaptation to new languages and domains, the language- and domain-specific knowledge sources were kept apart from the system's implementation. Also, machine learning techniques were used to train models for some components responsible for the linguistic analysis of the input text. The models – frequently encoded in the form of *Classification and Regression Trees* (CART) [5] – are then loaded the same way no matter what domain or language the system is dealing with.

1.4 Paper Organization

Although being a multi-lingual synthesis system, in this paper we will focus on the specific needs for speech synthesis in European Portuguese, namely, corpora and linguistic analysis. The paper is organized as follows. In section 2, we describe the corpora building procedures for limited domain and generic synthesis applications. Section 3 is reserved for describing the training of speaker-specific prosodic phrasing models, as well as the building of grapheme-to-phone (G2P) conversion models. In section 4, we present a detailed description of the system architecture. Conclusions and future work are presented in section 5.

2   Corpora Building

The quality of the synthetic speech produced by a corpus-based synthesizer depends, to a large extent, upon the suitability of the speech inventory to represent the variability of the language within the target application domain. While it is quite easy to design a set of limited domain sentences comprising units such as words in appropriate prosodic contexts, dealing with an unrestricted text task calls for another corpus design approach. In such a case, it is impossible to list all the target domain words, as unknown words can always appear in the TTS input. Hence, the representation of the language must be addressed by means of a finite set of linguistically motivated units (e.g. phonemes, diphones or syllables). Besides, several corpus design strategies can be chosen, as prompts can be manually designed or automatically selected from a a huge candidate sentence set. The design of limited domain and open domain speech inventories are described separately in this section, as they consist in problems of distinct nature.

2.1   Corpus Design

We followed a mixed approach for designing both the limited and open domain corpora. On the one hand, we automatically selected a set of sentences using a multi-leveled token search method described in [20]. On the other hand, a set of sentences were manually designed by a linguist, in order to cover a set of relevant linguistic units that could not be observed in the automatically selected sentences.

Open Domain. The design of speech corpora for our open domain voices was largely inspired by the language resource specification used in the TC-STAR project [4]. Thus, it started by automatically selecting candidate sentences from a large collection of newspaper articles in order to cover, as much as possible, a set of linguistically-motivated units,[2] so that even in the most unfavorable contexts, the TTS can render a speech signal with good enough quality.

The frequency that each linguistic unit can occur is highly dependent upon the text corpus. Moreover, in [17] the correlation coefficient between the frequency distributions of the triphones shared by two corpora was found not to reach values above 0.3. Therefore, covering only the most frequently observed units is not a solution, unless we are designing a corpus for a limited domain voice. Then our decision was to let the automatic selection algorithm cover all the units. However, since a complete coverage of such units cannot be achieved without a prohibitively large sentence set, that would take too much time to record and annotate, the search for relevant text prompts ends when a predefined coverage threshold is reached.

Another issue that must be addressed is the covering of some lexical items that are only observed in specific domains. Even though, such domains (*e.g.* phone numbers, economy, currencies, computer science terms, frequently used foreign names and expressions, typical dishes, touristic attractions, or even countries and their capitals) are sometimes so relevant for the daily use of the language that, at least, their most frequent lexical items are likely to be typed by the end users. On the other hand, the naturalness of speech is highly dependent on a set of linguistic characteristics that combined with purpose and context may convey distinct effects. Therefore, the manually designed prompts should account for several features, namely, a list of the most frequent verbs in EP in both first and second persons. Hence, we started by computing the occurring frequency of each verb lemma in a corpus of around 1,600,000 newspapers' articles, based on the results of a morpho-syntactic tool described in [10]. The human-computer interactions strongly benefits from the use of spontaneous prompts frequently observed in our daily conversations. Therefore, previously recorded human dialogs in the CORAL corpus [16] were orthographically transcribed and subsequently recorded by the speakers.

Another set of prompts was built with the purpose of providing the speech inventory with additional linguistic events, namely filled pauses and prolongations of segmental material in predictable locations and with different functions (changing a subject, preparing the subsequent units, taking the floor and also as mitigating devices) [8], as well as conversational grants, (*e.g.* *hum*) with different values. The manually designed prompts also account for all types of interrogative sentences and declarative sentences with the same lexical material as a yes/no question, as it is well known that intonational contours varies distinctly according to the sentence type.

Limited Domain. The design of limited domain speech resources was carried out as follows. Firstly, we gathered a large set of domain-specific sentences.

---

[2] Diphones, triphones and syllables.

Then, we followed the frequency-dependent[3] approach used in the LC-STAR project [22] while deciding which words should be included in the word lists. After defining the set of words and word sequences to be covered by the automatic procedure, the sentence selection starts. Finally, additional sentences are manually designed in order to provide the corpora with the most relevant words in appropriate contexts, if such words and contexts were not found in the automatically selected prompts.

2.2 Phonetic Segmentation and Multi-level Utterance Descriptions

The phonetic segmentation of the databases is performed in three different stages. Firstly, the speech files are segmented by a hybrid speech recognizer (*Audimus* [9]) working in forced alignment mode. Next, such segmentations are used by the HTK programs [21] for training context-independent speaker-specific HMMs. The speaker-adapted models are subsequently provided to a phonetic segmentation tool based on weighted finite state transducers allowing for many alternative word pronunciations [11].

The spoken utterances were prosodically annotated following ToBI guidelines.[4] The utterance's orthographic transcriptions are then combined with the respective phonetic segmentations, using a procedure described in [12], in order to obtain a realistic and multi-leveled description of the spoken utterances. Moreover, those descriptions are enhanced by additional descriptions, such as F0 values of the speech signal and prosodic annotations. The F0 values are assigned to the respective phonetic segments based on the temporal inclusion criterion.

3 Linguistic Analysis

3.1 Speaker-Adapted Prosodic Models

The general quality and naturalness of synthetic voices crucially depends on the building of large databases annotated at multiple levels for the training and testing of prosodic models able to generate adequate rhythmic and intonational patterns. One of the most striking difficulties in the building of new voices in the present framework is that the type of annotation required is extremely time consuming and the models trained for one voice or speaking style are most often inadequate for another. Models trained on a laboratory corpus of read texts or elicited sentences by non-professional speakers, for instance, may hardly be used to build a voice based on new databases recorded by professional ones, as strong mismatches are found both for the phrasing and tonal assignment strategies used. This clearly affects the selection of the units to concatenate, as often adequate exemplars cannot be found and several discontinuities are introduced.

---

[3] In limited domain applications, the text prompts do not aim to be an abstract representation of the language use, thus, occurring frequencies carry relevant information for modeling the language in that specific domain.

[4] http://www.ling.ohio-state.edu/~tobi

It is worthwhile to note that most studies based on laboratory corpora recorded by non-professional speakers present EP as a language with sparse accentuation and just one level of phrasing (e.g. [19]) whereas it is clear from available data of spontaneous speech and professional reading that at least two levels of phrasing are needed to improve our results. Although the basic pitch accent inventory is in agreement with laboratory studies, it is also mandatory to be able to account for rather common pitch accents in our data that are absent or not well enough represented in laboratory corpora (*e.g.* L+H*, in nuclear position, the most frequently used in association with new information or old information that need to be reactivated or ∧H*, consistently used in repetitions for further specification, or to correct given or inferable information).

Most of our effort in what prosody is concerned, has thus been dedicated to accelerate the annotation process by training statistical models for the automatic feature extraction, in order to reduce as much as possible the need for manual intervention. So far, we have been mainly concerned with the improvement of prosodic phrasing models. Those are essential to achieve better results in what tonal scaling is concerned. On the other hand, as the annotation scheme is closed to the English ToBI one, and the phonetic correlates of each type of tonal event for EP are relatively well known, we expect to be able to reduce the number of errors in the automatic tagging of such events. Drawing on previous work in the line of [6], a new database for a professional speaker was automatically parsed with a CART trained and tested on text based annotations, only [18]. In spite of *break/no-break* decisions produced correct results in only 70% of the cases, the manual correction was considerably facilitated. The use of manually annotated data for training prosodic phrasing models accounting for the speaker-specific reading strategies has proved to be worthwhile, as the adapted *break/no-break* detection models reached substantially higher performances (precision=88.44%; recall=93.90%).

3.2   Grapheme to Phone

The current G2P component follows the same approach of the Festival system comprising a lexicon, an addenda and a set of classification trees, one for each symbol of the alphabet. To train the classification trees a rather large lexicon is required. The size of the lexicon depends on the language and on its regularity. In our case we used an EP lexicon with around 80,000 entries. Each lexicon entry includes the word orthography, a part-of-speech (POS) tag and the corresponding sequence of phones with a lexical stress mark on the central vowel of the stressed syllable. The orthography and phonetic transcription must be aligned so that each letter corresponds to a single phonetic symbol. This symbol can represent a single phone, a sequence of phones or no phone at all. The goal of a classification tree is to predict which is the phonetic symbol associated with a given letter of the alphabet in a specific context and for a word with a given POS tag. The context must have a finite length that needs to be optimized for each language. In the case of EP we achieved better results by using 3 letters to the left and 6 to right of the letter being transcribed. This technique produced

93.28% and 99.12% accurate transcriptions in the test set (10% of the full lexicon) at the word and grapheme levels, respectively. The performance measures for the full 80K lexicon were 3.71% and 0.48% for word and phone error, respectively. These results are slightly worse (around 2%) than the ones that we have achieved using phonetically motivated rewriting rules. This approach, however, has the advantage of being automatically trainable and thus easily extendable to other languages.

The system lexicon must include all the words that are not correctly transcribed by the classification trees. For performance reasons, however, we have decided to include also the most frequently used words in EP. The third element of the G2P component is the addenda. It works as an exception lexicon in which the user can override the way the system reads certain words. The addenda can be particularly useful for non-standard words like company names or even foreign words that are not correctly pronounced by the system.

4   System Architecture

4.1   Text Splitter

The input text of a speech synthesizer can have a wide range of variability. Moreover, one may assist to a dramatic degradation of the system's overall performance when long sentences, paragraphs and text documents are processed as a single unit. These large text chunks require longer processing that delay the generation of the audio output. The input text is split into sentences based on its punctuation in order to minimize that delay. However, punctuation marks can be mistakenly parsed (*e.g.* dots are not used solely for sentence breaks, they can also be used in abbreviations, numbers and even dates). A solution for this is to require that the full stop is followed by a space or a capitalized word. Such restrictions handle the most cases like numbers and dates but not the abbreviations, which are addressed as follows. A large abbreviation inventory is built in advance to enable the system to spot such tokens within the input text. The abbreviation identification can help in distinguishing the dot from a full stop, but the system must also take into account that some abbreviations can occur at the end of sentences (*e.g.* etc.).

4.2   Text Normalizer

The text normalizer is responsible for rearranging text in a normalized form, so that the following components can be more effective. It is a task that requires constant maintenance. Moreover, conventions are useless when it comes to deal with general normalization problems, since there are many writing conventions for similar contexts. For example, numbers in a certain language can have different convention domains, like economical and scientific domains. This is a strong hit in any attempt to design *general* text normalization method. Besides, addressing so many ambiguities can make the system inflexible in the presence of new paradigms.

Considering a general approach, Dixi addresses the normalization problem in two distinct steps. Firstly, text tokens are tagged according to their syntactic form. For example, the token *"1234,34"* is tagged as a number, whereas the token *"76-12-01"* is marked as a date (according to traditional European Portuguese standards). The real text normalization only takes place in the following step, since token to word rules are applied only there. Hence, with all tokens already tagged, specific modules are then applied to carry out the necessary conversions. However, this solution is just a course of action, it does not solve the problem as a whole. In order to increase the system flexibility, both identification and modification levels were categorized according to *Language*, *Region* and *Domain*. The categorization of the normalizer levels minimizes ambiguity, and enables the users to parameterize the normalizer so that it can meet their own demands.

4.3   Part-of-Speech Tagging

The fundamentals of the POS tagger currently used in Dixi was described in [13]. It consists of a lexicon comprising around 22,000 orthographic forms, containing a pair list in the form of *tag/probability* each. The lexicon is used along with a POS tri-gram grammar in order to find the most likely POS tag sequence, whose tags are subsequently assigned to the respective words.

4.4   Prosodic Phrasing

Voice-specific word break models enconded in the form of CARTs were trained as described in 3.1. In run time, the prosodic phrasing is performed making use of the model specifically trained for the currently selected voice.

4.5   Phonological and Phonetic Descriptions

As soon as the word list is grammatically tagged, the pronunciation generation is triggered. This procedure consists of the following steps. Given a particular word, a user-supplied lexicon addenda, if any, is searched for a that written form with a matching grammatical tag. When a search is well-succeeded, pronunciation generation procedure is finished and the respective phonetic sequence is used. If no such entry is found, the procedure resumes by searching for the word pronunciation in the lexicon. Finally if the word is still not present in the lexicon, not even with another grammatical tag, the pronunciation is generated by a set of CARTs, trained as described in 3.2. Up to this stage, pronunciations were generated for isolated words. However, post-lexical phonological processes play an important role in connected speech. Hence, a set of post-lexical rules is then applied to address that problem and produce more realistic utterance descriptions at this level.

4.6   Acoustic Synthesis

Unit Selection Synthesis. The waveform generation is based on a multi-level version of the *cluster unit selection* algorithm [2] and will be further described in a future paper. Our unit selection algorithm makes use of phone target durations to discard durational outliers in run time, rather than adjusting the durations of the selected units. Therefore, despite the local signal modifications carried out in order to soften the transitions at the concatenation points, the system uses the speaker-specific prosody, available in the recordings. Moreover, even though phonetic segments constitute the basic acoustic units used by *Dixi*, they are searched in a top-down fashion, in order to first search for candidate units coming from the most appropriate prosodic and phonetic contexts (e.g. phone belonging to a word in a specific position, or a syllable in a specific position, or triphone, or a diphone, etc.).

Parametric Synthesis. A parametric approach, based on the HMM synthesis, can also be used within the Dixi system. Such an approach automatically draws a correlation between acoustic features and a set of symbolic features derived from the input text. The training procedure of the HMMs is carried out by the HTS [15] Toolkit. Using tree-based context cluster HMM models, HTS extracts spectral information, average F0 and a voiced/unvoiced decision every 5ms. The features utilized by the HMM synthesizer as well as the context clustering questions, will be described in a future paper. The speech signal is generated using the MLSA[5] filter, proposed in [7].

5 Conclusions

We have described the development of a new text-to-speech system for EP. Besides, a strong emphasis was also put in the description of the language resources needed by the system, as well as the training methods used in the linguistic analysis of the input text. Finally, we described how the system can be parameterized to meet the user-specific requirements.

Acknowledgments

This work was funded by PRIME National Project TECNOVOZ number 03/165.

References

1. Black, A.W., Lenzo, K.A.: Flite: a small fast run-time synthesis engine. In: SSW4 (2001)
2. Black, A.W., Taylor, P.: Automatically clustering similar units for unit selection in speech synthesis. In: Eurospeech 1997 (1997)
3. Black, A.W., Taylor, P., Caley, R.: The Festival Speech Synthesis (2002)
4. Bonafonte, A., Hoge, H., Kiss, I., Moreno, A., Ziegenhain, U., Heuvel, H., Hain, H., Wang, X., Garcia, M.: TC-STAR: Specifications of language resources and evaluation for speech synthesis. In: LREC 2006 (2006)

---

[5] Mel Log Spectral Approximation.

5. Breiman, L., Friedman, J.H., Olshen, R.A., Stone, C.J.: Classification and Regression Trees. Chapman and Hall, Boca Raton (1984)
6. Hirschberg, J., Prieto, P.: Training intonational phrasing rules automatically for english and spanish text-to-speech. Speech Communication 18 (1996)
7. Imai, S.: Cepstral analysis synthesis on mel frequency scale. In: ICASSP- 1983 (1983)
8. Moniz, H., Mata, A.I., Viana, C.: On filled and prolongations in european portuguese. In: Interspeech 2007 (2007)
9. Neto, J.P., Meinedo, H.: Combination of acoustic models in continuous speech recognition hybrid systems. In: ICSLP 2000 (2000)
10. Oliveira, B., Pona, C., Matos, D., Ribeiro, R.: Utilização de xml para desenvolvimento rápido de analisadores morfológicos flexíveis. In: XATA 2006 - XML: Aplicações e Tecnologias Associadas (2006)
11. Paulo, S., Oliveira, L.: Generation of word alternative pronunciations using weighted finite state. In: Interspeech 2005 (2005)
12. Paulo, S., Oliveira, L.C.: MuLAS: A framework for automatically building multi-tier corpora. In: Interspeech 2007 (2007)
13. Ribeiro, R.D., Oliveira, L.C., Trancoso, I.M.: Using morphossyntactic information in tts systems: Comparing strategies for european portuguese. In: Mamede, N.J., Baptista, J., Trancoso, I., Nunes, M.d.G.V. (eds.) PROPOR 2003. LNCS, vol. 2721. Springer, Heidelberg (2003)
14. Taylor, P., Black, A.W., Caley, R.: Heterogeneous relation graphs as a formalism for representing linguistic information. Speech Communication 33 (2001)
15. Tokuda, K., Zen, H., Black, A.W.: An HMM-based speech synthesis system applied to english. In: 2002 IEEE SSW (2002)
16. Trancoso, I., Viana, C., Duarte, I., Matos, G.: Corpus de dialogo CORAL. In: PROPOR 1998 (1998)
17. van Santen, J.P.H., Buchsbaum, A.L.: Methods for optimal text selection. In: Eurospeech 1997 (1997)
18. Viana, C., Oliveira, L.C., Mata, A.I.: Prosodic phrasing: Machine and human evaluation. Speech Technology 6 (2003)
19. Vigário, M., Frota, S.: The intonation of standard and northern european portuguese. Journal of Portuguese Linguistics 2(2) (2003)
20. Weiss, C., Paulo, S., Figueira, L., Oliveira, L.C.: Blizzard entry: Integrated voice building and synthesis for unit-selection tts. In: Blizzard 2007 (2007)
21. Young, S., Evermann, G., Hain, T., Kershaw, D., Moore, G., Odell, J., Ollason, D., Povey, D., Valtchev, V., Woodland, P.: The HTK Book (for HTK Version 3.2.1) (2002)
22. Ziegenhain, U., Hoge, H., Arranz, V., Bisani, M., Bonafonte, A., Castell, N., Conejero, D., Hartikainen, E., Maltese, G., Oflazer, K., Rabie, A., Razumikin, D., Shammass, S., and Zong, C.: Specification of corpora and word lists in 12 languages. Report 1.3, Siemens AG (April 2003)

European Portuguese Articulatory Based Text-to-Speech: First Results

António Teixeira[1], Catarina Oliveira[1], and Plínio Barbosa[2]

[1] DETI/IEETA, Universidade de Aveiro, Portugal
[2] IEL, UNICAMP, Brazil
ajst@ua.pt

Abstract. In this paper we present recent work on the development of Linguistic Models, resulting in a first "complete" articulatory-based TTS system for Portuguese. The system, based on TADA system, integrates our past work in automatic syllabification and grapheme-phone conversion plus a first gestural specification of European Portuguese sounds. The system was integrated with SAPWindows, an articulatory synthesizer for Portuguese. A demonstration of the system capabilities and a first perceptual evaluation are presented.

1 Introduction

Articulatory synthesis produces speech using models of physical, anatomical and physiological characteristics of the human production system. This technique models the system directly, instead of modelling the signal or its acoustic characteristics. This type of synthesizer has not yet been used in technological applications because of the costly computations involved and the underlying unsolved theoretical and practical problems. The production of fricatives, for example, has not been fully understood. Articulatory synthesis has also to deal with the particular characteristics of each language and the scarcity of articulatory data.

Recent developments [1,2] show that articulatory synthesis is worth revisiting as a research tool and as part of TTS systems. Because articulatory synthesizers have parameters which can be conceptualized, even though a token turns out to be wrong, a lot can be learned from trying to fix it. The work to produce an usable system can also be a fruitful way of fostering linguistic knowledge.

We have been working in articulatory synthesis of Portuguese, with encouraging results[3]. Having as a long-term aim the development of a articulatory text to speech system for Portuguese, in this paper we present recent work on the development of several modules and their integration to create a first "complete" from word to sound articulatory based synthesis system for Portuguese. Such system must comprise models of linguistic processing and the conversion of the discrete linguistic variables to the continuously varying articulatory parameters.

The paper is structured as follows: section 2, presents a brief description of the adopted Articulatory Phonology (AP) framework; section 3, describes the

A. Teixeira et al. (Eds.): PROPOR 2008, LNAI 5190, pp. 101–111, 2008.

TADA application, a computational implementation of AP; section 4, presents the developed system, with some information on its components; last 2 sections, results (of a perceptive evaluation) and conclusions are presented.

2   Articulatory Phonology

In this section of the paper, we present a very brief overview of the Articulatory Phonology (AP) framework [4], as it forms the basis of our linguistic model.

In AP, the basic unit of speech is not the segment or the feature, but the articulatory gesture. This theory claims the units, the gestures, are dynamic actions. Gestures are essentially instructions to achieve the formation (and release) of a constriction at some place in the vocal tract (for example, an opening of the velum or a raising of the tongue body). At the same time, gestures are units of contrast that play a role in representations similar to that of the feature [5]. Segments in traditional phonology are considered under the framework of Articulatory Phonology to be, in general, a combination of gestures. Single gestures can be isomorphic to segments, such as [b].

Formally, gestures are specified using a set of tract variables that refer to both the location (CL) and the degree (CD) of constrictions in the vocal tract. Five tract variables are identified: Lips (L) Tongue Tip (TT), Tongue Body (TB), Velum (VEL) and Glottis (GLO). CL specifies the place of the constriction in the vocal tract and it takes the values of [labial], [dental], [alveolar], [postalveolar], [palatal], [velar], [uvular] and [pharyngeal]. For the CD variable the descriptors are: [closed] (for stops), [critical] (for fricatives), [narrow], [mid] and [wide] (approximants and vowels). Thus, a gestural specification for the alveolar stop [t] would be Tongue Tip [CD: closed, CL: alveolar]. This specification defines the 'rest position' (target).

Each tract variable is additionally specified in terms of stiffness and damping ratio. These values, combined with the equilibrium position, define the task dynamic regime of a gesture, modelled as a damped mass-spring equation [6]. The goal for a gesture is achieved by the coordinated action of a set of articulators.

Gestures are spatiotemporal units. Each gesture has a duration in time and an internal cycle. This cycle begins with the onset of movement, progresses to the point where the target is reached, then to the release, where the movement away from the constriction begins, and finally to the offset, the point where the articulator ceases to be under active control of the gesture.

Individual gestures, 'atoms', combine with each other to form larger combinations, 'molecules' (segments, clusters, syllables, etc.). These combinations are governed by phasing principles: a certain point in the trajectory of one gesture is phased with respect to a certain point in the trajectory of another gesture. The pattern of intergestural coordination along with the interval of active control for individual gestures is represented in a two-dimensional graphic, called gestural score. A gestural score is a set of idealized instructions to articulators, which require some interpretation and modification in order to be implemented.

3   TADA – TAsk Dynamics Application

The articulatory phonology approach has been incorporated into a computational model being developed at Haskins Laboratories [7,6]. Figure 1 portrays the components of the system: first, the **Linguistic Gestural Model** analyses the input into a set of discrete, concurrently active gestures and specifies a gestural score; second, the **Task Dynamic Model** calculates the articulatory trajectories given the gestural score; third, these articulator trajectories are input to the **Vocal Tract Model** which then calculates the resulting global vocal tract shape, area function, transfer function, and speech waveform and generates an acoustic signal.

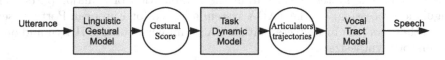

Fig. 1. Articulatory Phonology computational implementation

Recently, **TADA** (TAsk Dynamic Application), a new Matlab implementation of the computational model, was made available for research [8]. In this new version of the model, intergestural timing is determined by planning oscillators associated with each speech gesture [9].

The Gestural Model combines a coupled-oscillator model of inter-gestural planning, and a gestural-coupling model. The coupling model is based on syllable structure and, taking as input a text string it generates a graph that specifies the gestures composing the utterance (represented in terms of tract variable dynamical parameters and articulator weights) and the coupling relations among the gestures' timing oscillators. The coupled oscillator model of planning takes as input the coupling graph, and generates a gestural score, specifying gesture's activation intervals.

Gestural scores are the input to the task dynamic model [6], which generates the resulting time functions of the vocal tract constriction variables and articulator trajectories. The conversion from the gestural score to articulatory movements is assumed to be language independent. In TADA the articulators are those of CASY vocal tract model [10]. The articulator trajectories are then used to calculate the acoustic output in the vocal tract model. In the current version, TADA integrates a partial implementation of CASY [7], with several limitations: it does not include any treatment of nasality or any source control (friction, aspiration, f0). However, the output files, automatically produced by the system, can be used as control parameters for the HLsyn (pseudo-articulatory) synthesizer.

The TADA system has been designed to allow for the development of articulatory models in other languages through the use of language-specific dictionary files and gestural databases. It is also possible to add output for other external synthesizers.

In the current TADA models, the variable shape of constriction (CS) is not implemented. The possibility of controlling the cross-sectional shape would be essential for the correct treatment of lateral sounds. Furthermore, all gestures are assumed to be critically damped. However, the simulation of sounds such as taps and trills may require the revision of that assumption. These two problems constitute an important limitation to our objective of producing a full TTS system for EP. Nevertheless, by making available an interesting combination of a gestural model with a task dynamic model, the TADA system presented itself as the best choice for our purposes.

4   System Architecture and Strategies

Fig. 2 presents a block diagram of the main parts of the articulatory-based TTS system. The system results from the combination of 3 major parts: (1) Linguistic Processing; (2) the TADA system adapted for European Portuguese; (3) synthesizers (the incomplete Matlab CASY implementation, HLsyn and our articulatory synthesizer).

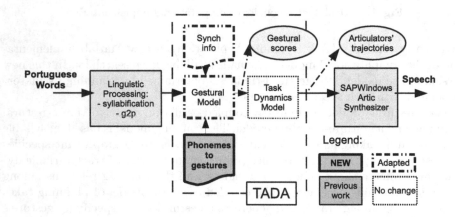

Fig. 2. General architecture of the developed articulatory-based TTS system

Some of the parts, as the Linguistic Processing, were entirely developed by the authors; others, as the Gestural Model, are adaptations to the original TADA system. The most important challenge in adapting TADA was the gestural definition of European Portuguese sounds. The precise definition of the articulatory characteristics of all EP segments is a considerable challenge, complicated by the scarce direct measures of EP production and the (many) limitations of all the models integrating the system.

4.1   Linguistic Processing

As the input for the Gestural Model in TADA system must provide information on the phones, stress and syllable structure for the words to be synthesized, the Linguistic Model must handle grapheme-to-phone (g2p) and syllabification from text. The syllable information is essential for the correct functioning of the Syllable-based Gestural Coupling Model.

Our work for this part of the system consisted in using our previous work on automatic syllabification from graphemes and previously developed g2p systems. Syllabification is performed by an adaptation of Mateus and d'Andrade algorithm [11] to graphemic input [12] Grapheme-phone conversion module uses a combination of a rule based system (implemented as a finite state transducer) with two machine learning systems (TBL - Transformation Based Learning - and MBL - Memory Based Learning) [13].

4.2 Gestural Model for European Portuguese

After converting the orthographic input to a syllabified broad phonetic transcription, the following step was the specification of gestural composition of each Portuguese segment. This information was coded in the Phonemes to Gestures dictionary (part of Fig. 2) containing all the Portuguese segments in SAMPA notation and the gestures associated with them.

The Portuguese gestural dictionary was created by an iterative process:

– descriptor values were, in a first approximation, estimated from phonetic literature on Portuguese and from articulatory data in general (since this kind of data is very scarce in EP, the main difficulty in all this process was to provide empirical support for an implementation of a Portuguese version of the Gestural Model);
– a specific 'target' value for location and degree of a constriction was estimated from examining MRI data. The MRI images were particularly useful for vowels;
– creation of test words and pseudo-words (e.g. CV or VCV sequences) and informal assessment of word and phone intelligibility and quality.

In the dictionary, the gestures are represented symbolically by four descriptors: Constricting Organ, Oscillator Type, Tract Variable, Constriction Type.

Oscillator Type identifies the timing oscillator that will be associated with the gesture. The oscillators 'clo', 'crt' and 'nar' correspond to the primary oral constriction for stops, fricatives and glides, respectively. Each of these may have a paired release oscillator ('rel'). Glottal gestures are associated with the 'h' oscillator, and velum lowering with the 'n' oscillator. Oscillators for vowels are 'v' (tongue constrictions) and 'v_round' (lip constrictions).

The *Tract Variables* currently implemented in the computational model were already presented in sec. 2. *Constriction Type* descriptors are similar to those defined in the theoretical proposal (sec. 2): CL for lips can be protruded (PRO) or dental (DENT); tongue tip CL can be dental (DENT), alveolar (ALV), alveolo-palatal (ALVPAL), palatal (PAL); CL descriptors for tongue body are palatal (PAL), velar (VEL), uvular (UVU), uvo-pharyngeal (UVUPHAR) and pharyngeal (PHAR). For the CD the descriptors are closed (CLO), critical (CRIT), narrow (NAR), wide (WIDE) and vocalic (V).

Vowels can be adequately characterized using one or two tract variables: Tongue Body alone or combined with Lips. TBCL can assume any of the values presented above; TBCD is generically defined as 'V' and the target specified in millimetres. By default the model assumes velum is closed and the glottis in the conditions to produced voiced sounds.

As an example, we present, in Table 1, the gestures for EP anterior oral vowels. The table shows the exactly content of the TADA configuration file for these 3 sounds. The 3 vowels are represented as having [palatal] constriction locations. To account for the different tongue positions of tongue in the anterior-posterior axis, constriction location was adjusted to 80 degrees for [i] and 90 degrees for [e], assuming [ɛ] the default value for palatal. The constriction degree was defined, based on MRI images measurements, as follows: 9 mm for [ɛ], 6 mm for [e] and 3 mm for [i].

The nasal vowels have not yet been included. As English doesn't include these sounds, there are some implementation "limitations" that must be solved, particularly regarding the synchronization of the velum with the consonantal and vocalic gestures of the syllable.

Table 1. Gestures for EP anterior oral vowels and 2 consonants, [b] and [m]

V	Org	Osc	TV	Const	Targ	Stif		C	Org	Osc	TV	Const	Targ	Stif
i	TB	v	TBCL	PAL	80	.		b	Lips	clo	LA	CLO	.	.
	TB	v	TBCD	V	3	.			Lips	rel	LA	REL	.	.
e	TB	v	TBCL	PAL	90	.			Velum	clo	VEL	CLO	.	.
	TB	v	TBCD	V	6	.		m	Lips	clo	LA	CLO	.	.
ɛ	TB	v	TBCL	PAL	.	.			Lips	rel	LA	REL	.	.
	TB	v	TBCD	V	9	.			Velum	n	VEL	WIDE	.	.

The criteria to define consonants are more heterogeneous. For stops and fricatives they consist essentially in defining constriction degree and location; for nasal consonants velum must also be defined; laterals demand the additional inclusion of a shape variable; taps and trills imply control of parameters such as stiffness. Sample definitions for 2 EP consonants are, also, presented in Table 1. The bilabials /b, m/ are specified here for the gestures Lips [closed]. Additionally, the nasal is specified for Velum [wide]. The correct definition of laterals would demand the inclusion of a constriction shape variable. Nevertheless, in the case of the alveolar lateral [l], the auditory effect was obtained by using a complex gestural constellation consisting of two coordinated oral gestures, TT [narrow, alveolar] and a TB [narrow, velar] [14]. In order to simulate the tap ([ɾ]) produced with a short duration TT gesture [closed, alveolar], we used stiffness to control the durational characteristics of TT movement associated with the gesture.

For intergestural coupling we used roughly the same phasing principles proposed for American English, due to shortage in such descriptions and data for EP.

The complete proposal for a gestural definition of EP will be presented elsewhere.

4.3   Synthesizer

Integrating the synthesizer into the system was considered a high priority. The main anticipated advantages are the potentialities for synthesizing nasal sounds and the possibility to change and update all the models. Even with limitations (it was essentially developed to handle vowels and nasals), this way we can have a complete open TTS system.

Briefly, our articulatory synthesizer [3] contains three major blocks: a source model; an articulatory model; and an acoustic model. Model articulators are tongue body, tongue tip, jaw, lips, velum and hyoid. In our implementation, targets for each articulator can be defined independently.

The work in the interface between TADA an our synthesizer comprised two types of procedures:

1. addition of a new output format to TADA. For these, after understanding CASY implementation, the necessary mappings between CASY parameters and our synthesizer parameters were implemented.
2. some adjustments to our articulatory synthesizer models. It was necessary to adjust some of the fixed structures sizes, as tongue radius, to achieve a better match between both models. The control of Lip opening was also updated to work as in CASY, where the parameter controls directly opening area independently of jaw movement. As our synthesizer prevented the passage of tongue through the tract walls and TADA uses this fictitious passage to create non-pontual occlusions, the restriction was disabled. Some adaptations to the source model were also performed to have a first treatment of unvoiced consonants, but work must be continued.

4.4   Synthesis Example

As an example of system functioning, in this section, we present the output of the several system models for the sample word "banana".

For our example, the Linguist Processing module produces the following information (using SAMPA): `banana (b-60_)(n-a1_)(n-60_)`. For each syllable, onset, nucleus and coda are separated. In the first syllable of our example, the onset is occupied by the stop consonant [b], nucleus by vowel [ɐ] and the coda is empty. The number following the vowel represents stress.

The output of the Gestural Model is presented in Fig. 3. It illustrates the use of LA tract variable in the first syllable to produce the bilabial [b], the effect of

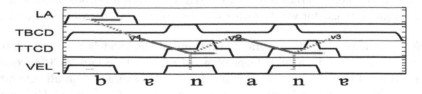

Fig. 3. Gestural score produced by the system for word banana

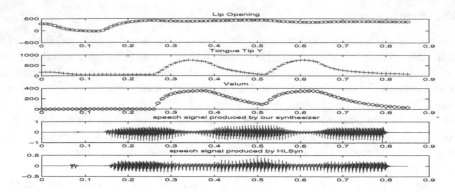

Fig. 4. Trajectories for articulator parameters Lip Opening, Tongue tip, Velum, and the speech signal produced by our articulatory synthesizer and HLSyn

the 3 vowels (v1,v2,v3) in TBCD and the use of TTCD for the two [n]. The lines between the several tract variables or inside the same variable represent gestural phasing.

As final simulation results, TADA system produces articulators' trajectories and creates output files for use with external synthesizers. The trajectories for 3 of the articulators, as well as the speech signal produced by our synthesizer, are presented in Fig. 4.

5   Identification Test

To put the Gestural Model to the test and also to obtain information on the main problems of the system, synthesized tokens were used in an open set identification test. The test was administered individually to 9 subjects in a quiet office using headphones. The listeners were asked to report the words or sounds they were able to identify.

For the test, 50 words were randomly selected from the "Português Fundamental" corpus [15], excluding all the words that presently the system is not capable to process (words with nasal vowels, the palatal lateral [ʎ] and the trill [R]). The words were distributed taking in consideration the number of syllables (50% with 2, 30% with 3 and 20% of 4 or more syllables) and the syllable structure (80% for words with only open syllables). Syllable and phone information were obtained manually, avoiding errors from the Linguistic Processing modules.

All the words were synthesized with HLSyn. Present limitations of our synthesizer prevented its use for words with fricatives; only 19 words were synthesized.

Results: As expected, the several limitations of the models and articulatory synthesizers resulted in a high level of word error rate. Only 25.3 % correct identification of the complete word was achieved. On average, stimuli generated by our synthesizer obtained better results in this metric (33.3 % vs 22.2 %), being the difference statistically significative ($p = 0.002$ in a paired sampled t-test).

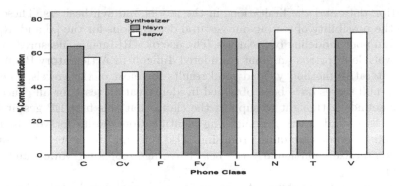

Fig. 5. Identification by class: stops (C), fricatives (F), laterals (L), nasal cons. (N), taps (T) and vowels (V). 'v' for voiced. Gray for HLsyn, white for our synthesizer.

Results for individual test participants ranged from 17.4 % to 40.6 %. Higher identification scores were obtained for: "mínimo" (minimum)['minimu] (94.4 %), "camisola" (shirt) [kɐmi'zɔlɐ] (77.8 %), "olá" (hello) [ɔ'la] (77.8 %), "sofá" (sofa) [su'fa] (77.8 %), "chocolate" (chocolate) [ʃuku'latɨ] (77.8 %), "pato" (duck) ['patu] (77.8 %), and "mapa" (map) ['mapɐ] (66.7 %). The most problematic were words such as "virar (turn) [vi'raɾ], "vosso" (yours) ['vɔsu] or "seco (dry) ['seku].

The identification percentage for the classes of phonemes included in our test are presented in Fig. 5. Best results were obtained for vowels and nasal consonants. For laterals the results depend on the synthesizer, being the HLsyn results similar to the ones obtained for the two better identifiable classes. Worst identification results, not surprisingly, were obtained for voiced fricatives and taps. A relevant result is the one obtained for tap [ɾ], with our synthesizer outperforming HLSyn average identification. On average, our synthesizer results are also good or even better than the ones obtained for HLsyn for vowels, nasal consonants and voiced stops. Paired samples t-tests didn't confirm as significative the effect of synthesizer for all the classes ($p > 0.05$).

For individual phones, best results were obtained for [f], [u] and [m], all above 80 %. The 5 worst results, all below 25 %, were obtained for [ɲ] [b], [v], [ʒ] and [e]. The poor results for [e] can be attributed to the non-inclusion of stress effect in the words containing the sound. Three of the other problematic phones point to a problem in producing voiced consonants.

6   Conclusions

We presented a new, and to best of our knowledge first, articulatory-based text-to-speech system for European Portuguese. One of the most important tasks performed to create the system concerned the development of the actions required to convert segments into gestures. The system constitutes a valuable tool to develop and test Articulatory descriptions of Portuguese. The major difficulties in creating the Gestural Model for EP came from the lack of enough production data and the limitations in the model and synthesizers. These prevented the possibility of testing our gestural descriptions for the [ʎ] and [R] and resulted in poor modeling of fricatives. The use of MRI images measures to support vowels descriptions - a point considered difficult in Articulatory Phonology - in our iterative method yielded good results for most of the vowels. The fact that non-null scores have been obtained in identification tests for all concerned kinds of sounds is thought to support the claim that this first EP gestural description is a valid starting point. The identification results for vowels, nasal consonants and stops are quite promising. Also relevant are the results obtained by our synthesizer, particularly the better average scores for vowels and nasal consonants.

Future work must contemplate the addition of nasal vowels, extension of our synthesizer models, proper treatment of the shape variable for laterals and addition of prosodic information to the system. For all these developments, more production data must be obtained, particularly for laterals, taps and trills.

Acknowledgement

The work presented was supported by FCT Project HERON (POSC/PLP/57680/2004) and PhD scholarship SFRH/BD/18058/2004 of Catarina Oliveira. We thank all participants in the identification test, our colleagues involved in the creation of MRI database and TADA developers for making it available.

References

1. Whalen, D.H.: Articulatory synthesis: Advances and prospects. In: ICPhS, pp. 175–177 (2003)
2. Birkholz, P.: Control of an articulatory speech synthesizer based on dynamic approximation of spatial articulatory targets. In: Interspeech, pp. 2865–2868 (2007)
3. Teixeira, A., Martinez, R., Silva, L., Jesus, L., Príncipe, J.C., Vaz, F.: Simulation of human speech production applied to the study and synthesis of European Portuguese. EURASIP Journal of Applied Signal Processing (2005)
4. Browman, C.P., Goldstein, L.: Gestural specification using dynamically-defined articulatory structures. Journal of Phonetics 18, 299–320 (1990)
5. Hall, N.E.: Gestures and Segments: Vowel intrusion as overlap. Phd thesis, University of Massachusetts (2003)
6. Saltzman, E., Munhall, K.: A dynamic approach to gestural patterning in speech production. Ecological Psychology 1/3, 333–382 (1989)
7. Rubin, P., Saltzman, E., Goldstein, L., McGowan, R., Tiede, M., Browman, C.P.: CASY and extensions to the task-dynamic model. In: Proc. 1st ESCA ETRW on Speech Production Modelling, Autrans, France (1996)
8. Nam, H., Goldstein, L., Browman, C., Rubin, P., Proctor, M., Saltzman, E.: TADA (TAsk Dynamics Application) manual. Manual, Haskins Laboratories (2006)
9. Goldstein, L., Byrd, D., Saltzman, E.: The role of vocal tract gestural action units in understanding the evolution of phonology. In: Arbib, M. (ed.) Action to Language via the Mirror Neuron System, pp. 215–249. CUP (2006)

10. Rubin, P., Baer, T., Mermelstein, P.: An articulatory synthesizer for perceptual research. J. Acoust. Soc. America 70(2), 321–328 (1981)
11. Mateus, M.H., d'Andrade, E.: Phonology of Portuguese. OUP (2000)
12. Oliveira, C., Moutinho, L., Teixeira, A.: On European Portuguese automatic syllabification. In: InterSpeech (2005)
13. Teixeira, A., Oliveira, C., Moutinho, L.: On the use of machine learning and syllable information in European Portuguese grapheme-phone conversion. In: Vieira, R., Quaresma, P., Nunes, M.d.G.V., Mamede, N.J., Oliveira, C., Dias, M.C. (eds.) PROPOR 2006. LNCS (LNAI), vol. 3960, pp. 212–215. Springer, Heidelberg (2006)
14. Browman, C.P., Goldstein, L.: Gestural syllable position effects in American English. In: Bell-Berti, F., Raphael, L.J. (eds.) Producing Speech: Contemporary Issues, for Katherine Safford Harris, pp. 19–33. AIP Press (1995)
15. Nascimento, F., Marques, L., Segura, L.: Português Fundamental: Métodos e Documentos. INIC-CLUL, Lisboa (1987)

Statistical Machine Translation of Broadcast News from Spanish to Portuguese

Raquel Sánchez Martínez, João Paulo da Silva Neto,
and Diamantino António Caseiro

L²F - Spoken Language Systems Laboratory, INESC ID Lisboa
R. Alves Redol, 9, 1000-029 Lisboa, Portugal
{raquel.sanchez,dcaseiro}@l2f.inesc-id.pt,
Joao.Neto@inesc-id.pt
http://www.l2f.inesc-id.pt/

Abstract. In this paper we describe the work carried out to develop an automatic system for translation of broadcast news from Spanish to Portuguese. Two challenging topics of speech and language processing were involved: Automatic Speech Recognition (ASR) of the Spanish News and Statistical Machine Translation (SMT) of the results to the Portuguese language. ASR of broadcast news is based on the AUDIMUS.MEDIA system, a hybrid ANN/HMM system with multiple stream decoding. A 22.08% Word Error Rate (WER) was achieved in a Spanish Broadcast News task, which is comparable to other international state of the art systems. Parallel normalized texts from European Parliament database were used to train the SMT system from Spanish to Portuguese. Preliminary non-exhaustive human evaluation showed a fluency of 3.74 and sufficiency of 4.23.

Keywords: Automatic Speech Recognition, Broadcast News Transcription, Acoustic Model, Language Model and Statistical Machine Translation.

1 Introduction

One of the main motivations beyond this research work was the opportunity to expand an existing and optimized Portuguese broadcast news recognition system to process Spanish broadcast news context and consequently to calculate it performance in different languages domain. In the best of our knowledge, is the first broadcast news machine translation system for the Spanish to Portuguese language pair, what did an appealing target.

A great focus has been placed in ASR research area due to emerging demands, for example, from people with hearing disabilities. This have driven an elevated research level and generated a great variety of services and commercial applications. Technological advances in recent years as digital signal processors, faster and affordable memories and increased capacity have also contributed in the evolution of ASR system.

A. Teixeira et al. (Eds.): PROPOR 2008, LNAI 5190, pp. 112–121, 2008.

The SMT research has regain focus research, after a few years where it was left aside in favor of linguistic knowledge representation, mainly due to not be comparable the results and the necessary effort for the latter area.

The costs involved in manual translation by a professional translator have also driven the companies to use SMT as an attractive solution.

A large amount of data is available for English Broadcast News enabling the development of concurrent ASR systems for this task. The current state-of-art WER is less than 16% [1] in real-time (RT) operations and under 13% [2] with 10 times RT. There are Spanish Broadcast News Recognition Systems based on reference English Recognition Systems developed by research centers CMU [3] and BBN [4] [5], cofinanced by DARPA, and IBM, LIMSI and RWTH within a project co-founded by the European commission [6].

For the development of the Spanish system the data available at LDC[1] was used in a total of 30 hours of Latin America Broadcast News audio and different Spanish newspapers corpus.

Table 1. Vocabulary and corpora text dimension, in number of words (Mw: Million words; Kw: Thousands of words) and the respective WER for the different system

	LIMSI_04	IBM_04	RWTH_04	BBN_97	CMU_97	BBN_98
TOTAL(Mw)	400	210	140	157	157	157
VOCABULARY(Kw)	65	47	50	40	40	40
WER(%)	17.8	23.3	17.8	19.9	23.3	21.5

The table 1 presents the total number of words contained in text corpora, the vocabulary size used to develop the language model and the WER of each system. RWTH_2004 has the best WER of 17.8% [6]. CMU_1997 and IBM_2004 have 23.3% WER [3] [6] which are the worst values in the systems under study. These values were obtained with the test set "1997 Hub-4 BN Evaluation Non-English Test Material" by LDC. It corresponds to one-hour Latin America Broadcast News audio, with the same acoustic conditions as those used in the acoustic model training.

The work involved in the development of Statistical Machine Translation of Broadcast News from Spanish to Portuguese has started with the study of a platform that was already being applied to Portuguese Broadcast News task [7]. Then audio and text were selected in order to create a lexicon, acoustic models and language models for the Spanish recognition system. This system was evaluated showing comparable results to other international state-of-the-art systems. Parallel normalized texts of both languages were used to train the translation probabilities and to develop the SMT system.

This paper is organized in the following way: in the section 2 it is described the selection and transformation process of necessary corpora; section 3 explains the changes made in the existing Portuguese Recognizer to adapt it to Spanish and in section 4 it is presented the translation system from Spanish to Portuguese. The last section present the conclusions and future work.

---

[1] http://www.ldc.upenn.edu/

2 Corpora Description

In order to allow the comparison of this work with the studied state-of-the-art systems, the same corpora sources were applied, whenever possible.

2.1 Audio Corpora

We use 30 hours of Latin America Broadcast News audio made available by the LDC, which represents mostly Cuba and Mexico dialect. Their corresponding transcriptions were normalized and transformed to AUDIMUS.MEDIA [7] [8] system.

The acoustic files are divided in 80 NIST SPHERE format files, without compression. The data are 16-bit linear PCM, 16-KHz sample frequency, single channel. Most files contain 30 minutes of recorded material and some contain 60 or 120 minutes. The sampling format requires roughly 2 megabytes (MB) per minute of recording, so the file sizes are typically around 60 MB, with some files ranging up to 120 or 240 MB. The transcripts are in SGML format, using the same markup conventions.

This corpus is divided in 23 hours, corresponding to 63 files, for training set (75%), 4 hours, 10 files, for development set (15%) and 3 hours, 7 files, for test set (10%). The audio selection process for building each set tried to give similar coverage to each phone in the different dialects, creating a more robust acoustic model to the dialectal variability. On the other hand, news of older dates were used in training set and news of more recent dates were used in development and test set, avoiding the context-dependence between news of near dates in each different set.

2.2 Text Corpora

A statistical language model requires a large quantity of data that should be adapted to the task to obtain proper probabilities. We had available a corpus with audio corpora training set transcriptions, they are totally adapted to Latin America Broadcast News task, but with a total of 300,000 words, being an insufficient dimension to generate a statistical language model. We created another corpus using a newspapers set from LDC, namely "Spanish Gigaword First corpus Edition", adding newspapers from previous editions of "English News Text" and "Spanish Newswire Text", which were not included in the last edition. They constitute a large data set of about 720 Million words, necessary to generate the statistic language model, despite the newspapers grammatical constructions are more formal than in Broadcast News. Text corpus was divided in training (75%), development (15%) and test (10%) sets, following the same rules than audio corpora. It was normalized by removing labels, punctuation and special symbols. Other normalization step expanded abbreviations, numbers and acronyms.

2.3 Parallel Text Corpora

The parallel text corpus consists of proceedings of the European Parliament session. This corpus was assembled by Philipp Koehn [9] and has been extensively used by researchers in Statistical Machine Translation. The language used in this corpus is more formal than Broadcast News, and consists of approximately 1.3 Million sentence pairs.

It was necessary to normalize this parallel text, deleting formatting tags and punctuation, and expanding abbreviations. For the Spanish side, we used the same tools as used for preparing the language model corpus. For the Portuguese side, an existent normalizer tool was used. Finally, we removed sentences deemed too long or nonexistent in one of the languages, obtaining approximately 700,000 sentence pairs to train the SMT system.

3 Spanish Broadcasts News Recognizer

3.1 Introduction

The automatic broadcast news recognition is still a challenge due to not resolved questions, since the almost frequent and unpredictable changes, in the speaker, type of speech, the topic, vocabulary, and the record and channel conditions, between others. Then a very important work in this research area is the obtaining of big quantities of audio and text resources with these characteristics included. Language model, acoustic model, vocabulary and lexicon to Spanish task did not exist previously to our work, being necessary to develop a complete system with specific language tools.

3.2 Reference Platform

The AUDIMUS.MEDIA system [7] is a hybrid speech recognition system that combines the temporal modeling capabilities of Hidden Markov Models (HMMs) with the pattern discriminative classification capabilities of multilayer perceptrons (MLPs).

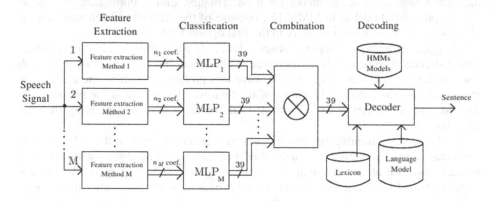

Fig. 1. AUDIMUS.MEDIA recognition system and processing stages [7]

In figure 1 is represented AUDIMUS recognition system. The first stage of processing comprises speech signal feature extraction where three different methods are used PLP [10], log RASTA [10] and MSG [11]. In the MLPs classification stage posteriori probabilities are generated to the 39 possible context-independent phones, 38 for the Portuguese language, and 1 representing the silence. The individual MLP processing results are combined in the next stage.

It use Weighted Finite-State Transducers Technology (WFST) [12] in the decoding stage, where combined phone stream is processed to produce the resulting sentences based on a finite vocabulary and a language model that represents the task restrictions set.

3.3 Vocabulary and Lexical Model

A similar dimension of vocabulary compared with state-of-the-art systems was desired. First we selected 70,500 most common words from the newspapers corpus. Then, we added some not included words from transcriptions training set. Finally, we filtered foreign words and uncommon acronyms to obtain 64,198 words vocabulary.

It was used an automatic grapheme to phone transcription system similar to [13] to generate the phonetic transcriptions. The lexicon uses symbolic representation from SAMPA, plus [N] and [z] phones. The total phone set comprises 32 phones, including a silence phone. In addition, we made manual lexicon corrections of typical foreign words. It was also created, a program based on regular expressions that detects abbreviations and transcribes them using a set of rules.

3.4 Alignment and Training of Acoustic Model

We use generic acoustic model without speaker-dependence, due to the high number of speakers in the corpus audio training set. In order to create this acoustic model, it is necessary an initial phone/audio alignment.

One of the options available was to train the model with small Spanish audio corpora with good acoustic conditions, as in CMU [3]. However it was decided to obtain the initial Spanish acoustic model by transformation of the Portuguese Broadcast News optimized model as in LIMSI [6], because of the difficulty to generate manual alignments at phone level, necessary in the first option.

We generate a phone mapping between the 23 Spanish phones (22 sound phones, plus a silence phone), and the 39 Portuguese phone set. For the remaining 9 Spanish phones there is not direct mapping. To solve this problem, we chose the Portuguese phone set with the most similar sound of the Spanish phone. Having the phone transformed, and parameters optimized, the acoustic model was trained applying an iterative process.

Firstly acoustic training set was aligned by the model optimized for Portuguese Broadcast News, and then we transformed the corresponding results with mapping phones described above, obtaining initial Spain Broadcast News targets. We made 4 alignments and MLP training with the new Spanish network. Since after these alignments was no significant change in the recognition results, we stopped the training process.

3.5 Language Model

In order to create language model we used the transcriptions and newspapers text corpora. It was not possible to separately use them, since individually did not gather the necessary characteristics to create a robust language model. The transcription text corpus is adapted to the Broadcasts News task, but it has a small dimension (300,000 words), and Newspaper corpus has a big dimension (700 Millions words) but their form is not a good representation of spontaneous speech of Broadcast News. After studying several alternatives [14], it was decided to first generate a language model for each individual corpus with the same vocabulary and then interpolate them, reducing the perplexity that models have separately.

Table 2. Newspaper Corpora LM Perplexity and dimension for different cut-off

Cut-off	n-gram				Total	PPL
	1	2	3	4		
50-50-50	64,198	109,880	128,942	135,008	438,028	305.78
25-25-25	64,198	216,904	288,024	302,562	871,688	250.59
4-4-4	64,198	1,378,934	2,819,181	3,153,520	6,993,616	158.76
2-4-4	64,198	2,849,752	2,819,181	3,153,520	8,886,651	153.27
2-3-4	64,198	2,849,750	3,994,811	3,153,520	10,062,279	150.93
2-2-3	64,198	2,849,750	7,599,741	5,025,741	15,539,430	146.48

Table 3. Transcription Corpora LM Perplexity and dimension for 3-4-grams

	n-gram				Total	PPL
	1	2	3	4		
order 3	64,198	99,883	168,284	----------	332,365	327.53
order 4	64,198	99,883	168,284	184,690	517,055	359.07

Table 4. Interpolated LM perplexity and dimension

Cut-off	n-gram				Total	PPL
	1	2	3	4		
50-50-50	64,198	172,306	267,548	303,937	807,989	170.47
25-25-25	64,198	266,685	413,675	464,146	5,968,687	151.77
4-4-4	64,198	1,400,967	2,903,685	3,289,301	7,658,151	110.09
2-4-4	64,198	2,865,844	2,903,685	3,289,301	9,123,028	106.86
2-3-4	64,198	2,865,844	4,073,443	3,289,301	13,848,416	105.41
2-2-3	64,198	2,865,844	7,669,114	5,156,557	15,755,713	102.60

For the newspapers corpus, we generated 4-gram language models through SRLIM tools[2], using discounting of Kneser-Ney [15]. We also conducted experiences with different cut-off values [16]. Table 2 shows the perplexity (PPL) for each cut-off experiment and different n-gram orders on the development set. The obtained PPL can be considered a good representation of the adaptation of the language model to Broadcasts News task, since OOV rate is approximately 1%.

For the transcriptions corpus, we generated two N-gram language models of 3rd and 4th order. Cut-off were not apply, due to the limited corpora dimensions. The PPL results are presented in Table 3. PPL was calculated on the audio corpora development set. This value was high because of the small data size. It was chosen a 4-gram language model in order to have a greater representation in the interpolation, despite having a greater perplexity than 3-gram language model.

---

[2] http://www.speech.sri.com/projects/srilm/

Finally both models were interpolated with the SRILM tool, giving greater weight to the transcription language model (λ=0.68). In Table 4 is observed an improvement of perplexity in the interpolation results in relation to individual models. In the end, it was chosen a language model with cut-off of 2-3-4 and perplexity 105.41. The model with lower perplexity had a greater dimension, penalizing overall real time system's performance and increasing the write error probability in the n-gram selection.

3.6 Evaluation

Two different evaluations sets were used. WER was calculated in audio corpora test set, corresponding to 7 files and 3 hours in total. In the table 5, it is represented the different files names of the test set and their individual WER. It is observed that for su97612.sph WER is larger than the other audio files because their acoustic conditions are worse than the others. The total mean was also calculated, obtaining a value of approximately 25.62% WER.

Also the WER was calculated with the same test set (called h4ne97sp.sph, made available by the LDC) as state-of-the-art systems. In table 5 it is observed a value of 22.08% WER. This is a comparable value to those obtained in the systems studied previously.

Table 5. Test set WERs: It is represented the recognition evaluation with the test set selected from audio corpora and test set eval97 by LDC

Audio Name	WER%
se97406.sph	23.02
su97610.sph	22.16
su97610.sph	27.31
su97611.sph	27.84
su97612.sph	33.12
sv97725b.sph	20.70
sv97725c.sph	27.04
TOTAL	25.62
h4ne97sp.sph	22.08

4   Machine Translation

We decided to use a statistical approach to machine translation, as the phrase-based SMT system for Spanish to English [17]. This approach has advantages relative to others systems [18], namely, it is a language independent technology, does not require linguistic experts, allows fast creation of prototypes, and the statistical framework is compatible with the statistical techniques used in automatic speech recognition.

The corpus was based on parallel texts from European Parliament session transcriptions. The SMT system was based on Weighted Finite-State Transducers [17], and consisted of a cascade of transducers each representing the knowledge source in the SMT system, including the phrase-table and the target language model.

It was used a bootstrapping process where word-based translation models of increasing complexity and accuracy are trained and used to align each sentence pair in the corpus at the word level. This word alignment was then refined by combining automatic Spanish to Portuguese word alignment. Finally, all possible phrases were extracted from the combined alignment.

This training process was done using available tools. In particular, word level alignments used IBM 4 model [19] as implemented in GIZA ++, and the phrase-table was extracted using the MOSES[3] software package.

The phrase-table generated from the European Parliament corpus was extremely large (approximately 155 Millions phrases). In order to reduce its size, all phrases containing Spanish words not included in the speech recognition vocabulary were removed.

The resulting system was still too large for on-line use, thus an off-line system was developed which, given an input text, selects the relevant phrases from the phrase-table prior to translation. A WFST based decoder was developed for translation, which consists of a WFST representing a phrase-table. In this transducer, each simple path between the initial and final states corresponds to a particular phrase, the input labels corresponding to Spanish words and the output one to Portuguese words. Decoding is done by:

1. Converting the input sentence to a linear automaton.
2. Compose the automaton with the phrase table transducer.
3. Search the best path in the composition.

This decoder is monotonic in the sense that input and output phrases are produced in the same order, although word reordering is allowed inside each phrase. We believe that this limitation is not very important given the proximity of the two languages. Furthermore, this monotonous prevents long delays that are not desirable in a near future on-line implementation.

An initial effort to assess the translation quality of the system was done using a non-exhaustive human evaluation. Seven evaluators scored 15 translated sentences, yielding a result of 3.74 fluency and 4.23 sufficiency (in a 1 to 5 scale). These are good results in which the similarity of the two languages plays an important role.

5 Summary and Future Work

In this work we built a Statistical Machine Translation System of Broadcast News from Spanish to Portuguese. The fusion of two wide research fields was necessary.

The hybrid real-time recognition system AUDIMUS.MEDIA [7] was used as the recognition engine. After creating the acoustic models, language models, lexicon and vocabulary for the Spanish Broadcast News and carry out successive trainings, we obtained a 22.08% WER for the test eval97 to Latin America Broadcast News. This is a comparable WER value to the one produced by state-of-the-art systems, which are based on HMM models and realize several passages in the decoding stage.

The SMT strategy adapted is phrase-based translation. The MOSES software and normalized parallel texts select from European Parliament collection available were used to train the translation probabilities and models. First, a large phrases-table was

---

[3] http://www.statmt.org/moses/

created, and later was reduced by smaller language models adapted to the Broadcast News. In the last, there was realized a not-exhaustive human evaluation, obtaining a result of 3.74 fluency and 4.23 sufficiency.

For improvements and futures implementation we will generate Spanish and Portuguese language models with the same translated vocabulary to the SMT system and we will adapted the models to European Spanish Broadcast News.

Acknowledgments

The authors would like to acknowledge the work by Isabel Trancoso on the grapheme to phoneme transcription system to Spanish. This work was funded by PRIME National Project TECNOVOZ number 03/165.

References

[1] Matsoukas, S., Prasad, R., Laxminarayan, S., Xiang, B., Nguyen, L., Schwartz, R.: The 2004 BBN 1xRT Recognition Systems for English Broadcast News and Conversational Telephone Speech. In: Proceedings INTERSPEECH, Lisbon, Portugal (2005)

[2] Nguyen, L., Abdou, S., Afify, M., Makhoul, J., Matsoukas, S., Schwartz, R., Xiang, B., Lamel, L., Gauvain, J., Adda, G., Schwenk, H., Lefevre, F.: The 2004 BBN/LIMSI 10xRT English broadcast news transcription system. In: Proceedings INTERSPEECH, Lisbon, Portugal (2005)

[3] Huerta, J.M., Thayer, E., Ravishankar, M.K., Stern, R.: The Development of the 1997 CMU Spanish Broadcast News Transcription System. In: Proceedings of the DARPA Broadcast News Transcription and Understanding Workshop, Lansdowne, VA (1998)

[4] Kubala, F., Davenport, J., Jin, H., Liu, D., Leek, T., Matsoukas, S., Miller, D., Nguyen, L., Richardson, F., Schwartz, R., Makhoul, J.: The 1997 BBN byblos system applied to broadcast news transcription. In: Proceedings DARPA Broadcast News Transcription and Understanding Workshop, Lansdowne, VA (1998)

[5] Matsoukas, S., Nguyen, L., Davenport, J., Billa, J., Richardson, F., Siu, M., Liu, D., Schwartz, R., Makhoul, J.: The 1998 BBN Byblos primary system applied to English and Spanish broadcast news transcription. In: Proceedings DARPA Broadcast News Workshop, Herndon, VA (1999)

[6] Westphal, M.: TC-STAR Recognition Baseline Results, TC-STAR Deliverable n° D6 (2004), http://www.tc-star.org/documents/deliverable/deliverable_updated14april05/D6.pdf

[7] Meinedo, H., Caseiro, D., Neto, J., Trancoso, I.: AUDIMUS.MEDIA - A Broadcast News speech recognition system for the European Portuguese language. In: Proceedings PROPOR, Faro, Portugal (2003)

[8] Neto, J., Martins, C., Meinedo, H., Almeida, L.: AUDIMUS - Sistema de reconhecimento de fala contínua para o Português Europeu. In: Proceedings PROPOR IV, Évora, Portugal (1999)

[9] Koehn, P.: Europarl: A Parallel Corpus for Statistical Machine Translation. MT Summit 2005 (2005)

[10] Hermansky, H., Morgan, N., Baya, A., Kohn, P.: RASTA-PLP speech analysis technique. In: Proceedings ICASSP, San Francisco, USA (1992)

[11] Kingsbury, B.E., Morgan, N., Greenberg, S.: Robust speech recognition using the modulation spectrogram. Speech Comunication 25, 117–132 (1998)
[12] Caseiro, D., Trancoso, I.: Using Dynamic WFST Composition for Recognizing Broadcast News. In: ICSLP, Denver, CO (2002)
[13] Caseiro, D., Trancoso, I., Oliveira, L., Viana, C.: Grapheme-to-Phone Using Finite-State Transducers. In: IEEE Workshop on Speech Synthesis, Santa Monica, CA (2002)
[14] Souto, N., Meinedo, H., Neto, J.: Building language models for continuous speech recognition systems. In: Ranchhod, E., Mamede, N.J. (eds.) PorTAL 2002. LNCS (LNAI), vol. 2389. Springer, Heidelberg (2002)
[15] Kneser, R., Ney, H.: Improved backing-off for m-gram language modeling. Proceedings ICASSP 1, 181–184 (1995)
[16] Jelinek, F.: Self-organized language modeling for speech recognition. Speech Recognition 1, 450–506 (1990)
[17] Caseiro, D.: The INESC-ID Phrase-based Statistical Translation System. In: TC-STAR OpenLab, Trento, Italy (2006)
[18] Callison-Burch, C., Koehn, P.: Introduction to Statistical Machine Translation. ESSLLI Summer Course on SMT (2005)
[19] Mohri, M., Pereira, F., Riley, M.: Weighted Finite-State Transducers in Speech Recognition. Computer Speech and Language 16(1), 69–88 (2002)

Combining Multiple Features for Automatic Text Summarization through Machine Learning

Daniel Saraiva Leite and Lucia Helena Machado Rino

Departamento de Computação, UFSCar
CP 676, 13565-905 São Carlos, SP, Brazil
Núcleo Interinstitucional de Lingüística Computacional
{daniel_leite, lucia}@dc.ufscar.br

Abstract. In this paper we explore multiple features for extractive automatic summarization using machine learning. They account for SuPor-2 features, a supervised summarizer for Brazilian Portuguese, and graph-based features mirroring complex networks measures. Four different classifiers and automatic feature selection are explored. ROUGE is used for assessment of single-document summarization of news texts.

1 Introduction

Extractive Automatic Summarization (eAS) aims at producing a condensed version of a text by copying and pasting relevant text segments from the source text into the final extract. Statistical methods are usually employed to compute features that may be taken into account to rank those segments. Several features are acknowledged to play an important role in such a process, usually addressing different types of information, such as grammatical or position features [9], title words (e.g., [13]), or even functional information conveyed by the source text, e.g. signaling nouns [9]. Edmundson [9] and others (e.g., Mani, [22]) suggest a general algebraic formula for combining those features that may be embedded in a reasoning model to determine salient information for eAS. Roughly, it adds up all the features weighted either according to their influence in context or to the summarizing model itself (Equation 1). Defining weights to adequately rank features is the main bottleneck in this strategy because features may be genre- or domain-dependent and the more features considered, the more complex their combination is. An additional drawback arises if we consider a linear combination of features. This is usually barely suitable for a general eAS reasoning model.

$$Salience\ (s) = w_1 \times F_1(s) + ... + w_n \times F_n(s) \tag{1}$$

To tune our eAS system and reduce the above difficulties we adopted a corpus-based approach following Kupiec et al.'s one [13]. They use a Naïve-Bayes classifier for Machine Learning (ML) and calculate the relevance of a sentence through its likelihood of inclusion in the extract. Five binary features are considered. In spite of its simplicity, such a model is still complex. Besides the Naïve-Bayes classifier, many others (e.g., [25]; [34]) assume that all features are equally important and, thus, have the same influence on deciding which sentence must be selected. The question that arises, thus, is if employing less features may yield better results than employing all. Kupiec et al., e.g., show that using only three features is

A. Teixeira et al. (Eds.): PROPOR 2008, LNAI 5190, pp. 122–132, 2008.

better than using five. Other works corroborate this (e.g., [28]; [17]). On top of that, the Naïve-Bayes classifier assumes that all the probabilities of the features are independent. As suggested by Leite and Rino [17] this may be even more problematic if the features are binary, as they are in the Kupiec et al.'s approach.

In this paper we address the above drawbacks of eAS by exploring other ways to combine multiple features through ML. We employ fully automated methods to make eAS as general as possible, taking advantage of an automatic feature selection method. Feature subsets are then used together with 4 classifiers to define our eAS models. Two sets of features were explored: 11 features of SuPor-2 [17], a supervised extractive summarizer for Brazilian Portuguese, and 26 features derived from Complex Networks Theory measures (e.g., [1]). These are used to rank sentences through graph-based relations. Both are described in Section 2 and Section 3, respectively. In Section 4 we present our filter to select subsets of features of those two sets. Section 5 briefly describes the employed classifiers, followed by their assessment in Section 6. Final remarks are presented in Section 7.

2 SuPor-2 Features

Amongst the eleven SuPor-2 numeric and mutinomial features (F1-F11), some of them depict full eAS methods and require language-dependent resources to pre-process the text (e.g., a thesaurus, tagger, stoplist, or lexicon). Others are totally language-independent. SuPor-2 features are the following (identifiers used hereafter): **F1** and **F2:** lexical chaining [3]; **F3:** sentence length [13]; **F4:** proper nouns [13]; **F5:** sentence location [9]; **F6** and **F7:** word frequency [20]; **F8** and **F9:** relationship mapping [30]; **F10** and **F11:** importance of topics [15]. Some of these are well-known and need no further explanation. The ones that are language-dependent are F1 and F2, F6 and F7, and F8 and F9. Features unfolded in two address two distinct ways of processing the text. The Lexical Chaining method uses the WordNet [24] to identify relations between lexical items. Heuristics used to pinpoint candidate sentences for an extract are the same as suggested in [3]: every sentence that contains the first occurrence of a member of a strong lexical chain is chosen (H1); similar to H1, but considering only the representative members of a strong lexical chain (H2); only sentence that convey representative lexical chains of every topic of the source text are chosen (H3). F1 signals that differing text topics are delimited through paragraphs; F2 employs Text Tiling [11] instead. They are used to compute H3. Nominal values range over the set {'None', 'H1', 'H2', 'H3', 'H1H2', 'H1H3', 'H2H3', 'H1H2H3'}. Two distinct location indicators are also considered for F5, although they do not yield two distinct features. The position of a sentence in the paragraph and in the text are considered, resulting in the set {'II','IM',-'IF','MI','MM','MF','FI','FM','FF'}, first and second letters signaling respectively the position within a paragraph and within the text (Initial, Medium, or Final). For word frequency, language-dependence is due to stemming the words (F6) or generating their 4-grams (F7). Similarly, relationship mapping also applies such pre-processing methods, correspondingly yielding features F8 and F9. Also similar to Lexical Chaining is the way of combining the possibilities, now of traversing a text graph, namely, through the dense or bushy path (P1), the deep path (P2), or the segmented path (P3). P1 addresses paragraphs as if they were totally independent from each other and, thus, does not guarantee a cohesive extract. P2 aims at overcoming that by choosing paragraphs semantically inter-related. However, only one topic (even an irrelevant one) may be chosen and proper coverage of the source text may be lacked. P3 aims at solving both former problems. F8 and F9 range over the set {'None','P1','P2','P3', 'P1P2', 'P1P3', 'P2P3', 'P1P2P3'}. Finally, the importance of topics method is also unfolded in two features: F10 uses Text Tiling for topic detection, and F11 uses paragraphs themselves as topic units. In this case numeric features are provided by the harmonic mean between the sentence similarity to the centroid of the topic and the importance of that topic.

3 Features Based on Complex Networks

Complex systems modeled as graphs are known as Complex Networks (CNs) and have great influence on Statistical Mechanics and Graph Theory [1]. Texts may be represented by CNs in many distinct ways, depending on the interconnection between the nodes. For example, Schorochod'ko [31] suggests that nodes convey sentences and edges, their relationship. So, edges can be labeled according to words co-occurrence, after stemming and stopwords removal.

In our approach the feature set for CNs follows Antiqueira's [2], which follows Schorochod'ko's [31] in turn. Antiqueira used classical CN measures to propose 26 numeric features for eAS. His full system was used to verify the potential of putting all the features together for eAS. Table 1 lists all the measures, grouped by in the following way:

Degree [8]. Number of edges associated to a node. It signals how connected the node is to its neighbor nodes, or how representative a sentence is of a text, for eAS.

Clustering Coefficient [33]. It quantifies how close a node and its neighbors are from being a clique, a graph-theory kind of cluster in which every node is connected to the others, or how useful a sentence is to represent the cluster in an extract.

Minimal Paths [8]. For each node, all its shortest paths to other nodes are calculated and the average minimal path measure is drawn. This is used then to signal the sentence relevance for eAS. It relates to determining the gist of a text in that a node closer to others may convey its most important information.

Locality Index ([7], [2]). It also addresses clustering similar nodes, but considering all the connections of the other nodes that are in the cluster.

Matching Index [8]. It helps determining the strength of a connection between two nodes, i.e., the contribution of an edge to the representativeness of a node in the graph. It is used to select sentences that convey distinct groups of information, or varied topics of a text.

Dilation [6]. It conveys the node importance regarding its hierarchical relations within the graph. Rings of nodes around the focused node are depicted: rings of distance one amount to all the neighbors linked by one edge to that node; rings of distance two, by all the neighbors linked to it by two edges, etc. The rings aim at capturing the connectivity between nodes in neighborhoods that are more far-away from the focused node. The hierarchical degree of level h is then defined as the number of nodes between this level and the next level ring. Sentences with a high hierarchical degree are more likely to compose an extract because they mirror better the gist.

Hubs [2]. It also considers dilation, but aiming at giving preference to the node that is more connected in the graph, the so-called hub. A hub, in this case, would signal the gist of the text. Additionally, other sentences are chosen amongst the rings around the hub, aiming at addressing varied topics of the text or complementing information conveyed by the hub. Sentence priority is given accordingly to the order they have in the text (following [9]) or according to the degrees of the nodes connected to the hub.

K-cores [4]. A k-core is a subgraph whose nodes have a minimum degree of k. For eAS, the subgraph that signals relevant sentences is that with the greatest k.

W-cuts [2]. It aims at finding a cohesive group of sentences by considering interconnected sentences with high degrees.

Communities [5]. A community groups together those nodes that are more interconnected to each other. Nodes in distinct communities are not significantly interconnected. So, communities signal the density of connections in the graph. Within the same community sentences with high degrees are preferred for eAS.

According to the definitions given above by Antiqueira, some of CN features correlates to SuPor-2 F8 and F9 features. For example, by addressing degrees or minimal paths, a dense path (P1) may be withdrawn; by addressing locality indexes, a deep path (P2) may be outlined. Matching indexes and dilation seem to address segmented paths (P3), whilst hubs seem to convey both deep and segmented ones (P2 and P3). However, a more profound investigation is needed to certify this. A clearer correspondence is given with respect to the last three groups: K-cores do not necessarily address cohesion in that they refer to dense paths (P1). W-cuts aim at overcoming the former problem, similarly to P2. Communities address distinct topics of the text at once, thus, they depict segmented paths (P3).

Table 1. Antiqueira´s CN numeric features

F #	Measure	F #	Measure
12	Degree	22	Dilation (level 2)
		23	Dilation (level 2, cumulative)
13	Degree (weighted variation)	24	Dilation (level 3)
14	Clustering Coefficient	25	Dilation (level 3, cumulative)
		26	Dilation (level 2, weighted)
15	Clustering Coefficient (weighted variation)	27	Dilation (level 2, weighted, cumulative)
16	Minimal Paths	28	Dilation (level 3, weighted)
		29	Dilation (level 3, weighted, cumulative)
17	Minimal Paths (weights complement variation)	30	Hubs (sorted by degree)
		31	Hubs (sorted by locality)
18	Minimal Paths (weights inverse variation)	32	Hubs (sorted by locality, with degree cut)
19	Locality Index	33	K-Cores (sorted by locality)
		34	K-Cores (sorted by degree)
20	Locality Index (modified)	35	W-Cuts (sorted by locality)
		36	W-Cuts (sorted by degree)
21	Matching Index	37	Communities

4 Feature Selection

Our problem amounts to finding a feature subset that maximizes the system's ability to classify correct instances, or else, to maximize the determination of relevant sentences to include in extracts. As such problem is usually intractable, two usual approaches are suggested [34]:

(1) *Filter* approach operates independently from any induction algorithm and filters features before induction takes place. It usually produces a rank features through statistical measures such as Chi-Square, Information Gain and Gain Ratio (e.g., [34]); (2) *Wrapper* approach use an induction algorithm along with cross-validation to evaluate feature subsets. The result is usually a recommended feature subset instead of a ranking.

A filter approach is preferred here due to the high computational cost of the wrapper one. Besides, wrapper algorithms usually consider classifier error rates only to determine how worthwhile a feature subset is. For eAS, this approach is barely suitable. Although the filter approach does not specifically address eAS, it is based on statistical relevance measures that are usually adequate to most ML models. However, there is a serious drawback regarding the filtering algorithms: especially for those that produce a feature ranking, one must define manually a cutoff on the number of selected features. Another problem is to determine the most suitable measure to use. To illustrate this, Figure 1 shows a graph presenting normalized scores for three usual filter ranking measures applied to our 37 features (features 1-11 come from SuPor-2; features 12-37, from CN). Note that there are great divergences in some cases.

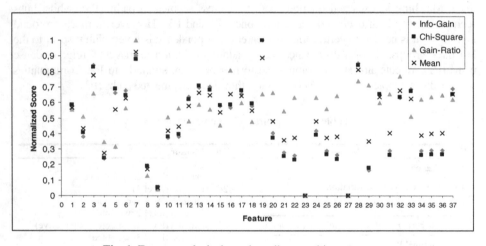

Fig. 1. Feature analysis through attribute ranking measures

To overcome the above, we chose to explore the Correlation Feature Selection (CFS) method [10]. It differs from standard filter algorithms in that (1) it does not need a previous definition of the amount of selected features, resulting in a recommended feature subset, instead of a rank; (2) it is quite fast; (3) it considers both feature redundancy and feature relevancy. So, CFS seems suitable for classifiers such as the Naïve-Bayes, which assumes statistically independent features.

CFS aims at finding features that are predictive of the class but that do not correlate with each other (i.e., non-redundant features). To judge the relevance of a feature subset, heuristics consider pair-wise feature correlations and individual feature relevance based on information gain. A heuristic search (such as Hill-Climbing) is then used to traverse the space of feature subsets. The subset with the highest measure is selected. We used CFS in the WEKA ML environment [34] in its default configuration.

5 Classifiers

Although we follow [13], our approach does not yield standard classification because the focus is not on the predicted class (e.g., 'Present in Extract' or 'Not Present in Extract'). Instead, a ranking that mirrors sentence importance is aimed at. In essence, this is accomplished by computing the likelihood of a sentence to belong to the class 'Present in Extract'. We explored four different classifiers running in WEKA with their default configurations. So, computing varies according to the classifier under focus. They are briefly described bellow, along with a description on sentence ranking for eAS.

Flexible-Bayes [12]. Variation of the Naïve-Bayes to handle numeric features (present on both SuPor-2 and CN feature sets). Sentence ranking follows Kupiec et al.'s.

C4.5 [27]. After decision trees are built for classification, each sentence probability is calculated through the relative frequency of the 'Present in Extract' class in the decision leaf.

SVM [32]. Support Vector Machines learn a decision boundary between two classes by mapping the training examples (labeled sentences) onto a higher dimensional space and determining the optimal separating hyperplane in that space. The likelihood of a sentence pertaining to the 'Present in Extract' class is then calculated based on the Euclidian distance between the hyperplane and the example.

Logistic Regression (e.g., [34]). Variation of standard regression, it is used when the dependent variable is binary ('Present in Extract' or 'Not Present in Extract'). The classifier model is depicted as follows (see Equation 2): p is the probability of the class 'Present in Extract', $X_1, \ldots, X_n$ are the values of the considered features and $\beta_0, \beta_1, \ldots, \beta_n$ are the regression coefficients, estimated through training. The log result is then transformed into the probability through a logistic function. Probabilities are thus used to rank the sentences.

$$\log\left(\frac{p}{1-p}\right) = \beta_o + \beta_1 X_1 + \ldots + \beta_n X_n \qquad (2)$$

6 Assessment of eAS Using Multiple Features

We explored distinct combinations of the above features for single-document summarization in three different ways: either features embedded in SuPor-2 or in CN systems were adopted, or features were classified altogether in varied ways. Such combinations of multiple features and classifiers yielded 24 automatic summarizers, which were compared using ROUGE-1 [19] at 95% confidence rate. This has been shown to agree the most with human judgments [19]. The most promising system was thus compared to other summarizers. TeMário corpus [26] was used in this assessment (100 Brazilian newswire articles along with their manual summaries) and 10 fold cross-validation was used for training and testing. TeMário manual summaries were used as reference data, with a 30% compression rate.

6.1 Determining the Best Feature Set and Classifier

Besides varying the classifiers and the source feature sets, we also explored the influence of CFS for feature selection. Table 2 shows ROUGE average recall indices for the resulting

TeMário extracts, for each system. Pair-wised systems are signaled with superscripted numbers. When 1st column = SuPor-2, 11 features are considered; 1st column = CN, 26 CN features are used instead; the union sums up all of them. CFS 'No' value indicates that all features were used. The confidence interval (last column) is also provided by ROUGE.

As shown, for all feature sets Logistic Regression and Flexible-Bayes outperform most C4.5 and all SVM models. This corroborates [17] in its claim that probability-based classifiers are better for eAS. This is certainly due to the accurate ranking of instances, or sentences in our case. The reason for lack of performance of C4.5 and SVM may be due to the following: (a) C4.5 estimates probabilities needed for sentence ranking using the frequency of positive training examples in a leaf. When the number of training examples associated to this leaf is small, observed frequencies may not be statistically reliable. Besides, the number of different probabilities is limited to the number of leaves in the tree; (b) WEKA employs in its default configuration a linear SVM. The results suggest that this was not suitable for delineating an eAS model dealing with multiple features. SuPor-2 features where weighted more expressively by this classifier, causing no difference using CN features or not – pairs (4) and (12). In fact, recent works suggest that SVM are adequate for eAS in some cases. For example, DUC[1] 2007 fifth-ranked summarizer [18] employed a linear SVM yielding good results. The difference is that SVMs were used for regression in that work, not for classification as we did here, and the number of features was only 6.

Table 2. ROUGE-1 recall measures

Source Feature Set	Classifier	CFS	Avg. Recall	95% conf. int.
SuPor-2[1]	Logistic Regression	No	0.5316	0.5208 - 0.5424
SuPor-2[1]	Logistic Regression	Yes	0.5288	0.5164 - 0.5407
SuPor-2[2]	Flexible-Bayes	Yes	0.5284	0.5160 - 0.5411
SuPor-2 $\cup$ CN[10]	Flexible-Bayes	Yes	0.5278	0.5142 - 0.5405
SuPor-2 $\cup$ CN[9]	Logistic Regression	Yes	0.5270	0.5146 - 0.5395
SuPor-2 $\cup$ CN[11]	C4.5	Yes	0.5253	0.5136 - 0.5374
SuPor-2 $\cup$ CN[10]	Flexible-Bayes	No	0.5249	0.5112 - 0.5382
SuPor-2[3]	C4.5	Yes	0.5238	0.5106 - 0.5356
CN[6]	Flexible-Bayes	No	0.5237	0.5104 - 0.5365
CN[6]	Flexible-Bayes	Yes	0.5236	0.5092 - 0.5372
CN[5]	Logistic Regression	No	0.5230	0.5098 - 0.5352
SuPor-2 $\cup$ CN[9]	Logistic Regression	No	0.5228	0.5088 - 0.5355
SuPor-2[2]	Flexible-Bayes	No	0.5227	0.5092 - 0.5361
SuPor-2[3]	C4.5	No	0.5212	0.5096 - 0.5330
CN[5]	Logistic Regression	Yes	0.5188	0.5053 - 0.5316
SuPor-2 $\cup$ CN[11]	C4.5	No	0.5184	0.5055 - 0.5316
CN[7]	C4.5	No	0.5167	0.5034 - 0.5294
SuPor-2 $\cup$ CN[12]	SVM	Yes	0.5158	0.5015 - 0.5294
SuPor-2 $\cup$ CN[12]	SVM	No	0.5158	0.5015 - 0.5294
SuPor-2[4]	SVM	Yes	0.5158	0.5015 - 0.5294
SuPor-2[4]	SVM	No	0.5158	0.5015 - 0.5294
CN[7]	C4.5	Yes	0.5157	0.5029 - 0.5287
CN[8]	SVM	Yes	0.5032	0.4897 - 0.5174
CN[8]	SVM	No	0.5032	0.4897 - 0.5174

---

[1] Document Understanding Conferences. http://duc.nist.gov (May/2008).

Regarding CFS, although the highest ranked system did not tune its feature set at all using it, the scores show that improving eAS through CFS depends on the source feature set. For example, when the union of both feature sets was considered, amounting to the 37 features, CFS improved the results for all models, except one – SuPor-2 $\cup$ CN pair (12). For this, the results were the same as using all features. However, considering only the CN source feature set, all but one classifier – CN pair (8) got worse when using CFS. Overall, using CFS improved the scores in 5 cases: see pairs (2), (3), (9), (10), and (11); not using it, it improved only in 4 cases – pairs (1), (5), (6), and (7). In conclusion, the results show that using CFS with single features sets (SuPor-2 or CN ones) does not improve recall. However, when all the features were considered, CFS improved the average recall rate for all the systems but one. This means that applying ML techniques to determine the most promising feature subset for classifying sentences for eAS may be worthwhile for some models. Still, it remains to be shown that our effort is justifiable, task reported next for the summarizer with the highest recall, i.e., that which uses only SuPor-2 feature set and the classifier based on Logistic Regression. Hereafter such system is named 'SuPor2-LogistRegr'.

Such results shall be analyzed more deeply in the future to see how features interact with each other, i.e., if they are independent or not, or how they contribute to eAS better. In fact, the CFS measure should account for the commonalities between CN and SuPor-2 features (see Section 3 for details), more specifically, pinpointing redundancy, as we stated in Section 4. However, the above recall rates for pair-wise systems do not confirm that.

6.2 Comparison to Other Summarizers

We compared the 'SuPor2-LogistRegr' system to four other summarizers , one of them being a baseline, as follows: (1) the original SuPor-2 system [17] which is actually depicted in the gray line in Table 2; (2) TextRank [23], which is also a graph-based eAS method; (3) a single-feature summarizer based on CNs proposed by Antiqueira [2]; and, finally, (4) the baseline, which just selects the topmost sentences of the source text.

TextRank represents a text similarly to the process described in Section 3, but its ranking measure is based upon Google™ PageRank algorithm instead. It was used here in its backward configuration. Mihalcea [23] showed that TextRank is a language-independent summarizer, presenting results for English and Brazilian Portuguese. Mihalcea actually used the very same TeMário corpus as we did here.

Antiqueira's summarizer uses only feature #13 – 'Degree - weighted variation' (see Table 1), which was the best CN single-feature summarizer pinpointed by Antiqueira (hereafter named 'BestCN'). Such summarizer is unsupervised and ranks sentences according to feature #13 lowest values: sentences which score lower are the most important ones to consider for eAS.

The comparison setting was identical to that in [23] and [2], in order for us to compute only our new data and fully reproduce their scores in comparison to the new ones just reported above. In other words, we did not run those systems again. Table 3 shows that all the systems outperformed the baseline. Adding to previous observations, the supervised summarizers – SuPor2-LogistRegr and SuPor2 – yielded the best results, being the former slightly better than the latter.

Table 3. ROUGE-1 recall average rates - summarizers

Summarizer	Avg. Recall
SuPor2-LogistRegr	0.5316
SuPor-2	0.5227
TextRank	0.5121
BestCN	0.5020
Baseline	0.4984

7 Final Remarks

The combination of multiple features for eAS is clearly a complex issue. Our assessment shows that eAS may be improved provided that adequate feature classifiers and source feature sets are used.

Despite the fact that CN features did not outperform SuPor-2 features in most cases, we believe that exploring them may be still worthwhile. SuPor-2 features depict full eAS methods and require many language-dependent resources. In contrast, graph-based methods usually do not rely on analyzing linguistic information conveyed by the text. They tend to be more language-independent. This is one of the main reasons for the recent interest on them. Actually, if we focus only on graph-based methods, CN features combined with ML (CN$^{(6)}$ in Table 2) outperforms TextRank best configuration.

Although many features we used are language-dependent (especially those of SuPor-2) the approach presented here to combine multiple features is portable to other languages and domains.

As future work, other techniques and methods for combining features may also be explored, e.g., that suggested by Lee et al. [16], which still uses filtering, but outperforms CFS in some cases, as shown by Lee et al.

Acknowledgments

The authors are grateful to the Brazilian agencies CNPq and CAPES for supporting this work and to Lucas Antiqueira for extracting CN features for TeMário corpus.

References

1. Albert, R., Barabási, A.L.: Statistical mechanics of complex networks. Rev. Mod. Phys. 74, 47–97 (2002)
2. Antiqueira, L.: Development of Complex Networks Techniques for Extractive Automatic Sumarization. MSc. Dissertation. ICMC, USP (2007) (in Portuguese)
3. Barzilay, R., Elhadad, M.: Using Lexical Chains for Text Summarization. In: Mani, I., Maybury, M.T. (eds.) Advances in Automatic Text Summarization, pp. 111–121. MIT Press, Cambridge (1999)
4. Batagelj, V., Zaversnik, M.: Partitioning approach to visualization of large networks. In: Kratochvíl, J. (ed.) GD 1999. LNCS, vol. 1731, pp. 90–98. Springer, Heidelberg (1999)
5. Clauset, A., Newman, M.E.J., Moore, C.: Finding community structure in very large networks. Phys. Rev. E 70, 66–111 (2004)

6. Costa, L.F., da Rocha, L.E.C.: A generalized approach to complex networks. Eur. Phys. J. B 50, 237–242 (2006)
7. Costa, L.F., Kaiser, M., Hilgetag, C.: Beyond the average: detecting global singular nodes from local features in complex networks. Physics, 0607272 (2006a)
8. Costa, L.F., Rodrigues, F.A., Travieso, G., Villas Boas, P.R.: Characterization of complex networks: A survey of measurements. cond-mat/0505185 (2006b)
9. Edmundson, H.P.: New methods in automatic extracting. Journal of the Association for Computing Machinery 16, 264–285 (1969)
10. Hall, A.M.: Correlation-based Feature Selection for Discrete and Numeric Class Machine Learning. In: Langley, P. (ed.) Proc. of 17th International Conference on Machine Learning, pp. 359–366. Morgan Kaufmann, San Francisco (2000)
11. Hearst, M.A.: TextTiling: A Quantitative Approach to Discourse Segmentation. Technical Report 93/24. University of California, Berkeley (1993)
12. John, G., Langley, P.: Estimating continuous distributions in Bayesian classifiers. In: Besnard, P., Hanks, S. (eds.) Proc. of the 11th Conference on Uncertainty in Artificial Intelligence, Quebec, Canada, pp. 338–345 (1995)
13. Kupiec, J., Pedersen, J., Chen, F.: A trainable document summarizer. In: Fox, E.A., Ingwersen, P., Fidel, R. (eds.) Proc. of the 18th ACM-SIGIR Conference on Research & Development in Information Retrieval, Seatlle, WA, pp. 68–73 (1995)
14. Larocca Neto, J., Freitas, A.A., Kaestner, C.A.A.: Automatic text summarization using a machine learning approach. In: Bittencourt, G., Ramalho, G.L. (eds.) Proc. of 16th Brazilian Symposium on Artificial Intelligence (SBIA 2002). LNCS (LNAI), vol. 2057, pp. 205–215. Springer, Heidelberg (2002)
15. Larocca Neto, J., Santos, A.D., Kaestner, C.A.A., Freitas, A.A.: Generating Text Summaries through the Relative Importance of Topics. In: Monard, M.C., Sichman, J.S. (eds.) SBIA 2000 and IBERAMIA 2000. LNCS (LNAI), vol. 1952, pp. 300–309. Springer, Heidelberg (2000)
16. Lee, H.D., Monard, M.C., Wu, F.C.: A Fractal Dimension Based Filter Algorithm to Select Features for Supervised Learning. In: Sichman, J.S., Coelho, H., Rezende, S.O. (eds.) IBERAMIA 2006 and SBIA 2006. LNCS (LNAI), vol. 4140, pp. 278–288. Springer, Heidelberg (2006)
17. Leite, D.S., Rino, L.H.M.: Selecting a Feature Set to Summarize Texts in Brazilian Portuguese. In: Sichman, J.S., Coelho, H., Rezende, S.O. (eds.) IBERAMIA 2006 and SBIA 2006. LNCS (LNAI), vol. 4140, pp. 462–471. Springer, Heidelberg (2006)
18. Li, S., Ouyang, Y., Wang, W., Sun, B.: Multi-document Support Vector Regression. In: Proc. of Document Understanding Conference - DUC 2007, Rochester, NY (2007)
19. Lin, C., Hovy, E.H.: Automatic Evaluation of Summaries Using N-gram Co-occurrence Statistics. In: Proceedings of Language Technology Conference (HLT-NAACL 2003), Edmonton, Canada (2003)
20. Luhn, H.: The automatic creation of literature abstracts. IBM Journal of Research and Development 2, 159–165 (1958)
21. Mani, I., Maybury, M.T.: Advances in Automatic Text Summarization. MIT Press, Cambridge (1999)
22. Mani, I.: Automatic Summarization. John Benjamin's Publishing Company (2001)
23. Mihalcea, R.: Language Independent Extractive Summarization. In: Proc. of the 43th Annual Meeting of the Association for Computational Linguistics, Companion Volume (ACL 2005), Ann Arbor, MI (June 2005)
24. Miller, G.A., Beckwith, R., Fellbaum, C., Gross, D., Miller, K.: Introduction to WordNet: An On-line Lexical Database. International Journal of Lexicography 3(4), 235–244 (1990)

25. Mitchel, T.M.: Machine Learning. McGraw Hill, New York (1997)
26. Pardo, T.A.S., Rino, L.H.M.: TeMário: A corpus for automatic text summarization (in Portuguese). NILC Tech. Report NILC-TR-03-09 (2003)
27. Quinlan, J.R.: C4.5 Programs for machine learning. Morgan-Kaufman, San Mateo (1993)
28. Rino, L.H.M., Módolo, M.: SuPor: An environment for AS of texts in Brazilian Portuguese. In: Vicedo, J.L., Martínez-Barco, P., Muñoz, R., et al. (eds.) Advances in Natural Language Processing. LNCS, vol. 3230, pp. 419–430. Springer, Heidelberg (2004)
29. Rino, L.H.M., Pardo, T.A.S., Silla Junior, C.N., Kaestner, C.A.A., Pombo, M.: A Comparison of Automatic Summarizers of Texts in Brazilian Portuguese. In: Bazzan, A.L.C., Labidi, S. (eds.) SBIA 2004. LNCS (LNAI), vol. 3171, pp. 235–244. Springer, Heidelberg (2004)
30. Salton, G., Singhal, A., Mitra, M., Buckley, C.: Automatic Text Structuring and Summarization. Information Processing & Management 33, 193–207 (1997)
31. Skorochod'ko, E.F.: Adaptive method of automatic abstracting and indexing. In: Freiman, C.V. (ed.) Proceedings of the IFIP Congress, vol. 71, pp. 1179–1182 (1971)
32. Vapnik, V.: The Nature of Statistical Learning Theory. Springer, Heidelberg (1995)
33. Watts, D.J., Strogatz, S.H.: Collective dynamics of 'small-world' networks. Nature 393, 440–442 (1998)
34. Witten, I.H., Frank, E.: Data Mining: Practical machine learning tools and techniques, 2nd edn. Morgan Kaufmann, San Francisco (2005)

Some Experiments on Clustering Similar Sentences of Texts in Portuguese

Eloize Rossi Marques Seno and Maria das Graças Volpe Nunes

NILC-ICMC University of São Paulo
CP 668P, 13560-970 São Carlos – SP, Brazil
{eloize,gracan}@icmc.usp.br

Abstract. Identifying similar text passages plays an important role in many applications in NLP, such as paraphrase generation, automatic summarization, etc. This paper presents some experiments on detecting and clustering similar sentences of texts in Brazilian Portuguese. We propose an evalution framework based on an incremental and unsupervised clustering method which is combined with statistical similarity metrics to measure the semantic distance between sentences. Experiments show that this method is robust even to treat small data sets. It has achieved 86% and 93% of F-measure and Purity, respectively, and 0.037 of Entropy for the best case.

Keywords: Sentence Similarity, Sentence Clustering, Statistical Metrics.

1 Introduction

Identifying similar text passages plays an important role in many Natural Language Processing (NLP) applications, such as paraphrase generation [1], automatic summarization [4] [5] [6], ontology building [11], digital library systems [13], dialogue systems [15], etc. In this paper, we present experiments on identifying and clustering similar sentences from one or multiple documents written in Brazilian Portuguese. Sentence clustering is performed as a primary step towards aligning and fusing common information (e.g., paraphrases and synonyms) among semantically similar sentences.

We propose an evaluation framework named SiSPI – *Similar Short Passages Identifier*, which is based on an incremental and unsupervised clustering method. The incremental method is particularly appealing since it is not based on learning and, therefore, it does not require a great training data set.

In order to compute semantic distance between a sentence and a cluster, SiSPI implements three different statistical similarity measures. The first measure, called Word Overlap [16], is based on the total of words in common between a sentence and a cluster. The two latter are the well-known TF-IDF (Term Frequency Inverse *Document* Frequency) measure from Information Retrieval [10] and the TF-ISF (Term Frequency Inverse *Sentence* Frequency) measure [3], which is an adaptation of the TF-IDF (see Section 3).

Aiming at identifying sets of highly semantically-related sentences from a collection of documents, a key concept to SiSPI is the notion of similarity. In this study, we

A. Teixeira et al. (Eds.): PROPOR 2008, LNAI 5190, pp. 133–142, 2008.
© Springer-Verlag Berlin Heidelberg 2008

follow Hatzivassiloglou et al.'s similarity definition [5], which has been proposed for the same task of detecting similar sentences. Thus, we regard two sentences as similar if they refer to the same object or event and i) the object either accomplishes the same action in both units, or ii) is the subject of the same description. Next, we present three sentences on the same event, the domestic bomb explosion, extracted from the experimental corpus (see Section 4)[1]. Despite all sentences refer to the same fact, sentences (a) and (b) focus on the explosion in Ministério Público, while sentence (c) focuses on the explosion in Secretaria de Estado da Fazenda. Therefore, only sentences (a) and (b) are considered similar.

(a) Uma bomba caseira foi atirada contra a sede do Ministério Público (MP).
(b) Uma bomba caseira foi jogada contra o prédio do Ministério Público, na capital do estado.
(c) Uma bomba caseira atingiu o prédio da Secretaria de Estado da Fazenda, localizado na avenida Rangel Pestana, ao lado do Poupatempo Sé.

The remainder of this paper is organized as follows. Some related works are described in Section 2 and the proposed clustering framework is described in Section 3. An experimental evaluation using SiSPI is presented in Section 4, and some final remarks are presented in Section 5.

2   Related Work

Various methods for detecting similar short passages (e.g. sentences and paragraphs) have been proposed in the literature recently. Most of them are based on machine learning techniques and rely on statistics of words in common [11] [15]. In general, they make use of the Salton et al.'s vector space model [10] and of some statistical similarity measure to identify similar passages. In [11], for example, the TF-IDF model, which is widely used for document clustering (e.g., [3] [8]) is combined with a non-hierarquical clustering algorithm in order to cluster sentences and paragraphs for ontology enhancement. No evaluation result for the clustering process in specific is presented by the authors.

Despite those works treat short passages, our concept of similarity is more restrict than the one used in those works. The concept of similarity used in this work is similar to the one used in Hatzivassiloglou et al. [5] (see Section 1). The differences rely on the fact that they utilize a supervised approach based on linguistic knowledge to classify paragraph pairs of documents written in English as similar or non-similar. More specifically, those authors make use of a rule induction method, called RIPPER, which combines 43 linguistics features. Such features include morphological, syntactic and semantic information. RIPPER has been trained with a corpus of 10.345 manually-classified paragraph pairs and obtained 45.6% F-measure. In a subsequent experiment, reported by [6], a log-linear regression model was based on a more refined set of those features. In addition, they have used a co-reference resolution component that allows comparing multiple forms of the same name. This model resulted in a performance increase of 51.0% F-measure compared with RIPPER. In [6] an

---

[1] The sentences have been kept in Brazilian Portuguese in order to avoid noise in the translation.

experiment using a variation of the TF-IDF model which treats paragraphs rather than documents is also presented. By using the same data set used by RIPPER and by the regression model, such model has obtained 36.7% F-measure on average.

In spite of machine learning techniques being widely used, they usually require a great data set of similar passage instances, which is hard to obtain. Trying to solve this, we employ an incremental clustering method which does not require training. Our hypothesis is that with an incremental clustering approach it is possible to achieve satisfactory results even using statistical similarity metrics only.

3 The Clustering Framework

SiSPI is composed by two main processing modules named Sentence Splitting and Sentence Clustering (Figure 1). The former splits each document of a collection into sentences. The latter identifies and clusters similar sentences. During this process, SiSPI makes use of a stemmer [2] and a stoplist. The output is a set of sentence cluster files.

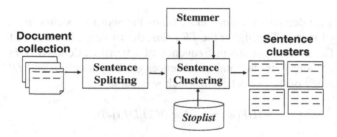

Fig. 1. SiSPI architecture

SiSPI is domain independent, for it is based only on lexical information. It is also weakly language-dependent, for it does not use any deeper linguistic knowledge (e.g., syntactic and semantic information).

The Sentence Splitting is performed by a textual-segmentation tool called SENTER [7], which is based on a list of abbreviations and some sentence delimiters. SiSPI could manage longer passages, as paragraphs, by just substituting this tool.

The Sentence Clustering module uses the incremental clustering method Single-pass [14], an effective and widely used algorithm for document clustering ([8]).

Single-pass requires a single sequential pass over the set of sentences to be clustered. The first cluster is created by selecting the first sentence of the first document. At each iteration, the algorithm decides on whether a new input sentence should be inserted in an existing cluster or should originate a new one. This decision is based on a condition specified by the similarity function employed, that is, a similarity threshold.

In this study, two different similarity functions are evaluated. The first one is based on the Word Overlap metric [16], which calculates the number of common words between a sentence S and a cluster C, normalized by the total of words of S plus C (Formula 1). According to (1), the similarity threshold is a value that ranges from 0 to 0.5, which is derived experimentally (see Section 4). The larger the similarity value, the more similar the sentence and that cluster are. Notice that in SiSPI each sentence belongs to a single cluster.

$$Wol\ (S,C) = \#CommonWords(S,C)\ /\ (|S| + |C|). \tag{1}$$

The second similarity function is the cosine coefficient [10], which is applied to the term frequency vector of a sentence and to the vector that represents the most important terms of a cluster, named centroid. According to this function, the similarity threshold is a value in the range of 0 to 1. The larger the similarity value between the vectors, the more similar the sentence and the cluster are.

The determination of a cluster centroid is based on the relevance of the corresponding words of that cluster, computed by two different metrics. The first metric is a slightly modified version of TF-IDF (Term Frequency Inverse Document Frequency) [10]. The TF-IDF value of a word w of a cluster c, denoted $TF\text{-}IDF(w,c)$, is given by Formula 2.

$$TF\text{-}IDF(w,c) = TF(w,c) * IDF(w). \tag{2}$$

where $TF(w,c)$ depictes the number of times the word w occurs in cluster c, *i.e.*, the frequency of w in c. The higher the *TF value*, the more representative the word w is of cluster c. The inverse document frequency of a word w, denoted $IDF(w)$, is given by Formula 3, where C is the total of sentences of the collection and $DF(w)$ is the sentence frequency of the collection in which w occurs.

$$IDF(w) = 1 + \log (|C| / DF(w)). \tag{3}$$

According to (3), the *IDF* value is high if the word w occurs in few sentences of a collection, meaning that w has a great document-discriminating power. On the other hand, the *IDF* value is low if the word w occurs in many sentences of the collection, indicating that w has a little document-discriminating power.

The second metric used is TF-ISF (Term Frequency Inverse Sentence Frequency) [3]. The TF-ISF measure is similar to (1), but we compute the inverse sentence frequency for a specific cluster rather than for the document collection. The inverse sentence frequency of a word w, denoted $ISF(w)$, is given by Formula 4, where C is the total number of sentences in the current cluster, and $SF(w)$ is the sentence frequency of the cluster in which w occurs.

$$ISF(w) = 1 + \log (|C| / SF(w)). \tag{4}$$

For a word to be representative of a given cluster it must have both a high *TF* value and a high *ISF* (or *IDF*) value (therefore, a high *TF-ISF* (or *TF-IDF*) value). Thus, only the words with highest *TF-ISF* (or *TF-IDF*) scores are selected to represent the cluster centroid. The number of words to be selected is a given parameter, which was derived experimentally, as it will be explained in the next Section.

4 Experimental Evaluation

External or internal quality measures can be used to assess the quality of a clustering solution [12]. External quality measures evaluate how good the clusters are when compared with reference clusters (often manually classified clusters). So, this kind of evaluation can be carried out only if the class of each sentence is determined a priori. On the other hand, internal quality measures do not use any kind of external knowledge, and assess only the cohesiveness of a clustering solution, i.e., how similar the elements of each cluster are. If the purpose is to measure the goodness of a solution or the effectiveness of the clustering method, external measures are more appropriate. In this study, we use three external quality measures that are described in Section 4.2. Next, we describe the corpus used for the evaluation.

4.1 The Corpus

The corpus is composed by 20 collections of news articles, with 3.6 documents on average on the same topic per collection (one example of topic is the Virginia Tech massacre). This corpus has been manually collected from several web news agencies and totalizes 1.153 sentences in 71 documents.

Aiming at creating a reference clustering corpus, each sentence of each document collection has been manually classified (i.e. associated with a cluster name) by the first author of this work, according to the similarity definition presented in Section 1. In cases when there were more than one possible cluster for a single sentence, only one has been chosen. Decisions about the best cluster to be chosen were based on semantic similarity (that is, the cluster which was most semantically similar to that sentence) or randomly, in cases where clusters were considered equally similar to that sentence. Henceforth, we will refer to manual classifications as *classes* and automatic clustering as *clusters*.

4.2 The Evaluation Measures

The accuracy of the produced clustering solution has been assessed by using the well-known Precision and Recall metrics, redefined in the cluster domain (see [4] and [12]).

Let N be the total number of sentences to be clustered, K the set of classes, C the set of clusters and n_{ij} the number of sentences of the class $k_i \in K$ that are present in cluster $c_j \in C$. The Precision and Recall for k_i and c_j, denoted $P(k_i, c_j)$ and $R(k_i, c_j)$, respectively, are computed by formulas 5 and 6. Precision is given by the number of sentences of cluster c_j that belong to the class k_i, thus measuring the homogeneity of cluster c_j with respect to class k_i. Similarly, Recall is given by the number of sentences of class k_i that are present in cluster c_j, thus measuring how complete cluster c_j is with respect to class k_i. We also measure the quality of cluster c_j in describing the class k_i, by calculating the harmonic mean between Recall and Precision of cluster c_j regarding class k_i (Formula 7). This is also known as F-measure.

$$P(k_i, c_j) = \quad n_{ij} / |c_j|. \tag{5}$$

$$R(k_i, c_j) = \quad n_{ij} / |k_j|. \tag{6}$$

$$F(k_i, c_j) = (2 * R(k_i, c_j) * P(k_i, c_j)) / (R(k_i, c_j) + P(k_i, c_j)). \tag{7}$$

The F-measure for each class over the entire data set is based on the cluster that best describes each class k_i, i.e., the one that maximizes $F(k_i, c_j)$ for all j. Thus, the overall F-measure of a clustering solution S, denoted $F(S)$, is calculated by using the weighted sum of such maximum F-measures for all classes, according to Formula 8. $F(S)$ values range from 0 (worse) to 1 (best).

$$F(S) = \sum_{k_i \in K} \frac{|k_i|}{N} \ max \ c_j \in C \ \{F(k_i, c_j)\}. \tag{8}$$

The second metric employed is Entropy [12]. It measures how well each cluster is organized, i.e., how the various classes of sentences are distributed in each cluster. A perfect clustering solution will be the one in which all clusters contain sentences from a single class only. In this case the Entropy is zero. The calculation of Entropy is based on the class distributions in each cluster. This is exactly what is done by Precision metric. In fact, Precision represents the probability of a sentence randomly chosen from cluster c_j to belong to class k_i. Hence, the Entropy of a cluster c_j, denoted $E(c_j)$, can be calculated by Formula 9.

$$E(c_j) = -\sum_{k_i} P(k_i, c_j) \ log \ P(k_i, c_j). \tag{9}$$

The Entropy of a whole clustering solution S, denoted $E(S)$, is given by the sum of the individual cluster entropies weighted by the size of the cluster, (Formula 10). $E(S)$ values are always positive. The smaller the $E(S)$, the better the clustering solution is.

$$E(S) = \sum_{c_j} \frac{|c_j|}{N} E(c_i). \tag{10}$$

The third metric used is Purity [9], which is given by the percentual of the most frequent class of a given cluster. Thus, the Purity of a cluster c_j, denoted $P(c_j)$, is defined by the class k_i that maximizes the Precision of that cluster (Formula 11).

$$P(c_j) = max \ k_i \ \{P(k_i, c_j)\}. \tag{11}$$

The overall Purity of a clustering solution, denoted $P(S)$, is given by a weighted sum of the individual cluster purities (Formula 12). $P(S)$ values range from 0 (worse) to 1 (best).

$$P(S) = \sum_{c_j \in C} \frac{|c_j|}{N} P(c_j). \tag{12}$$

It is interesting to note that the Entropy and Purity metrics evaluate the goodness of a clustering solution, while F-measure evaluates the effectiveness of the clustering method. In the next section we present the goodness and effectiveness results for SiSPI.

4.3 Experimental Results

Regarding TF-IDF and TF-ISF models, two parameters are relevant for evaluating a clustering solution: the centroid size and the similarity threshold. The first one is used to measure the similarity between a cluster and a candidate sentence to be added to it. The second one plays the role of a similarity limit, indicating when a sentence originates a new cluster.

The first experiment was carried out with four different configurations of centroids: 5, 10, 15 and 20 words. For this experiment, a similarity threshold of 0.4 (empirically determined) has been used. The average values obtained for each assessment measure for all collections are depicted in Table 1. The purpose of this experiment was to identify the centroid configuration that best describes our data set for each similarity measure.

Table 1. Average results obtained for TF-IDF and TF-ISF with 4 different centroid sizes

Centroid size in words	TF-IDF			TF-ISF		
	Entropy	F-measure	Purity	Entropy	F-measure	Purity
5	0.035	0.860	0.941	0.101	0.860	0.917
10	0.037	0.860	0.939	0.106	0.863	0.912
15	0.036	0.862	0.940	0.101	0.864	0.913
20	0.042	0.862	0.938	0.106	0.863	0.913

In general, the difference between the results of all configurations for both models is little. Regarding effectiveness (i.e. F-measure), the TF-IDF best performance was achieved using a 15 and a 20-word centroid, while the TF-ISF best performance was achieved using a 15-word centroid. However, regarding cluster goodness (measure in terms of Entropy and Purity), a 5-word centroid was the best configuration for both cases (except for TF-ISF whose Entropy values were the same for both configurations).

As F-measure is more complete than Entropy and Purity (those do not address the question of whether all elements of a given class are present in a single cluster), we preferred to use the configuration with the highest F-measure instead of the highest Entropy and Purity values. So, in the following experiments we have used a 15-word centroid. This value is close to the one used in document clustering, whose experiments show a 10-word centroid is enough to give a clear idea of what each cluster is about [8].

To identify the best similarity threshold, each similarity model has been assessed with several different threshold configurations that range from 0.1 to 1 (except Word Overlap that ranges from 0.1 to 0.5). The average values for all collections are shown in Table 2.

According to Table 2, in all cases, the Entropy values improve in a considerably way as the threshold increases. This also happens with F-measure and Purity values, but up to a given point, from which those values decrease smoothly. F-measure achieves its maximum at a threshold of 0.2, 0.3 and 0.4 for Word Overlap, TF-IDF and TF-ISF, respectively. Regarding Purity, the values increase until a similarity of 0.3 for Word Overlap, and of 0.5 for TF-IDF and TF-ISF models.

Table 2. Average results obtained for each similarity measure with different thresholds

	Similarity	0.1	0.2	0.3	0.4	0.5	0.6	0.7	0.8	0.9	1.0
TF-IDF	Entropy	0.843	0.287	0.096	0.037	0.016	0.005	0.004	0.003	0.002	0.001
	F-measure	0.603	0.814	**0.886**	**0.860**	0.841	0.828	0.812	0.799	0.775	0.736
	Purity	0.549	0.808	0.907	**0.934**	**0.945**	0.945	0.942	0.940	0.941	0.938
TF-ISF	Entropy	1.759	0.900	0.319	0.101	0.043	0.013	0.004	0.003	0.002	0.002
	F-measure	0.348	0.603	0.805	**0.864**	0.856	0.843	0.828	0.813	0.798	0.786
	Purity	0.315	0.564	0.804	0.913	**1.000**	0.950	0.954	0.953	0.952	0.951
Word Overlap	Entropy	0.572	0.079	0.010	0.000	0.001	-	-	-	-	-
	F-measure	0.695	**0.860**	0.838	0.809	0.786	-	-	-	-	-
	Purity	0.654	0.908	**0.946**	0.943	0.941	-	-	-	-	-

Specifically regarding Entropy and Purity values, they can be explained by the fact that whereas the threshold increases, the number of clusters also grows in a way that they become more homogeneous, i.e., the variety of classes in each cluster tend to decrease. Moreover, since the corpus contains many non-similar sentences, it is expected that those values increase even more, once many clusters contain only one sentence. With respect to F-measure, in spite of the cluster tendency to become more homogenous (increasing the precision), as the threshold increases, it becomes harder to identify those sentences that are semantically equivalent but lexically different (e.g. paraphrases). Hence, the recall values tend to decrease, damaging the model performance.

In terms of providing both good performance and cluster goodness, the TF-IDF model with a similarity of 0.4^2 (here TF-IDF-0.4), performed as the most appropriate for our purpose. Besides TF-IDF-0.4 has achieved a F-measure of 86.0% (the best F-measure was 88.6% (TF-IDF-0.3)), its Entropy and Purity values are good, mainly if they were compared with those obtained for TF-IDF-0.3, TF-ISF-0.4 and Word-Overlap-0.2. Moreover, the standard deviation obtained for TF-ISF-0.4 (0.07 for F-measure, 0.06 for Purity and 0.05 for Entropy) was smaller than that obtained for TF-IDF-0.3 (0.08 for F-measure, 0.07 for Purity and 0.10 for Entropy), TF-ISF-0.4 (i.e. 0.09 for F-measure, 0.08 for Purity e 0.09 Entropy) and Word Overlap (0.08 for F-measure, 0.06 for Purity and 0.07 Entropy). Figure 2 shows an example of sentence cluster built by using TF-IDF-0.4. According to the human classification, this cluster consists of 4 sentences and SiSPI found 3 of them (therefore, 85% F-measure, 100% Purity and 0 Entropy for this specific cluster).

---

[1] A polícia informou que o grupo já desviou R$ 70 milhões, desde 2004.
[2] O grupo é acusado de lesar os cofres públicos em cerca de R$ 70 milhões.
[3] Segundo divulgado pela PF, o grupo criminoso desviou desde 2004 cerca de R$ 70 milhões dos cofres públicos, por meio do pagamento de serviços, compras e obras superfaturadas.

Fig. 2. Example of a cluster generated by SiSPI with TF-IDF-0.4 version

---

[2] Coincidentally, this value is equal to the empirical value used in the first experiment.

5 Conclusions

We presented experiments using SiSPI, a sentence clustering framework which, for our best knowledge, is the first one proposed for Portuguese. SiSPI is domain independent and may be easily customized to other languages. Moreover, it can treat other similarity definitions just by adjusting the similarity threshold.

SiSPI's incremental clustering approach makes it robust even to treat small data sets. We believe that such approach will allow SiSPI to manage larger corpora with similar performance to that achieved using small corpus. Performance gains should be obtained by making use of, for instance, a synonym and/or paraphrase set, what may be useful to identify sentences with a lot of paraphrases.

Acknowledgements. We thank CNPq for financial support.

References

1. Barzilay, R., McKeown, K.: Sentence Fusion for Multi-document News Summarization. Computational Linguistics 31(3), 297–327 (2005)
2. Caldas Junior, J., Imamura, C.Y.M., Rezende, S.O.: Avaliação de um Algoritmo de Stemming para a Língua Portuguesa. In: 2nd Congress of Logic Applied to Technology, vol. 2, pp. 267–274 (2001)
3. Larocca Neto, J., Santos, A.D., Kaestner, C.A.A., Freitas, A.A.: Document Clustering and Text Summarization. In: 4th International Conference Practical Applications of Knowledge Discovery and Data Mining – PAAD 2000, pp. 41–55 (2000)
4. Fung, B.C.M., Wang, K., Ester, M.: Hierarchical Document Clustering using Frequent Itemsets. In: Barbará, D., Kamath, C. (eds.) 3rd SIAM International Conference on Data Mining, pp. 59–70 (2003)
5. Hatzivassiloglou, V., Klavans, J.L., Eskin, E.: Detecting Text Similarity over Short Passages: Exploring Linguistic Feature Combinations via Machine Learning. In: Empirical Methods in Natural Language Processing and Very Large Corpora – EMNL 1999, pp. 203–212 (1999)
6. Hatzivassiloglou, V., Klavans, J.L., Holcombe, M.L., Barzilay, R., Kan, M., McKeown, K.R.: SimFinder: A Flexible Clustering Tool for Summarization. In: Workshop on Automatic Summarization at NAACL 2001, pp. 41–49 (2001)
7. Pardo, T.A.S.: SENTER: Um Segmentador Sentencial Automático para o Português do Brasil. Technical Report NILC-TR-06-01, São Carlos-SP, Brazil, 6p (2006)
8. Radev, D.R., Hatzivassiloglou, V., McKeown, K.R.: A Description of the CIDR System as Used for TDT-2. In: DARPA Broadcast News Workshop (1999)
9. Rosell, M., Kann, V., Litton, J.: Comparing Comparisons: Document Clustering Evaluation Using Two Manual Classifications. In: Sangal, R., Bendre, S.M. (eds.) International Conference on Natural Language Processing, pp. 207–216. Allied Publishers Private Limited (2004)
10. Salton, G., Allan, J.: Text Retrieval Using the Vector Processing Model. In: 3rd Symposium on Document Analysis and Information Retrieval. In: 3rd Symposium on Document Analysis and Information Retrieval. University of Nevada, Las Vegas (1994)
11. Schaal, M., Müller, R.M., Brunzel, M., Spiliopoulou, M.: RELFIN - Topic Discovery for Ontology Enhancement and Annotation. In: The Semantic Web: Research and Applications. LNCS, pp. 608–622. Springer, Berlin (2005)

12. Steinbach, M., Karypis, G., Kumar, V.: A comparison of document clustering techniques. In: International Conference on Knowledge Discovery & Data Mining - KDD 2000 (2000)
13. Tombros, A., Jose, J.M., Ruthven, I.: Clustering Top-Ranking Sentences for Information Access. In: Koch, T., Sølvberg, I.T. (eds.) ECDL 2003. LNCS, vol. 2769, pp. 523–528. Springer, Heidelberg (2003)
14. Van Rijsbergen, C.J.: Information Retrieval, 2nd edn. Butterworths, Massachusetts (1979)
15. Ye, H., Young, S.: A Clustering Approach to Semantic Decoding. In: 9th International Conference on Spoken Language Processing – ICSLP, Pittsburgh, PA, USA (2006)
16. Radev, D., Otterbacher, J.: Zhang, Zhu.: Cross-document Relationship Classification for Text Summarization. In: Computational Linguistics (to appear, 2008)

Portuguese Part-of-Speech Tagging Using Entropy Guided Transformation Learning

Cícero Nogueira dos Santos[1], Ruy L. Milidiú[1], and Raúl P. Rentería[2]

[1] Departamento de Informática, Pontifícia Universidade Católica,
Rio de Janeiro, Brazil
nogueira@inf.puc-rio.br, milidiu@inf.puc-rio.br
[2] Fast Search & Transfer
raul.renteria@fast.no

Abstract. Entropy Guided Transformation Learning (ETL) is a new machine learning strategy that combines the advantages of Decision Trees (DT) and Transformation Based Learning (TBL). In this work, we apply the ETL framework to Portuguese Part-of-Speech Taggging. We use two different corpora: Mac-Morpho and Tycho Brahae. ETL achieves the best results reported so far for Machine Learning based POS tagging of both corpora. ETL provides a new training strategy that accelerates transformation learning. For the Mac-Morpho corpus this corresponds to a factor of three speedup. ETL shows accuracies of 96.75% and 96.64% for Mac-Morpho and Tycho Brahae, respectively.

1 Introduction

Part-of-Speech (POS) tagging is the process of assigning a POS or other lexical class marker to each word in a text [1]. POS tags classify words into categories, based on the role they play in the context in which they appear. The POS tagging is a key input feature for NLP tasks like phrase chunking and named entity recognition.

Since the last decade, many machine learning based POS taggers were proposed, such as Transformation Based Learning (TBL) [2], Maximum Entropy Models (MaxEnt) [3], Hidden Markov Models (HMM) [4], Decision Trees (DT) [5], Support Vector Machines (SVM)[6] and Cyclic Dependency Network using a rich feature set [7]. State-of-the-art POS taggers for English language achieve accuracies between 96.6% and 97.2%.

Some of the above refered machine learning techniques have been tested for the Portuguese language: Transformation Based Learning [8,9], Markov Models [10,11], Maximum Entropy Models [12] and Decision Trees [12]. The best reported results for Portuguese language achieve accuracies between 95% and 95.5%.

In this work, we apply Entropy Guided Transformation Learning (ETL) to Portuguese Part-of-Speech Taggging. ETL is a new machine learning strategy that combines the advantages of DT and TBL [13]. The ETL key idea is to use decision tree induction to obtain feature combinations (templates) and then use

A. Teixeira et al. (Eds.): PROPOR 2008, LNAI 5190, pp. 143–152, 2008.
© Springer-Verlag Berlin Heidelberg 2008

the TBL algorithm to generate transformation rules. ETL produces transformation rules that are more effective than decision trees. It also eliminates the need of a problem domain expert to build TBL templates, which is a very labor intensive task. One advantage of ETL over probabilistic methods is that its output is a set of transformation rules that can be converted into a deterministic finite-state transducer [14]. This yields to optimal time implementations of ETL POS taggers. In [14] it is showed that a transformation based English pos tagger converted into a finite-state tagger requires n steps to tag a sentence of length n, independently of the number of rules and the length of the context they require.

We evaluate the performance of ETL over two Portuguese corpora: Mac-Morpho [15] and Tycho Brahe [16]. We compare the ETL results with the ones of DT, TBL, and MXPOST taggers [3]. For both corpora, ETL shows the best results reported so far.

The remainder of this paper is organized as follows. In section 2, the ETL strategy is described. In section 3, we describe the ETL Part-of-Speech tagger. In section 4, the experimental design and the corresponding results are reported. Finally, in section 5, we present our concluding remarks.

2   Entropy Guided Transformation Learning

Entropy Guided Transformation Learning (ETL) is a new machine learning strategy that combines the advantages of Decision Trees (DT) and Transformation-Based Learning (TBL) [13]. The key idea of ETL is to use decision tree induction to obtain templates. Next, the TBL strategy is used to generate transformation rules. ETL has been successfully applied to the multilanguage phrase chunking task [17]. The ETL method is illustrated in the Fig. 1.

ETL method uses a very simple DT decomposition scheme to extract templates. The decomposition process includes a depth-first traversal of the DT. For each visited node, a new template is created by combining its parent node

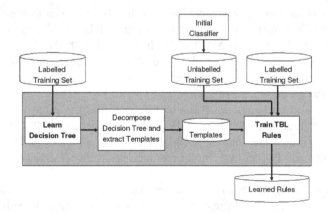

Fig. 1. ETL - Entropy Guided Transformation Learning

template with the feature used to split the data at that node. We use pruned trees in all experiments shown in section 4.

TBL training time is highly sensitive to the number and complexity of the applied templates. On the other hand, ETL provides a new training strategy that accelerates transformation learning. This strategy is based in an evolutionary template approach as described in [18]. The basic idea is to successively train simpler TBL models using subsets of the template set extracted from the DT. Each template subset only contains templates that include feature combinations up to a given tree level. In this way, only a few templates are considered at any point in time. Nevertheless, the descriptive power is not significantly reduced.

The next two sections briefly review the DT learning algorithm and the TBL algorithm.

2.1 Decision Trees

Decision tree learning is one of the most widely used machine learning algorithms. It performs a partitioning of the training set using principles of Information Theory. The learning algorithm executes a general to specific search of a feature space. The most informative feature is added to a tree structure at each step of the search. Information Gain Ratio, which is based on the data Entropy, is normally used as the informativeness measure. The objective is to construct a tree, using a minimal set of features, that efficiently partitions the training set into classes of observations. After the tree is grown, a pruning step is carried out in order to avoid overfitting.

One of the most used algorithms for induction of a DT is the C4.5 [19]. We use Quinlan's C4.5 system throughout this work.

2.2 Transformation-Based Learning

Transformation Based error-driven Learning (TBL) is a successful machine learning algorithm introduced by Eric Brill [2]. It has since been used for several Natural Language Processing tasks, such as part-of-speech (POS) tagging [2], English text chunking [20,21], spelling correction [22], Portuguese appositive extraction [23], Portuguese named entity extraction [24] and Portuguese nounphrase chunking [25], achieving state-of-the-art performance in many of them.

The TBL algorithm generates an ordered list of rules that correct classification mistakes in the training set, which have been produced by an initial classifier. The requirements of the algorithm are:

- two instances of the training set, one that has been correctly labeled, and another that remains unlabeled;
- an initial classifier, the *baseline system*, which classifies the unlabeled training set by trying to apply the correct class for each sample.
- a set of rule templates, which are meant to capture the relevant feature combinations that would determine the sample's classification.

The TBL algorithm can be depicted as follows:

1. Starts applying the baseline system, in order to guess an initial classification for the unlabeled version of the training set;
2. Compares the resulting classification with the correct one and, whenever a classification error is found, all the rules that can correct it are generated by instantiating the templates. Usually, a new rule will correct some errors, but will also generate some other errors by changing correctly classified samples;
3. Computes the rules' scores (errors repaired - errors created). If there is not a rule with a score above an arbitrary threshold, the learning process is stopped;
4. Selects the best scoring rule, stores it in the set of learned rules and applies it to the training set;
5. Returns to step 2.

When classifying a new sample item, the resulting sequence of rules is applied according to its generation order.

3   Part-of-Speech Tagging Using ETL

Our POS modeling approach follows the two stages strategy proposed by Brill [2]. First, we apply the morphological stage. Next, the contextual stage is applied.

Morphological stage, where we learn rules to classify unknown words. These rules contain morphological information such as:

- The prefix/suffix of the word up to c characters (we use c=5)
- Does the word contain a specific character
- Adding/subtracting a prefix/suffix of c characters results in a word in vocabulary

Contextual stage, where we use contextual information (neighbors words and pos tags) to learn rules to classify known words.

We use the ETL strategy for the learning of contextual rules only. The features and template set used in the morphological stage are the same used in [2].

The used base line system (BLS) assigns to each word the POS tag that was most frequently associated with that word in the training set. If captalized, an unknown word is tagged as a proper noun, otherwise it is tagged as a common noun.

The DT learning works as a feature selector and is not affected by irrelevant features. For POS tagging, we have tried several context window sizes when training the ETL tagger. Some of the tested window sizes would be very hard to be explored by a domain expert using TBL alone. The corresponding huge number of possible templates would be very difficult to be managed by a template designer.

4   Experiments

This section presents the experimental setup and results of the application of ETL to Portuguese POS Tagging. ETL results are compared with the results of DT, TBL using hand-crafted templates, Hidden Markov Models (HMM), Variable Length Markov Chain (VLMC) and Maximum Entropy Models (MaxEnt).

The following experimental setup provided us our best results.

ETL In the ETL learning, we use the features *word* and *POS*. In order to overcome the sparsity problem, we only use the 200 most frequent words to induce the DT. In the DT learning, the pos tag of the word is the one applied by the initial classifier (BLS). On the other hand, the pos tag of the neighbor words are the true ones. We report results for ETL trained with all the templates at the same time and also using template evolution.

TBL The results for the TBL approach refer to the contextual stage trained using the lexicalized templates set proposed in [2]. These template set uses combinations of words and pos tags in a context window of size 7.

DT the best result for the DT classifier is shown. The features *word* and *POS* in a context window of size 7 are used to generate the DT classifier. The pos tag of a word and its neighbors are the ones guessed by the initial classifier. Using only the 200 most frequent words gives our best results.

The MaxEnt tagger tested is the MXPOST [3]. We train and test the MX-POST tagger using the same corpus partitions used for training ETL. MXPOST does not have parameters to be customized. We compare ETL with HMM and VLMC only for the Tycho Brahe corpus, the HMM and VLMC results are the ones reported in [11].

In all experiments, the term WS=X subscript means that a window of size X was used for the given model. For instance, $ETL_{WS=3}$ corresponds to ETL trained with window of size three, that is, the current token, the previous and the next one.

Next, we present the results for each of the two corpora: Mac-Morpho and Tycho Brahe.

4.1   Mac-Morpho Corpus

The Mac-Morpho corpus [15] contains 1.2 million manually tagged words. Its tagset contains 22 POS tags and 10 more tags that represent additional semantic aspects. We divide the corpus into an 1M words training set and a 200K words test set. We carry out tests with and without using the 10 additional tags.

In Table 1, we compare the performance of ETL with DT, TBL and the MXPOST tagger for the POS tagging of the Mac-Morpho corpus without the

Table 1. POS Taggers performance on the Mac-Morpho Corpus

Classifier	Accuracy (%)	# templates
BLS	90.71	–
DT	85.58	–
MXPOST	96.30	–
TBL	96.60	26
$ETL_{WS=3}$	96.60	21
$ETL_{WS=5}$	96.72	52
$ETL_{WS=7}$	**96.75**	72
$ETL_{WS=9}$	96.74	82

Condition	Rule Change
word=*m	N → V
',' ∈ word	N → NUM
word=*ar	N → V
word=*ndo	* → V
word=*do	N → PCP
word+'m'= known word	* → V
word=*u	N → V
word=*da	N → PCP
word=*dos	N → PCP
word=*das	N → PCP

(a) Morphological Rules

Condition	Rule Change
pos[1]=ART	ART → PREP
pos[-1]=V	PRO-KS-REL → KS
word[0]=de ∧ pos[-1]=NPROP ∧ pos[1]=NPROP	PREP → NPROP
pos[1]=V ∧ word[0]=a	ART → PREP
pos[-1]=N ∧ pos[-2]=ART	N → ADJ
word[1]=de	ART → PROSUB
pos[1]=ART	VAUX → V
pos[1]=PCP	V → VAUX
pos[1]=PRO-KS-REL	ART → PROSUB
pos[1]=PREP ∧ pos[-1]=ART	ADJ → N

(b) Contextual rules

Fig. 2. Top 10 morphological (left) and contextual (right) rules for POS Tagging of the Mac-Morpho. For the morphological rules, *word=*m* is short for the test *word ends in m*, *'x' ∈ word* is short for *word contains the character 'x'*.

10 additional tags. We also report the total number of templates for each case. We can see that ETL, even using a small window size, produces better results than DT and MXPOST. ETL slightly outperforms TBL when using a window size larger than three. The accuracy of the $ETL_{WS=7}$ classifier is 0.45% higher than the one of MXPOST, 11.17% higher than the one of the DT classifier, and 0.15% higher than the one of TBL. The $ETL_{WS=7}$ accuracy, 96.75%, is the best one reported so far for the Mac-Morpho corpus. Figure 2(b) displays the top 10 contextual rules[1] learned by the $ETL_{WS=7}$ classifier. The first morphological rule states that unknown words ending with m and tagged as N (noun), must have their tags changed to V (verb). The first contextual rule states that if a word and its right neighbor are tagged as ART (article) the word must have its tag changed to PREP (preposition).

In Table 2, we show the performance of ETL using template evolution. The template evolution strategy reduces the average training time in approximately 68% with no loss in the tagger efficacy. This is a remarkable reduction, since we

Table 2. ETL with template evolution performance on the Mac-Morpho Corpus

Classifier	Accuracy (%)	Training time reduction (%)
$ETL_{WS=3}$	96.57	55.6
$ETL_{WS=5}$	96.71	70.5
$ETL_{WS=7}$	**96.74**	72.6
$ETL_{WS=9}$	96.73	74.0

---

[1] Where: N=Noun, NPROP=Proper noun, V=Verb, VAUX=Auxiliary verb, ADJ=Adjective, NUM=Number, PCP=Past participle or adjective, ART=Article, PREP=Preposition, KS=Coordinating conjunction, PRO-KS-REL=Relative subordinating pronoun, PROSUB=Non-subordinating pronoun as a noun phrase nucleus.

Table 3. POS Taggers performance for unknown words on the Mac-Morpho Corpus

Classifier	Accuracy (%)
BLS	63.01
Morpho. stage (TBL)	85.89
TBL	87.64
ETL$_{WS=7}$	87.95
MXPOST	**88.87**

use an implementation of the *fastTBL* algorithm [26] that is already a very fast TBL version.

In the test set, 4.18% of the words are unkown, what means that these words do not appear in the training set. In table 3, we show the performance of the taggers in the classification of unkown words. MXPOST has the best performance for this part of the test set. ETL and TBL accuracies correspond to the ones obtained after the application of their respective contextual rules. Figure 2(a) displays the top 10 rules learned in the morphological stage.

In Table 4, we compare the performance of ETL with DT, TBL, MXPOST and TreeTagger [5] for the POS tagging of the Mac-Morpho corpus with the 10 additional tags. The TreeTagger result is the one reported in [12]. Again, ETL outperforms the other taggers.

Table 4. POS Taggers performance on the Mac-Morpho Corpus with 10 additional tags

Classifier	Accuracy (%)	# templates
BLS	85.58	–
DT	85.26	–
TreeTagger	94.16	–
MXPOST	95.84	–
TBL	96.32	26
ETL$_{WS=3}$	96.25	21
ETL$_{WS=5}$	96.33	53
ETL$_{WS=7}$	**96.37**	74
ETL$_{WS=9}$	96.36	87

4.2   Tycho Brahe Corpus

The Tycho Brahe corpus [16] contains a total of 1,035,593 manually tagged words. It uses a set of 383 tags and is composed of various texts from historical Portuguese. In our experiments, we use exactly the same split of [11], which divide the corpus into a training set containing 775,602 words, and a testing set containing 259,991 words. Hence, our results can be directly compared with the ones of [11].

In Table 5, we compare the performance of ETL with DT, TBL, HMM, VLMC and the MXPOST tagger for the POS tagging of the Tycho Brahe corpus. The

Table 5. POS Taggers performance on the Tycho Brahe Corpus

Classifier	Accuracy (%)	# templates
BLS	91.12	–
DT	83.25	–
HMM	93.48	–
VLMC	95.51	–
MXPOST	96.32	–
TBL	96.63	26
$ETL_{WS=3}$	96.53	16
$ETL_{WS=5}$	96.62	36
$ETL_{WS=7}$	**96.64**	43
$ETL_{WS=9}$	96.63	49

results for HMM and VLMC are the ones reported in [11]. We can see that ETL, even using a small window size, produces better results than DT, HMM, VLMC and MXPOST. The accuracy of the $ETL_{WS=7}$ classifier is 13.39% higher than the one of the DT, 3.16% higher than the one of the HMM, 1.13% higher than the one of the VLMC and 0.32% higher than the one of MXPOST. ETL and TBL achieve similar results. The $ETL_{WS=7}$ accuracy, 96.64%, is the best one reported so far for the Tycho Brahe corpus.

In Table 6, we show the performance of ETL using template evolution. The template evolution strategy reduces the average training time in approximately 59% with no loss in the tagger efficacy.

Table 6. ETL with template evolution performance on the Tycho Brahe Corpus

Classifier	Accuracy (%)	Training time reduction (%)
$ETL_{WS=3}$	96.50	46.5
$ETL_{WS=5}$	96.59	62.2
$ETL_{WS=7}$	96.60	62.6
$ETL_{WS=9}$	**96.61**	63.3

Table 7. POS Taggers performance for unknown words on the Tycho Brahe Corpus

Classifier	Accuracy (%)
BLS	17.85
Morpho. stage (TBL)	72.87
TBL	76.49
$ETL_{WS=7}$	76.42
MXPOST	**79.45**

In the test set, 3.42% of the words are unkown, which means that these words do not appear in the training set. In table 7, we show the performance of the taggers in the classification of unkown words. MXPOST has the best performance for this part of the test set. ETL and TBL accuracies correspond to the ones obtained after the application of their respective contextual rules.

5 Conclusions

In this paper, we approach the Portuguese Part-of-Speech task using Entropy Guided Transformation Learning (ETL). We carry out experiments with two corpora: Mac-Morpho and Tycho Brahae. ETL achieves the best results reported so far for Machine Learning based POS tagging of both corpora. ETL provides a new training strategy that accelerates transformation learning. For the Mac-Morpho corpus this corresponds to a factor of three speedup.

One advantage of ETL over probabilistic methods is that its output is a set of transformation rules that can be interpreted by humans and can be extremely fast to apply [14]. The main ETL advantage over TBL is that we do not need a problem domain expert to construct rule templates. Therefore, ETL makes easy to incorporate new input features such as the ones provided by a lemmatizer. As a future work, we plan to use a richer feature set, as in [7], to improve the ETL Portuguese POS tagger performance.

Acknowledgments

The authors would like to thank Fábio Kepler and Marcelo Finger for providing a clean and well formated version of the Tycho Brahe corpus.

References

1. Jurafsky, D., Martin, J.H.: Speech and Language Processing. Printice Hall (2000)
2. Brill, E.: Transformation-based error-driven learning and natural language processing: A case study in part-of-speech tagging. Comput. Linguistics 21, 543–565 (1995)
3. Ratnaparkhi, A.: A maximum entropy model for part-of-speech tagging. In: Brill, E., Church, K. (eds.) Proceedings of the Conference on Empirical Methods in Natural Language Processing, Somerset, New Jersey, pp. 133–142. Association for Computational Linguistics (1996)
4. Brants, T.: Tnt – a statistical part-of-speech tagger. In: ANLP, pp. 224–231 (2000)
5. Schmid, H.: Probabilistic part-of-speech tagging using decision trees. In: International Conference on New Methods in Language Processing, Manchester, UK (1994)
6. Giménez, J., Màrquez, L.: Fast and accurate part-of-speech tagging: The svm approach revisited. In: RANLP, pp. 153–163 (2003)
7. Toutanova, K., Klein, D., Manning, C.D., Singer, Y.: Feature-rich part-of-speech tagging with a cyclic dependency network. In: HLT-NAACL (2003)
8. Aires, R.V.X., Aluísio, S.M., e Silva Kuhn, D.C., Andreeta, M.L.B., Osvaldo, N., Oliveira, J.: Combining classifiers to improve part of speech tagging: A case study for brazilian portuguese. In: IBERAMIA-SBIA, pp. 227–236. ICMC/USP (2000)
9. Finger, M.: Técnicas de otimização da precisão empregadas no etiquetador tycho brahe. In: Proceedings of PROPOR, São Paulo, pp. 141–154 (2000)
10. Kepler, F.N., Finger, M.: Part-of-speech tagging of portuguese based on variable length markov chains. In: Vieira, R., Quaresma, P., Nunes, M.d.G.V., Mamede, N.J., Oliveira, C., Dias, M.C. (eds.) PROPOR 2006. LNCS (LNAI), vol. 3960, pp. 248–251. Springer, Heidelberg (2006)

11. Kepler, F.N., Finger, M.: Comparing two markov methods for part-of-speech tagging of portuguese. In: IBERAMIA-SBIA, pp. 482–491 (2006)
12. The lacio web project (accessed in January 23, 2008), http://www.nilc.icmc.usp.br/lacioweb/ferramentas.htm
13. dos Santos, C.N., Milidiú, R.L.: Entropy guided transformation learning. Technical Report 29/07, Departamento de Informática, PUC-Rio (2007)
14. Roche, E., Schabes, Y.: Deterministic part-of-speech tagging with finite-state transducers. Comput. Linguist. 21, 227–253 (1995)
15. Aluísio, S.M., Pelizzoni, J.M., Marchi, A.R., de Oliveira, L., Manenti, R., Marquiafável, V.: An account of the challenge of tagging a reference corpus for brazilian portuguese. In: Mamede, N.J., Baptista, J., Trancoso, I., Nunes, M.d.G.V. (eds.) PROPOR 2003. LNCS, vol. 2721, pp. 110–117. Springer, Heidelberg (2003)
16. IEL-UNICAMP, IME-USP: (Corpus anotado do português histórico tycho brahe (accessed in January 23, 2008), http://www.ime.usp.br/~tycho/corpus/
17. Milidiú, R.L., dos Santos, C.N., Duarte, J.C.: Phrase chunking using entropy guided transformation learning. In: Proceedings of ACL 2008, Columbus, Ohio (2008)
18. Curran, J.R., Wong, R.K.: Formalisation of transformation-based learning. In: Proceedings of the ACSC, Canberra, Australia, pp. 51–57 (2000)
19. Quinlan, J.R.: C4.5: programs for machine learning. Morgan Kaufmann Publishers Inc., San Francisco (1993)
20. Ramshaw, L., Marcus, M.: Text chunking using transformation-based learning. In: Armstrong, S., Church, K., Isabelle, P., Manzi, S., Tzoukermann, E., Yarowsky, D. (eds.) Natural Language Processing Using Very Large Corpora. Kluwer Academic Publishers, Dordrecht (1999)
21. dos Santos, C.N., Milidiú, R.L.: Probabilistic classifications with tbl. In: Proceedings of Eighth International Conference on Intelligent Text Processing and Computational Linguistics – CICLing, Mexico City, Mexico, pp. 196–207 (2007)
22. Mangu, L., Brill, E.: Automatic rule acquisition for spelling correction. In: Proceedings of The Fourteenth ICML. Morgan Kaufmann, San Francisco (1997)
23. Freitas, M.C., Duarte, J.C., dos Santos, C.N., Milidiú, R.L., Renteria, R.P., Quental, V.: A machine learning approach to the identification of appositives. In: Proceedings of Ibero-American AI Conference, Ribeirão Preto, Brazil (2006)
24. Milidiú, R.L., Duarte, J.C., Cavalcante, R.: Machine learning algorithms for portuguese named entity recognition. In: Proceedings of Fourth Workshop in Information and Human Language Technology, Ribeirão Preto, Brazil (2006)
25. dos Santos, C.N., Oliveira, C.: Constrained atomic term: Widening the reach of rule templates in transformation based learning. In: Bento, C., Cardoso, A., Dias, G. (eds.) EPIA 2005. LNCS (LNAI), vol. 3808, pp. 622–633. Springer, Heidelberg (2005)
26. Ngai, G., Florian, R.: Transformation-based learning in the fast lane. In: Proceedings of North Americal ACL, pp. 40–47 (2001)

Learning Coreference Resolution
for Portuguese Texts*

José Guilherme C. de Souza[1], Patricia Nunes Gonçalves[2], and Renata Vieira[2]

[1] Universidade do Vale do Rio dos Sinos (UNISINOS)
[2] Pontifícia Universidade Católica do Rio Grande do Sul (PUCRS)

Abstract. This work presents the implementation of an automatic coreference resolution system based on supervised machine learning that is capable of processing any type of noun phrases for Portuguese. The system was trained and tested in a journalistic corpus formed by 50 texts with a total of 5047 markables. Both the induced classifier and the anaphoric clustering algorithm were evaluated using appropriate metrics. The clustering eveluation was performed using the MUC and B^3 scorers.

1 Introduction

The Natural Language Processing (NLP) area focuses on the construction of applications capable of interpreting or generating information provided in Natural Language (NL)[1]. In Information Extraction (IE), the target data must be found in a set of texts. In a text, the target information (or objects of interest) are linked in different ways in different places. The problem of determining which references point to which objects is one of the several challenges of the process. This is the problem studied in this work.

In this work we present an approach based on supervised machine learning to perform the automatic coreference resolution for Portuguese. The coreferential data may used to help the resolution of other NLP problems such as automatic translation and summarization. The use of coreferential data might enrich the results of these applications by making them more cohesive and intelligible[2].

The system presented here bases its ideas on the work of Soon et al. [3], one of the first works using the machine learning approach for English. The process consists in deriving subsets of text expressions (the coreference chains) first identifying pairs of expressions that are anaphoric and then grouping the pairs that relate to each other thus forming the coreference chains.

2 Related Work

A large number of works for the English language restrict the problem to the resolution of pronominal anaphora (e.g. [4], [5], [6], [7] and [8]). A portion of

---

* This work received partial support of CNPq and CAPES.

A. Teixeira et al. (Eds.): PROPOR 2008, LNAI 5190, pp. 153–162, 2008.

these works ([4], [6] and [8]) use a rule-based approach to determine pronominal anaphoricity that makes use of syntactic information.

There are also several works for resolving anaphora and coreference in English. Part of these works use the corpus based machine learning approach (e.g. [9], [10], [11], [12], [13], [3]) and others adopt statistical approaches to solve the same problem (e.g. [4], [14], [15], [6], [7]). Some other works ([16], [17] and [18] - the latter ones for Portuguese) aim to solve a related problem, the referential expressions classification. The expressions are classified as new or old (the latter being anaphoric or coreferential expressions) in the text.

Soon et al.[3] was one of the first works to use machine learning together with data provided by annotated corpora for English. Besides, Soon et al. does not have restrictions on the type of noun phrases (it uses all types of noun phrases) and has presented results comparable or even better than other works that don't use the machine learning approach. The system uses 12 features to help the determination of whether a given pair of expressions is anaphoric or not. The features are based on positional, syntactic, morphological and seman-tic (with the help of the WordNet[19]) information. The system was evaluated with the datasets and metrics provided and defined by Message Understanding Conferences 6 (MUC-6)[20] and MUC-7[21].

There aren't many works for resolving anaphora and coreference for Por-tuguese. We can cite the works of Coelho and Carvalho[22], a Lappin and Leass[6] algorithm adaptation for Portuguese; a multi-agent implementation developed by Paraboni[23] and a Mitkov's algorithm[24] adaptation for Portuguese[2].

All these three works make use of linguistic knowledge. Coelho and Carvalho[22], for example, needs the text's syntactic tree. Paraboni[23] requires morphological, syntactic and pragmatic information. In the work of Chaves[2], an annotated cor-pus with morphological and syntactic information is used to help the process of resolving pronominal anaphora. These systems are restricted to pronominal noun phrases and don't use machine learning nor semantic knowledge.

The machine learning corpus based approach has been having, in the worst case, as good results as the approaches that are not based on machine learning. Furthermore, there's a tendency for the utilization of semantic information to help the task of anaphoricity and coreference resolution. Ng presents a study[25] that explicitly shows that better results may be achieved when using semantic data. This is reinforced by the results of a series of other works ([9], [13], [3], [26], [27], [28]).

3   A Coreference Resolution Approach for Portuguese

The system's final goal is to automatically extract the coreference chains of texts written in Brazilian Portuguese. The solution implemented uses machine learning to learn a classifier which receives as input noun phrases pairs as well as properties (features) about them.

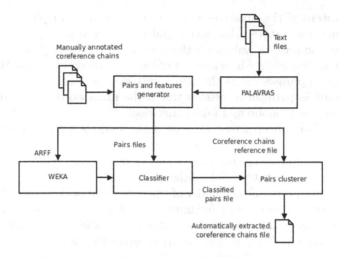

Fig. 1. The system's architecture

The system is composed of three modules: the pairs and features generator, the classifier and the anaphoric pairs clustering module. The relationship of these modules can be seen in Fig. 1.

The pairs and features generator module is the main part of the system. It receives input files with linguistic information of the texts being processed provided by the PALAVRAS[29] parser and the manually annotated coreference chains of these texts. The objective of this module is to generate pairs of noun phrases as well as some features about them. This pairs list is then used to induce a classifier using a decision tree algorithm.

Since we use a supervised machine learning algorithm, we must provide positive and negative instances for the classifier's induction. The positive and negative pairs generation is implemented following the algorithm proposed by Soon et al.[3]. The positive pair generation consists in forming pairs of adjacent expressions of the manually annotated chains of each text. For example, if the manually annotated chain is (A, B, C, D), the positive instances are (A, B), (B, C) and (C, D). The negative pair generation was conceived with the idea that between the two members of each antecedent-anaphor pair, there are other expressions that don't belong to any coreference chain or belong to other chains. To form the negative instance, each one of these expressions is paired with the anaphor. If the expressions x, y and z appear between the pair (A, B), according to the algorithm, the negative instances generated would be: (x, B), (y, B) and (z, B).

For each pair, negative or positive, a total of 10 features are processed. These features are based on positional, syntactic, morphological and semantic information. The syntactic, morphological and semantic information are obtained using the PALAVRAS parser. The semantic data provided by PALAVRAS relies on the concept of semantic prototype introduced by Bick[29][30]. The features processed by this module were all developed for Portuguese. They are:

1. **cores-match:** if the antecedent and the anaphor noun phrase's core are equal, then this feature value is true (false if different).
2. **distance:** the possible values for this feature are integers greater than 0. It determines the distance in sentences of each expression forming the pair. If the two noun phrases are in the same sentence, the distance is 0.
3. **antecedent-is-pronoun:** receives true if the main word that composes the antecedent is a pronoun and false otherwise.
4. **anaphora-is-pronoun:** receives true if the main word that composes the anaphor is a pronoun and false otherwise.
5. **both-proper-names:** if both the antecedent and the anaphor are proper names, this feature is true (false otherwise).
6. **gender-agreement:** if the gender of the core word of the expressions is the same (i.e. both masculine or feminine), this feature receives the value 1. If the value of one of the core words is unknown, this feature receives the value 2. If the gender is different this feature receives the value 0.
7. **number-agreement:** if both expressions agree on number (i.e. singular or plural), this feature is true (false otherwise).
8. **both-subject:** this feature is true if both noun phrases are subject of the sentences they belong to (false otherwise).
9. **semantic-agreement:** if both expression's core words are different (i.e. the core-matches feature is false) and they have the same semantic tags, this feature is true. False otherwise.
10. **same-semantic-group:** if both the cores are different and the semantic tags belong to the same semantic group, the value of this feature is true (false otherwise).

The pairs and features generator creates three output files: the Attribute-Relation File Format (ARFF), the pairs file and the manually annotated coreference chains file. The ARFF is the input for the Waikato Environment for Knowledge Analysis (WEKA)[31] application. WEKA is a collection of machine learning algorithms for data mining tasks. The pairs file is a list of the antecedent-anaphora pairs generated along with its features. The last file is just a reference file which contains all the manually annotated coreference chains. These three output files contain information for all the texts being processed. If 10 texts are being processed, these files contain data from all of them.

WEKA is used to induce a classifier using the ARFF file. With the induced classifier, the classifier module is built. The classifier module is then used to determine whether each pair of the pairs file (the input file for this module) is anaphoric or not. The classifier module outputs one list with all the pairs antecedent-anaphora classified.

The next step is to use the pairs clusterer module with the classified pairs list and the manually annotated coreference chains file to group the anaphoric pairs that relate to each other into groups (the coreference chains). The pairs clusterer algorithm works as follows: for each anaphoric antecedent-anaphora pair, if both the antecedent and the anaphora weren't processed yet, a new chain with the antecedent and the anaphora is created. If only the the anaphora wasn't processed, it is likely that the chain to which the pair belongs to has already been created. If the antecedent of the current pair matches the last element of a chain already created, the anaphora is added to this chain. This process continues until the end of the list of pairs. The pairs clusterer module generates a file which contains all the automatically extracted coreference chains of all texts processed by the system.

4   Evaluation

In order to evaluate the system's perfomance we have used a brazilian Portuguese corpus with manually annotated coreference information. The Summ-It corpus[32] is formed by 50 texts from the Ciências section of the Folha de São Paulo newspaper. These texts are a fraction of the PLN-BR[1] corpus. The Summ-It is the first proposal of a corpus annotated with coreference data for Portuguese since there isn't anything like the MUC[2] or the Automatic Content Extraction (ACE)[3] for Portuguese.

Table 1. Summ-it corpus coreference annotation

Classification	# (%)
new	1428 (60,05%)
associative	183 (7,68%)
other	17 (0,69%)
direct	407 (17,12%)
old indirect	291 (12,24%)
other	53 (2,21%)
Total	**2377 (100%)**

Each document in the corpus corresponds to a text file with size between 1 and 4KB (ranging from 127 to 654 words). There is a total of 5047 markables. The great part of them (95,15%) are noun phrases with core names and pronouns are only 4,82% of the corpus. From 2377 definite descriptions (noun phrases with a definite article, e.g. "the chair"), 1428 (60,05%) are new, confirming a high number of new references in the texts. The definite descriptions classified as old represent 31,57% of the corpus. The direct (direct anaphoras) subclass of the old class comprises 17,12% of the markables and the subclass indirect (indirect anaphora) represents 12,24% of the corpus. These data are summarized in Table 1. More information about the Summ-It corpus can be found in Coelho et al.[33] and Collovini et al.[32].

For the evaluation, the corpus was divided in two sets: a set for training the decision tree algorithm (31 texts) and a set for testing the system (19 texts).

---

[1] http://www.nilc.icmc.usp.br:8180/portal/
[2] http://www-nlpir.nist.gov/related_projects/muc/
[3] http://www.nist.gov/speech/tests/ace/

The pair and features generator module has created a database of 7293 instances using the training set and 1207 instances using the testing set. The number of negative instances is greater in both sets (6387 and 955 respectively). This converges with the fact that even in the manual annotation the number of anaphoric and coreferent noun phrases is inferior when compared to the number of new expressions in the texts (see Table 1).

The antecedent-anaphora pairs were used in conjuction with WEKA[31] to induce a classifier. The algorithm used was the J48, a C4.5[34] implementation in Java. The J48 algorithm was invoked with the default parameters and 10-fold cross-validation using the training set. The classifier used only five of the ten features: cores-match, number-agreement, gender-agreement, antecedent-is-pronoun and anaphora-is-pronoun. Figure 2 shows the induced tree.

```
cores-match = yes
|   number-agreement = yes: yes (354.0/39.0)
|   number-agreement = no
|   |   gender-agreement = yes
|   |   |   antecedent-is-pronoun = yes: no (16.0/4.0)
|   |   |   antecedent-is-pronoun = no: yes (40.0/15.0)
|   |   gender-agreement = no
|   |   |   both-proper-names = yes: yes (3.0/1.0)
|   |   |   both-proper-names = no: no (17.0)
|   |   gender-agreement = unknown: no (0.0)
cores-match = no
|   anaphora-is-pronoun = yes
|   |   number-agreement = yes: yes (9.0)
|   |   number-agreement = no
|   |   |   gender-agreement = yes
|   |   |   |   antecedent-is-pronoun = yes: no (5.0/1.0)
|   |   |   |   antecedent-is-pronoun = no: yes (118.0/54.0)
|   |   |   gender-agreement = no: no (85.0/18.0)
|   |   |   gender-agreement = unknown: no (0.0)
|   anaphora-is-pronoun = no: no (6646.0/468.0)
```

Fig. 2. The tree induced by J48 using the training set

With the testing set, 1037 instances (85,91%) were correctly classified. The precision for both the positive and negative class is high, 80,6 and 86,6% respectively. The recall is higher for negative cases (97,3%) than for positive cases (42,9%). The F-Measure for the positive examples is of 56% and for the negative 91,6%. This data is summarized in Table 2.

We noticed that the classifier tends to generalize towards the more frequent class, failing to find many anaphoric pairs. The number of true positives for the anaphoric samples is not elevated, only 108 cases or 42,9%. Also, there is an elevated number of false positives for the non-anaphoric class (929 cases or 57,1%). These data may be viewed in Table 3.

Table 2. Detailed accuracy by class using the testing set

Class	TP (%)	FP (%)	Precision (%)	Recall (%)	F-Measure (%)
Anaphoric	42,9	0,027	80,6	42,9	56
Non-anaphoric	97,3	57,1	86,6	97,3	91,6

Table 3. Confusion matrix for the testing set

Class / Prediction	Anaphoric	Non-anaphoric
Anaphoric	108	144
Non-anaphoric	26	929

We have also evaluated perfomance of the pairs clusterer module. We have used the class `ClusterScore` implemented in the LingPipe[4] library to help with the evaluation. This library provides scorers that can be used to evaluate response partitions in function of reference partitions. A reference partition is a set of coreference chains that are used as a key. The reference partition contains the chains considered to be the "truth". The response partition is a set of coreference chains that one wish to evaluate in relation to a reference partition. Here, the reference partition is the set of manually annotated coreference chains and the response partition the list of coreference chains outputted by the pairs clusterer module.

The LingPipe library implements the MUC scorer proposed and explained by Vilain et al.[35]. This scorer is a MUC initiative to evaluate automatic coreference resolution systems in terms of precision, recall and F-Measure. Besides the MUC scorer, the library also implements the metric proposed by Bagga et al.[36]. We used both metrics to evaluate our automatic extracted coreference chains.

Table 4. Coreference chains evaluation

	MUC			B^3		
System	Precision	Recall	F-Measure	Precision	Recall	F-Measure
Our system	97,47	36,18	51,33	99,24	54,30	69,66
Soon et al.	67,3	58,6	62,6	75,3-78,4	53,4-58	63,5-65,6

In Table 4 we present our system's accuracy results using the MUC and the B^3 scorers. We evaluated the pair clusterer module in the testing set (19 texts). Our baseline is the work of Soon et al.[3] because it is a reference in the area and also because there isn't any work that performs coreference resolution in Portuguese. It's important to state that the corpus used to evaluate the work of Soon et al. is not the same as our corpus. The numbers of the MUC metric were taken from Soon et al.[3] and the numbers for the B^3 were taken from Ng (2005)[27]. The latter work uses three different testing sets provided by ACE[5].

We are aware that we cannot strictly compare our system's results with those of Soon et al. since they are experiments over different languages and data sets. Our comparison here is just to see the performance of our system in regards to a much referred baseline in the area. We have a lower recall (36,18%) than the one of Soon et al (58,6%). Because of the low recall, our F-Measure is also inferior

---

[4] http://www.alias-i.com/lingpipe/
[5] http://www.itl.nist.gov/iad/894.01/tests/ace/

(51,33% and 62,6%). When using the B^3 metric, however, the results seems more alike. The F-Measure of our system is even superior to the F-Measure obtained by Soon et al. This is explained by the fact that the B^3 algorithm takes into account the singleton chains, i.e., coreference chains composed by only one expression. A thorough study about the differences between the MUC and B^3 scorers are found in the works of Luo[37] and Baldwin[38].

5   Final Remarks

We have presented a solution to automatic coreference resolution of noun phrases based on a corpus data-driven supervised machine learning approach for Portuguese. The implementation shown in this work is domain free, i.e., it may be applied to texts of any domain. The system was trained and tested using a journalistic corpus of 50 texts with 5047 markables. Both the classifier and the clustering algorithm were evaluated using metrics appropriated for these tasks. It is important to notice that this is the first proposal for automatic coreference resolution of noun phrases of any type for domain independent texts for Portuguese. Related work present solutions restricted to the pronominal anaphora resolution. The results demonstrated to be encouraging for a first proposal.

As future work we believe that there is room to improve the set of features used to determine whether a given pair is anaphoric. Despite of the presence of two semantic features, none of them were used in the final induced tree. With this, cases of indirect anaphora (referentiation by lexical substitution) are not being covered. The testing instances generation scheme of this work described in the section 3 does not use the scheme proposed by Soon et al.[3]. This scheme could improve the final results of the system. Besides, another point that could be improved is the algorithm used to generate positive and negative instances for training. Ng(2002)[26] has improved the work of Soon et al.[3] in this aspect and has achieved better results. The clustering algorithm could be also improved with other schemes proposed in the literature for English coreference resolution ([26],[28]).

References

1. Vieira, R., de Lima, V.L.S.: Lingüística computacional: princípios e aplicações. In: As Tecnologias da informação e a questão social, Ana Teresa Martins and Díbio Leandro Borges (2001)
2. Chaves, A.R.: A resolução de anáforas pronominais da língua portuguesa com base no algoritmo de mitkov. Master's thesis, Universidade Federal de São Carlos (Julho 2007)
3. Soon, W.M., Ng, H.T., Lim, D.C.Y.: A machine learning approach to coreference resolution of noun phrases 27(4), 521–544 (2001)
4. Hobbs, J.R.: Pronoun resolution. SIGART Bull. 61, 28 (1977)
5. Dagan, I., Itai, A.: A statistical filter for resolving pronoun references. Artificial Intelligence and Computer Vision, 125–135 (1991)

6. Lappin, S., Leass, H.J.: An algorithm for pronominal anaphora resolution. Computational Linguistics, 535–561 (December 1994)
7. Mitkov, R.: Robust pronoun resolution with limited knowledge. In: Proceedings of 17th Internaltional Conference on Computational Linguistics, pp. 869–875 (1998)
8. Palomar, M., Moreno, L., Peral, J., Mun, R., Ferrández, A., Martínez-Barco, P., Saiz-Noeda, M.: An algorithm for anaphora resolution in spanish texts. Comput. Linguist. 27(4), 545–567 (2001)
9. Aone, C., Bennett, S.W.: Evaluating automated and manual acquisition of anaphora resolution strategies. In: Proceedings of the 33rd annual meeting on Association for Computational Linguistics, pp. 122–129. Association for Computational Linguistics, Morristown (1995)
10. McCarthy, J.F., Lehnert, W.G.: Using decision trees for coreference resolution. In: Proceedings of the 14th IJCAI, Montreal, Canada, pp. 1050–1055 (1995)
11. Fisher, D., Soderland, S., Feng, F., Lehnert, W.: Description of the umass system as used for muc-6. In: MUC6 1995: Proceedings of the 6th conference on Message understanding, pp. 127–140. Association for Computational Linguistics, Morristown (1995)
12. Mccarthy, J.F.: A trainable approach to coreference resolution for information extraction. PhD thesis, Director-Wendy G. Lehnert (1996)
13. Cardie, C., Wagstaff, K.: Noun phrase coreference as clustering. In: Proceedings of the Joint SIGDAT Conference on Empirical Methods in Natural Language Processing and Very Large Corpora, University of Maryland, MD, pp. 82–89. Association for Computational Linguistics (1999)
14. Baldwin, B.: Cogniac: High precision coreference with limited knowledge and linguistic resources. In: Proceedings of the ACL Workshop on Operational Factors in Practical, Robust Anaphora Resolution for Unrestricted Texts, pp. 38–45 (1997)
15. Kameyama, M.: Recognizing referential links: An inforation extraction perspective. In: Proceedings of the ACL Workshop on Operational Factors in Practical, Robust Anaphora Resolution for Unrestricted Texts, pp. 46–53 (1997)
16. Uryupina, O.: High-precision identification of discourse new and unique noun phrases. In: Proceedings of the ACL Student Workshop, Sapporo (2003)
17. Collovini, S., Coelho, J.C.B., Vieira, R.: Classificação automática de expressões anafóricas em textos da língua portuguesa. In: Proceedings of ENIA 2005 (2005)
18. de Abreu, S.C.: Análise de expressões referenciais em corpus anotado da língua portuguesa. Master's thesis, UNISINOS, São Leopoldo, RS (2005)
19. Fellbaum, C.: WordNet: An Electronical Lexical Database. MIT Press, Cambridge (1998)
20. MUC-6: Coreference task definition. In: Proceedings of the Sixth Message Understanding Conference (MUC-6), San Francisco, CA, 8 September 1995, vol. 2.3, pp. 335–344 (1995)
21. MUC-7: Coreference task definition. In: Proceedings of the Seventh Message Understanding Conference (MUC-7), San Francisco, CA (13 July 1997)
22. Coelho, T.T., Carvalho, A.M.B.R.: Uma adaptação do algoritmo de lappin e leass para resolução de anáforas em português. In: Anais do XXV Congresso da Sociedade Brasileira de Computação (III Workshop em Tecnologia da Informação e da Linguagem Humana - TIL 2005, São Leopoldo, RS, pp. 2069–2078 (2005)
23. Paraboni, I.: Uma arquitetura para a resolução de referências pronominais possessivas no processamento de textos em língua portuguesa. Master's thesis, PUCRS, Porto Alegre (1997)
24. Mitkov, R.: Anaphora Resolution. Longman (2002)

25. Ng, V.: Shallow semantics for coreference resolution. In: Proceedings of IJCAI-2007, pp. 1689–1694 (2007)
26. Ng, V., Cardie, C.: Improving machine learning approaches to coreference resolution. In: Proceedings of the 40th Annual Meeting of the Association for Computational Linguistics (ACL), pp. 104–111 (2002)
27. Ng, V.: Machine learning for coreference resolution: from local classification to global ranking. In: ACL 2005: Proceedings of the 43rd Annual Meeting on Association for Computational Linguistics, pp. 157–164. Association for Computational Linguistics, Morristown (2005)
28. Ponzetto, S.P., Strube, M.: Exploiting semantic role labeling, wordnet and wikipedia for coreference resolution. In: Proceedings of the Human Language Technology Conference of the NAACL, Main Conference, June 2006, pp. 192–199. Association for Computational Linguistics,New York (2006)
29. Bick, E.: The Parsing System PALAVRAS - Automatic Grammatical Analysis of Portuguese in a Constraint Grammar Framework. PhD thesis, Department of Linguistics, University of Århus, DK (2000)
30. Bick, E.: Noun sense tagging: Semantic prototype annotation of a portuguese treebank. In: Proceedings of TLT 2006 (2006)
31. Witten, I.H., Frank, E.: Data Mining: Practical machine learning tools and techniques, 2nd edn. Morgan Kaufmann, San Francisco (2005)
32. Collovini, S., Carbonel, T.I., Fuchs, J.T., Coelho, J.C., Rino, L., Vieira, R.: Summ-it: um corpus anotado com informações discursivas visando sumarização automática. In: TIL 2007 (2007)
33. Coelho, J.C.B., Collovini, S., Vieira, R.: Instruções para anotação de relações anafóricas e referência dêitica (2006)
34. Quinlan, J.R.: C4.5: programs for machine learning. Morgan Kaufmann Publishers Inc., San Francisco (1993)
35. Vilain, M., Burger, J., Aberdeen, J., Connolly, D., Hirschman, L.: A model-theoretic coreference scoring scheme. In: Proceedings of the 6th Message Understanding Conference (MUC6), pp. 45–52. Morgan Kaufmann, San Francisco (1995)
36. Bagga, A., Baldwin, B.: Algorithms for scoring coreference chains. In: Proceedings of the First International Conference on Language Resources and Evaluation Workshop on Linguistic Coreference (1998)
37. Luo, X.: On coreference resolution performance metrics. In: HLT 2005: Proceedings of the conference on Human Language Technology and Empirical Methods in Natural Language Processing, pp. 25–32. Association for Computational Linguistics, Morristown (2005)
38. Baldwin, B., Morton, T., Bagga, A., Baldridge, J., Chandraseker, R., Dimitriadis, A., Snyder, K., Wolska, M.: Description of the university of pennsylvania camp system as used for coreference. In: Proceedings of the 7th Message Understanding Conference (MUC-7) (1998)

Domain Adaptation of a Broadcast News Transcription System for the Portuguese Parliament

Luís Neves[1], Ciro Martins[1,2], Hugo Meinedo[1], and João Neto[1]

[1] L2F – Spoken Language Systems Lab – INESC-ID/IST
Rua Alves Redol, 9, 1000-029 Lisboa, Portugal
[2] Department Electronics, Telecomunications & Informatics/IEETA
Aveiro University, Portugal
{Luis.Neves,Ciro.Martins,Hugo.Meinedo,
Joao.Neto}@l2f.inesc-id.pt

Abstract. The main goal of this work is the adaptation of a broadcast news transcription system to a new domain, namely, the Portuguese Parliament plenary meetings. This paper describes the different domain adaptation steps that lowered our baseline absolute word error rate from 20.1% to 16.1%. These steps include the vocabulary selection, in order to include specific domain terms, language model adaptation, by interpolation of several different models, and acoustic model adaptation, using an unsupervised confidence based approach.

Keywords: Vocabulary selection, model adaptation, domain adaptation, Portuguese Parliament, transcription systems.

1 Introduction

In the last decade Broadcast News (BN) transcription systems have been subject to a large effort of investigation and development by several international laboratories [1] [2] [3]. In our group we have been working specially on subtitling systems. This development allowed the construction of robust transcription systems, with high vocabulary coverage, low transcription word error rates and, in some cases, real time performance and online operation.

For the Portuguese language particular case, there have been great improvements, allowing the use of transcription systems in practical applications with diverse and complex context. An example of this is the large vocabulary transcription system currently being used to generate subtitles in the RTP1 evening news, working below real time on online mode.

There are several applications that could benefit from a large vocabulary automatic transcription system. The acknowledgment of this fact led us to adapt our BN transcription system to other domains of application, and in this particular work, for parliament meetings.

Automatic speech recognition of European Parliament Plenary Sessions has been one of the tasks of the TC-STAR project and one of the components involved in the translation process, with several participants like LIMSI [4], IBM [5], and NOKIA [6]

A. Teixeira et al. (Eds.): PROPOR 2008, LNAI 5190, pp. 163–171, 2008.

submitting their speech transcription systems for evaluation. This project was only focused in three different languages: European English, European Spanish, and Mandarin Chinese.

In the Portuguese Parliament plenary meetings there is the specific need to produce a journal that can be viewed by the general public, according to Parliament's Rules of Procedure. Our transcription system can be used to produce the journal entries, or to generate an initial transcription to be manually corrected, reducing the time demand of this process. The parliament plenary meetings are also broadcasted in television and online web video stream. Our transcription system can be used to produce subtitles allowing hearing impaired persons to follow these programs.

Section 2 summarizes the first task of the project, corpus collection, by retrieving previous plenary meetings available on the web, and recording video streams from the parliament's television channel. Section 3 describes our baseline Broadcast News transcription system and the corresponding results achieved without adaptation to the Parliament meetings task. Section 4 is dedicated to the adaptation of the transcription system's modules to the new domain. Finally we present some conclusions.

2   Corpora Collection

In this section we describe the corpora that were collected and processed to perform the work of domain adaptation.

2.1   Textual Corpora

In order to accomplish the speech transcription task in a given domain, it is necessary to obtain information about terms that are frequently used, and the way they appear in a sentence. This kind of information can be found in text material related to the target domain, and it is used to build the transcription system's vocabulary and language model. The system's performance is highly dependent of the quantity and quality of the text material. It was desirable to find manually transcribed plenary meetings, because they were most representative of the speech that would be recognized by our system. This text material was found in the Portuguese Parliament online site http://www.parlamento.pt, under The Journal of the Assembly - 1st Series section. Each document had the transcription of one parliament session; there were available 287 documents from the X Legislature and 281 from the IX Legislature, as shown in table 1. All of them were available as pdf files.

It was necessary to make the conversion from the pdf files to plain text. The format conversion was followed by a normalization process which eliminated punctuation, converted all text to lowercase, expanded abbreviations, spelled numbers and deleted speaker tags. This normalization was made with the same system used to normalize the text corpora from the broadcast news.

The X Legislature 3rd session was reserved as the text corpus development set, as shown in table 1. This set was required for the linear interpolation between language models, described in the domain adaptation section.

Table 1. Collected documents organization

Legislature	Series	Time interval of the meetings	Number of documents	Set
X Legislature	1st series	2005-03-11 to 2006-09-08	149	Training
	2nd series	2006-09-16 to 2007-09-07	110	Training
	3rd series	2007-09-20 to 2007-12-14	26	Development
IX Legislature	1st series	2002-04-06 to 2003-09-04	146	Training
	2nd series	2003-09-18 to 2004-09-03	108	Training
	3rd series	2004-09-16 to 2005-01-27	24	Training

This way we had two different text corpora sets. The training set with 907,281 sentences and around 17M words, and the development set with 13,429 sentences and 205,795 words.

2.2 Audio Corpora

We have collected two video streams from the parliament channel's website, in 9 January and 10 January 2008.

For both video programs the audio stream was extracted to mp3 format, using open source tools. It was necessary to perform the conversion of the compressed audio to raw format at 16 KHz sampling rate, 16 bits per sample, which is currently one of the audio formats supported by our transcription system.

There were saturation levels in the 9 January program's audio, because the microphone recording level of the plenary session participants' was extremely variable. Usually this audio signal saturation increases the transcription system's error rate.

The total duration of the audio signal collected was 3 hours and 36 minutes. In order to evaluate the transcription system's performance, we selected a 21 minutes and 40 seconds audio segment, which was transcribed manually and used as the audio corpora evaluation set. This set has five male speakers and one female speaker, all of them with Lisbon accent, totalling 19 minutes and 12 seconds of net speech. The manual transcription of the evaluation set has 248 sentences with 2,850 words.

3 Baseline Transcription System

Our baseline large vocabulary transcription system was trained for Broadcast News in European Portuguese, entitled AUDIMUS [7]. It uses hybrid acoustic models that try to combine the temporal modeling capabilities of hidden Markov models with the pattern classification capabilities of MLPs (Multi-Layer Perceptrons).

The models have a topology where context independent phone posterior probabilities are estimated by three MLPs given the acoustic parameters at each frame. The MLPs were trained with different feature extraction methods: PLP (Perceptual Linear Prediction), Log-RASTA (log-RelAtive SpecTrAl) and MSG (Modulation Spectro-Gram). The two first referred above incorporate local acoustic context via an input window of 13 frames, with the energy algorithm are extracted 12 coefficients and their first derivative, totaling a 26 elements vector. The last method uses an input

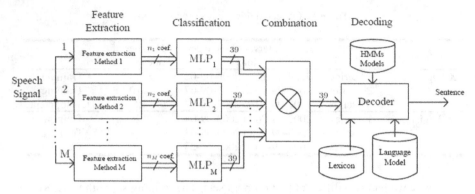

Fig. 1. Baseline transcription system's architecture [11]

window of 15 frames being extracted 14 coefficients. These are submitted to two filters (high-pass and band-pass), producing a 28 elements vector. The resulting network has two non-linear hidden layers with over 2000 units and 39 softmax output units (38 phones plus silence). The phone probabilities generated at the output of the MLPs classifiers are combined using an appropriate algorithm [8] to be used in the decoding process.

The decoder used in this system is based on a weighted finite-state transducer (WFST) approach to large vocabulary speech recognition [9]. In this approach, the decoder search space is a large WFST that maps observation distribution to words. The transcription system's full architecture is described in figure 1.

The transcription system vocabulary and language model were built from two training corpora, newspapers texts collected from the WWW since 1991 until the end of 2003 with 604M words, and broadcast news transcripts with 531K words. The vocabulary was created selecting the 100,000 (100K) more frequent words from both corpora. The baseline language model (LM) [10] combines a backoff 4-grams LM trained on the newspapers corpus, and a backoff 3-grams LM estimated on the transcripts corpus. The two models were combined by means of linear interpolation, generating a mixed model.

The acoustic model used by the baseline system was trained with 46 hours of manual transcribed broadcast news programs, recorded during October 2000, and afterwards adapted using 332 hours of automatically transcribed material [11].

For the BN evaluation set corpus [11], the out-of-vocabulary (OOV) word rate is 1.4%, the average word error rate (WER) is 21.5% for all conditions and 10.5% for F0 conditions (read speech in studio).

Using our evaluation set, described in the audio corpora section, this baseline system achieved a WER of 20.1% for all acoustic conditions.

4   Domain Adaptation

The following subsections describe the adaptation stages to the Parliament's domain of the lexical, language and acoustic models.

4.1 Vocabulary and Lexical Model

In the adaptation to a new domain the vocabulary selection is extremely important. The specific frequent terms from the domain must be included, in order to the transcription system to recognize them.

To build the vocabulary and language model, we had available three different corpora. Two of them had been used training the broadcast news system, as described in section 3, and the third was our training textual corpora, described in section 2.1. The corpora collected for the broadcast news system, because of its size and generic characteristics, gave us the terms that were frequently used in the Portuguese language, while our textual corpora of manually transcribed plenary meetings gave us the specific terms of the domain.

We have decided to build our 100,000 words vocabulary based on word relative frequency. We have started by calculating the relative frequency value of each word in the three corpora, added these values for equal words, and selected the 100,000 words with the highest value. This extremely simple solution revealed itself effective, but there are other solutions for this problem, like morpho-syntactic analysis [12]. This selection method added 6,549 parliament transcriptions words that weren't in the initial broadcast news vocabulary. For our text development set the out-of-vocabulary (OOV) word rate was reduced from 2.0% to 1.1%.

The pronunciation lexicon was built by running the vocabulary through an automatic grapheme-to-phone conversion module for European Portuguese [13]. This module has the ability to produce multiple SAMPA pronunciations for each word, generating a pronunciation lexicon with 107,784 entries.

4.2 Language Model

It is important to introduce rules that can describe linguistic restrictions present in the language. This is accomplished through the use of a language model in the system. A language model represents a grammar which is a set of rules that regulate the way the words of the vocabulary can be arranged into groups and form sentences. Usually the grammar of a large vocabulary transcription system is a stochastic model based on probabilities for sequences of words. To create this kind of models it is required to use large amounts of training data as to obtain valid statistics that allow the construction of robust stochastic language models. This need for large amounts of training data lead us to build a mixed model, by linear interpolation of the broadcast news models and the model created with our manual transcriptions of plenary meetings, as shown in figure 2.

The process of creation, interpolation and quality analysis (perplexity calculation) of the language models was performed with the SRILM Toolkit [14].

Our first step was to create a language model for each textual corpus available using our previous selected 100K vocabulary. To do this we selected the order and discount method that minimizes the perplexity of the model. This way we created a backoff 4-grams LM using absolute discounting trained with the newspapers texts, a backoff 3-grams LM using unmodified Kneser-Ney discounting trained with the broadcast news transcriptions, and a backoff 3-grams LM using unmodified Kneser-Ney discounting trained with the parliament transcriptions. The perplexity values, obtained for each one of these models for our development set, can be viewed in table 2.

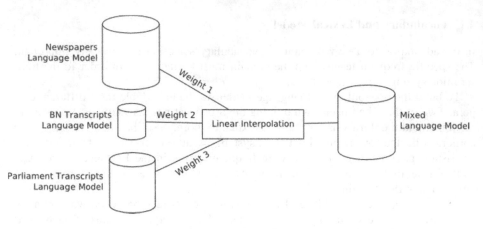

Fig. 2. Language Model linear Interpolation schematic

Table 2. Language models parameters and perplexity

Language Model	Order	Discounting	Perplexity
Newspapers	4	Absolute	140.2
BN Transcripts	3	Kneser-Ney	436.6
Parliament Transcripts	3	Kneser-Ney	71.8

To perform the linear interpolation between the three models, it is necessary to compute the interpolation weights with regard to the development set. The resulting model will have the minimum perplexity possible for the development set using a mixture of those three models [15]. The interpolation weights were set to 0.190 0.002 0.808 for the newspapers, BN transcripts and Parliament Transcripts LM's respectively. The result was a single backoff 4-grams language model.

After interpolation the perplexity value decreased to 50.9 with an OOV rate of 1.1% for the development set. The absolute WER for our evaluation set using the new language model decreased to 16.3%, resulting in a relative WER reduction of 18.9%.

4.3 Acoustic Model

Usually the adaptation of the acoustic model requires a large amount of manual transcribed audio to adapt the Multi-Layer Perceptron (MLP) network weights effectively. Unfortunately, for this work in particular, the only manually transcribed material available was the evaluation set. To solve this problem we have used a multiple decoding stage approach.

In a first stage a first transcription of the entire audio corpora can be obtained with the base system acoustic model and the adapted language model. Then the transcribed targets from the first decoding stage are pruned according to the degree of confidence of the transcription obtained.

The transcription system can provide a confidence measure that compared to a threshold allow rejecting transcriptions potentially erroneous. To determine the value of the threshold, we have created the ROC curve based on the evaluation set, as shown in figure 3, and determined an appropriate working point. It was more important to have a low false alarm (erroneous transcription that was accepted) than a high detection (correct transcription that was accepted) percentage to assure the quality of the transcribed targets. This way we have selected a confidence threshold value of 0.915 which produced 12% false alarm and 66% detection.

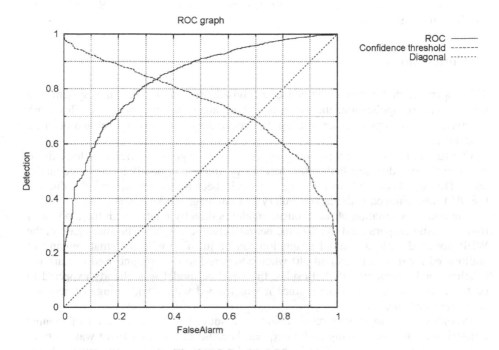

Fig. 3. ROC curve of the test set

Table 3. Transcription word error rate (WER) in the second stage decoding, without cross-validation set in the neural network adaptation

Training Iteration	Without Cross-validation
1	16.1%
2	16.4%
3	16.6%
4	18.6%
5	17.3%

Using the pruned transcribed targets and the corresponding audio segments was possible to adapt the acoustic model to the new domain. We have conducted two different adaptations of the Multi-Layer Perceptron (MLP) network weights. In the first one we used 10% of the selected targets as a cross-validation set and the rest of them as the training set. The cross-validation classification error defined the number of training iterations of the neural network. There were performed 6 training steps, producing a WER result of 16.7% in the evaluation set. In the second kind of adaptation we have used all the previous selected targets to adapt the weights. The results, available in table 3, showed that this was the best solution using all the selected targets to make just one neural network training iteration, to avoid over-adaptation.

The final result from the second stage transcription of the evaluation set, with the adapted acoustic model, was 16.1% WER, resulting in a relative WER reduction of 19.9% to the baseline system.

5 Conclusions

This paper reported our work on adapting a broadcast news transcription system to a new domain of application, the Portuguese Parliament plenary meetings. This work involved several steps in order to adapt the vocabulary, language model and acoustic model used.

Our first impressions of this work, suggested that the greater difference that existed between the two domains lied in the vocabulary and consequently in language model used. This was later confirmed during our work, because the greater WER reduction (18.9%) was achieved with the vocabulary and language model adaptations.

The correct adaptation of the acoustic model is directly related with the amount of training audio corpora used to adapt the neural network weights. The small gain in the WER obtained with our model adaptation can be justified with the small amount of audio used for this task (around 4h) when compared to the amount used to train the baseline model (around 348h). Besides, the baseline model was already expected to perform well in the new domain since it was trained with a wide range of acoustic conditions and speakers.

Probably a slightly better result could be achieved with the creation of manual transcriptions for the training audio corpora, because in this case there was no transcription error in the targets used in the neural network adaptation, but usually this is the most time demanding task. Our adaptation process allows us to eliminate this problem, thus, reducing drastically the time needed to deploy our transcription system in a new domain.

Acknowledgements

The authors would like to thank Alberto Abad, for many helpful discussions. This work was funded by PRIME National Project TECNOVOZ number 03/165.

References

1. Gales, M., Kim, D., Woodland, P., Mrva, D., Sinha, R., Tranter, S.: Progress in the CU-HTK Broadcast News Transcription System. IEEE Transactions on Audio Speech and Language Processing (2006)

2. Sinha, R., Gales, M., Kim, D., Liu, X., Sim, K., Woodland, P.: The CU-HTK Mandarin Broadcast New Transcription System. In: Proceedings ICASSP (2006)
3. Nguyen, L., Abdou, S., Afify, M., Makhoul, J., Matsoukas, S., Schwartz, R., Xiang, B., Lamel, L., Gauvain, J., Adda, G., Schwenk, H., Lefevre, F.: The 2004 BBN/LIMSI 10xRT English Broadcast News Transcription System. In: Proceedings DARPA RT 2004, Palisades, NY (November 2004)
4. Lamel, L., Gauvain, J., Adda, G., Barras, C., Bilinski, E., Galibert, O., Pujol, A., Schwenk, H., Zhu, X.: The LIMSI 2006 TC-STAR EPPS Transcription Systems. In: Proceedings of ICASSP, Honolulu, Hawaii, pp. 997–1000 (April 2007)
5. Ramabhadran, B., Siohan, O., Mangu, L., Zweig, G., Westphal, M., Schulz, H., Soneiro, A.: The IBM 2006 Speech Transcription System for European Parliamentary Speeches. In: ICSLP (September 2006)
6. Kiss, I., Leppanen, J., Sivadas, S.: Nokia's system for TC-STAR EPPS English ASR evaluation task. In: Proceedings of TC-STAR Speech-to-Speech Translation Workshop, Barcelona, Spain (June 2006)
7. Meinedo, H., Caseiro, D., Neto, J., Trancoso, I.: AUDIMUS.media: a Broadcast News speech recognition system for the European Portuguese language. In: Mamede, N.J., Baptista, J., Trancoso, I., Nunes, M.d.G.V. (eds.) PROPOR 2003. LNCS, vol. 2721. Springer, Heidelberg (2003)
8. Meinedo, H., Neto, J.: Combination of acoustic models in continuous speech recognition. In: Proceedings ICSLP 2000, Beijing, China (2000)
9. Mohri, M., Pereira, F., Riley, M.: Weighted finite-state transducers in speech recognition. In: ASR 2000 Workshop (2000)
10. Martins, C., Teixeira, A., Neto, J.: Language models in automatic speech recognition. Magazine of DET-UA. Aveiro 4(4) (2005)
11. Meinedo, H.: Audio pre-processing and speech recognition for Broadcast News. PhD thesis, IST (2008)
12. Martins, C., Teixeira, A., Neto, J.: Dynamic Broadcast News transcription system. In: ASRU 2007 (2007)
13. Caseiro, D., Trancoso, I., Oliveira, L., Viana, C.: Grapheme-to-phone using finite state transducers. In: Proc. 2002 IEEE Workshop on Speech Synthesis, Santa Monica, CA, USA (2002)
14. Stolcke, A.: Srlim - an extensible language modeling toolkit. In: Proc. ICSLP 2002, Denver, USA (2002)
15. Souto, N., Meinedo, H., Neto, J.: Building language models for continuous speech recognition systems. In: Ranchhod, E., Mamede, N.J. (eds.) PorTAL 2002. LNCS (LNAI), vol. 2389. Springer, Heidelberg (2002)

Automatic Classification and Transcription of Telephone Speech in Radio Broadcast Data

Alberto Abad, Hugo Meinedo, and João Neto

L^2F - Spoken Language Systems Lab
INESC-ID / IST, Lisboa, Portugal
{Alberto.Abad,Hugo.Meinedo,Joao.Neto}@l2f.inesc-id.pt
http://www.l2f.inesc-id.pt/

Abstract. Automatic transcription of telephone speech involves additional challenges compared to wideband data processing, mainly due to channel limitations and to particular characteristics of conversational telephone speech. While in TV speech recognition applications, such as automatic transcription of broadcast news, the presence of telephone data is nearly insignificant (less than 1 %), in most radio broadcast stations the presence of telephone speech grows significantly. Thus, transcription of telephone speech data deserves special attention in radio broadcast applications. In this work, we describe our initial efforts to tackle this particular problem. First, a telephone channel classifier is proposed to automatically detect telephone segments. Then, some strategies for increasing robustness of the automatic transcription system are investigated.

Keywords: Speech recognition, radio broadcast transcription, telephone speech processing, channel classification.

1 Introduction

Continuous advances in speech and language technology, and more concretely, in automatic speech recognition (ASR) have made possible the development of successful very large vocabulary continuous speech recognition systems in certain constrained conditions. Particularly, high quality speech – free of noise and reverberation – and planned non-spontaneous speaking style are usually required.

Due to the generally favorable speech data characteristics, automatic transcription of TV broadcast news has been one of the application fields that has received major attention by the research community. As a consequence, several research groups worldwide have developed their own high performance broadcast transcription system for different languages [1,2,3]. In the particular case of the European Portuguese language, the AUDIMUS.media system described in [4] is up to our knowledge the most successful one.

In the context of our actual research projects, we are currently investigating application of broadcast news transcription technology to the problem of automatic transcription of commercial radio broadcast stations.

A. Teixeira et al. (Eds.): PROPOR 2008, LNAI 5190, pp. 172–181, 2008.

Although the TV and radio broadcast transcription problems share many similarities, there are some major differences that make the radio broadcast problem more challenging. Mainly, there is a considerable increase in the amount of telephone data that is present in radio broadcast programs compared to TV shows, where most of the speech data is wideband data recorded in a free of noise environment.

On the one hand, the problems of speech recognition in telephone applications are very well-known. In addition to the inner limitations of narrow band speech, in most cases a considerable presence of environmental noise appears due the use of mobile telephones in adverse environments. Consequently, the performance of speech recognition systems well-matched to the clean wideband problem fail dramatically in these conditions.

On the other hand, increase of the amount of telephone speech in radio programs is usually related with the presence of live press conferences, interviews to personalities, audience calls and participation of journalists out of the studio. In general, a common characteristic of these telephone contributions is that they are highly spontaneous. This fact results in similar difficulties to the problems of conversational telephone speech recognition, which is known to be significantly more challenging than the transcription of broadcast news [5,6]. Actually, this problem has been extensively tackled in the context of the Switchboard [7] benchmark tasks for the English language.

In this work, automatic detection and transcription of excerpts of telephone speech in radio broadcast data is investigated and some directions for future improvements are drawn. For this purpose, a small corpora of one complete day of broadcast of a Portuguese commercial station was collected and telephone segments were manually transcribed.

With respect to the telephone/non-telephone speech detection, a channel classifier based on linear discriminant analysis (LDA) of logarithmic filter bank energies is proposed.

Regarding the adaptation of the TV broadcast news system for tackling the problem of conversational telephone speech, initial efforts have been focused on the acoustical missmatch problem. New phonetic classifiers for connectionist speech recognition system [8] have been trained using both downsampled TV broadcast news data and real telephone (fixed and mobile) speech data. A considerable improved performance was achieved compared to alternative use of phonetic networks trained only with TV broadcast news data (11.8 % relative word error reduction) or only with telephone data (28.5 % relative word error reduction).

The rest of this paper is organized as follows. The two baseline systems for TV broadcast news transcription and telephone speech recognition are described in next section. Corpora considered in the work is reported in Section 3. Sections 4 and 5 are respectively devoted to the description of the proposed telephone channel classifier and to the developed radio telephone transcription system. Some future work and challenges are also drawn at the end of Section 5 before the concluding remarks.

2   Baseline Transcription Systems

2.1   TV Broadcast News Transcription System

In this work, the broadcast news transcription (BNT) system for the European Portuguese language described in [4] is adapted to the particular needs of radio telephone broadcast speech. A block diagram of the BNT system is shown in Figure 2.1.

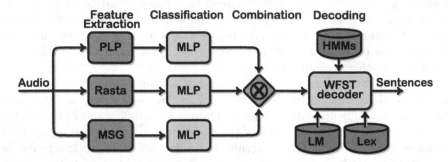

Fig. 1. Block diagram of the broadcast news transcription system after [4]

The system is based on the hybrid ANN/HMM paradigm for speech recognition [8]. This kind of recognisers are generally composed by a phoneme classification network, particularly a Multi-Layer Perceptron (MLP), that estimates the posterior probabilities of the different phonemes for a given input speech frame (and its context). These posterior probabilities are associated to the single state of context independent phoneme hidden Markov models (HMM). An appealing characteristic of the hybrid systems is that they are very flexible in terms of merging multiple input streams.

Concretely, the BNT system combines three network outputs trained with Perceptual Linear Prediction (PLP) features (13 static + first derivative), log-RelAtive SpecTrAl (log-RASTA) features (13 static + first derivative) and Modulation SpectroGram (MSG) features (28 static). In addition to the feature representation, MLP networks are characterized by the size of their hidden layers (2 hidden layers of 2000 units) and the size of the output layer (39 phonemes including silence pattern). The phonetic networks of the recognizer have been trained and adapted along years of speech recognition research with 57 hours of manually annotated data (46 train + 11 development) and more than 300 hours of automatically transcribed broadcast news data.

The decoder of the BNT system is based on weighted finite-state transducer (WFST) approach to large vocabulary speech recognition [9]. In this approach, the decoder search space is a large WFST that maps observation distributions to words. The language model (LM) in the one described in [10] with an active lexica size of 100K word. It is build based on a daily and unsupervised adaptation approach which dynamically adapts the active vocabulary and LM to the topic of the current news. Thus, a remarkable reduction of the out-of-vocabulary (OOV) words and of the word error rate (WER) is achieved.

In an evaluation set composed of 6 broadcast news programs recorded in 2007, the best WER performance achieved up to now with the BNT system is 20.6% for all conditions and 10.1% for F0 condition (read speech in studio). Current state of the art BN ASR systems for the English language have WER performances of less than 13% with 10x Real-Time [1] and less than 16% in real-time [6]. It is worth to notice that Portuguese BNT system results are in real-time performance.

2.2   The Telephone Speech Recognizer

A telephone speech recognizer (TSR) similar to the one described in [11] – known as AUDIMUS.telephone – has also been considered in this work for baseline comparison. The recognizer is particularly developed for both fixed and mobile telephone dedicated applications, such as automatic informational retrieval systems based on voice-command operated dialogs.

The architecture of the TSR system is the same multi-stream ANN/HMM paradigm shown in Figure 2.1. Main differences rely on the MLP networks and the corpora used for training them.

The phone classification networks were trained following the refrec 0.96 training procedure for SpeechDat [12] (without *garbage* model) using fixed telephone data (∼115 hours) and mobile telephone data (∼6 hours). Actually, the total amount of effective data was considerably reduced due to the unbalanced representation of some phone patterns and to the excessive amount of silence (approximately 36 % in fixed telephone and 50 % in mobile telephone data). Finally, networks with 7 window context and a unique hidden layer of 1500 units were trained. The different size with respect to previous BNT system is basically due to the different amount of available data.

In Table 1, WER results on SpeechDat II test sets proposed in [12] are shown together with the performace of some reference systems in other languages.

Table 1. WER results of the telephone speech recognizer (TSR) for Portuguese language compared to other language references after [12]. SpeechDat test categories are: isolated digits (I), yes/no (Q), application words (A), connected digits (BC), city names (O) and phonetically rich words (W).

Language	I	Q	A	BC	O	W
TSR system	0.4	0.1	1.8	8.2	5.6	6.8
Danish	0.0	0.3	1.9	2.4	13.8	46.2
English	3.5	0.0	0.8	4.4	6.0	30.8
German	0.0	0.0	1.7	2.8	5.3	7.1
Norwegian	3.5	0.0	2.8	5.3	14.9	22.1
Slovenian	5.2	1.2	3.5	4.7	7.3	15.9
Swiss German	0.2	1.0	0.6	2.5	9.2	25.0

3   Corpora Description

TV broadcast news data and fixed and mobile telephone speech data were used on the training and development of the telephone radio broadcast speech recognition system. Additionally, real radio data was collected for both development of a telephone classifier and for evaluation of this transcription system.

3.1   TV Broadcast News Corpus (TVBN)

The TV Broadcast News Corpus in an excerpt of the data collected from April 2000 to January 2001 to support the research and developments associated with automatic transcription of Portuguese BN. A total of 123 programs have been considered with an approximate duration of 57 hours. This corpus was divided in two sets: training (46 hours) and development (11 hours). Audio data is stored at 16 kHz sampling and 16 bits PCM encoding.

3.2   Fixed Telephone Corpus (FT)

A sub-set of the well-known Portuguese SpeechDat corpora have been used for training and development purposes. Concretely, the training data set consists of 24 hours of phonetically rich sentences and 12 hours of spontaneous speech from the SpeechDat II database. The development data set consists of 7 hours of phonetically rich sentences and 2 hours of spontaneous speech from the SpeechDat I database.

3.3   Mobile Telephone Corpus (MT)

Mobile telephone data of about 800 sessions recorded from the mobile GSM network in Portugal following the model of SpeechDat (yes/no categories, digit strings, application words...) was also considered. Non-spontaneous speaking style is dominant. Additionally, although being mobile data, there is not a significant amount of background noise. Data was also split into training ($\sim$ 11 hours) and development sets ($\sim$ 2 hours).

3.4   Radio Broadcast Corpus (RB)

One entire day, that is 24 hours, of a Portuguese commercial radio station was collected at 16 kHz sampling frequency. The data was used for developing the channel classifier described in next Section 4 in order to automatically detect the segments of telephone speech. The classified segments resulted in 116.6 minutes of telephone speech data. These telephone segments were orthographically transcribed to define the test data set of the speech recognition system for telephone radio broadcast data. Hereinafter, this test sub-set of only radio telephone speech will be referred to as RTB. For this corpus, the OOV word rate with the 100K words vocabulary is 0.33 %. Most of the OOV words are typical forms of conversational speech, such as clitics and some verb conjugations.

4   Detection of Telephone Segments in Radio Broadcast

Telephone channel is characterized by narrow band transmission in the frequency range from 300 Hz to 3400 Hz. Thus, a simple way for detecting telephone speech consists on computing energies in the different frequency bands and classify it.

Concretely, 15 logarithmic filter bank energies of 20 msec frames at 16 kHz sampling frequency are extracted with a time shift of 10 msec. The feature vectors are complemented with their first derivative. Then, each speech frame is classified with a binary LDA classifier into non-telephone or telephone classes.

In a first stage, the LDA classifier was initially trained with less than 4 minutes of telephone data and around 5 minutes of randomly selected non-telephone data (also including music and jingles) that were manually extracted from the RB corpora. A small portion of the training data was used for validation purposes. The rate of correct classified frames in the validation data set was of 99.8 %.

This initial classifier was then used to detect telephone segments in the whole RB corpus (24 hours). According to automatic frame classification, the segments with more than 1 second of duration and with a rate of telephone class labels above a fixed threshold were marked as telephone segments. Then, these automatically detected telephone segments were manually validated. In general, only few errors could be observed due to short telephone segments missed because of simple detection rule and mainly false positive detection corresponding to segments of music and jingles.

This new telephone segmentation (partially automatic) resulted in 116.6 minutes of telephone data. This sub-set constitutes the RTB test set. Notice that these almost two hours represents around 8 % of one complete day of radio broadcast, which is quite significant if it is taken into account that in the rest of the data there is a significant amount of non-speech acoustic events.

In a second stage, semi-automatically detected telephone segments together with around 4 hours of non-telephone data randomly selected and extracted from the same RB corpus were used to train a new LDA classifier. Again a short set of data was used for validation. In this case, the rate of correct classified frames in the validation data set was of 96.5 %. The drop in the classification rate is due to the higher variability in the data used for training. However, a generalized improved performance of the resulting classifier could be observed, particularly, when it is combined with a robust speech-non-speech detector that permits rejecting music and other non-speech acoustic events.

5   Automatic Transcription of Radio Telephone Speech

The object of detecting telephone speech segments in radio broadcast data is to apply a different processing to that applied to regular wideband data.

In the context of speech recognition area there is a countless number of robust techniques aimed to improve the performance of telephone speech recognition in different domains such as speech enhancement, robust feature extraction or acoustic model adaptation [13]. Additionally, attending to the fact that most of telephone data corresponds to spontaneous and even conversational speech, adaptation of the lexica and of the language model might also provide some additional benefits.

However, in this work we have focused only in the construction of robust acoustic phonetic classifiers adapted to the characteristics of conversational telephone data. The language model used in all the following experiments is the one described in Section 2.1 and the large lexicon of 100K words.

5.1 Baseline Systems Performace

Some initial experiments were carried out to confirm the need of developing new phonetic classification networks matched to the characteristics of telephone speech radio broadcast shows.

The well-trained system for transcription of Portuguese BN described in Section 2.1 and the set of well-trained phonetic classifiers for automatic recognition of telephone speech described in Section 2.2 were assessed with the RTB test set. In the case of the telephone dedicated system, the RTB test set was downsampled to 8 kHz.

Table 2 shows the performance of the two baseline systems. The TSR system, which is entirely trained with telephone speech data, is not well-matched to the problem of continuous speech recognition and obtains a poor WER of 72.0 % for all conditions. On the other hand, the BNT recognizer achieves a considerable error reduction with respect to the TSR system (WER of 58.4 % for all conditions). Despite the BNT system is trained with wideband data and suffers from channel missmatch problem, it is more appropriate for this concrete task. However, its performace in both planned and spontaneous speech is still far of the reference results provided in Section 2.1 obtained on TV broadcast news data. In general, it can be clearly stated that both systems fail to provide a reasonable performace independently of the speaking style. These observations are in well-accordance with [5], where a state of the art BN transcription system had a WER of around a 50 % in Switchboard data.

5.2 Robust Network Training

The TVBN corpus, the FT corpus and the MT corpus are used to develop a system in more accordance to the needs of telephone speech in radio broadcast shows. The combination of the three corpora results in a training set of approximately 93 hours (46 TVBN + 36 FT + 11 MT) and a development set of 22 hours (11 TVBN + 9 FT + 2 MT). The development data is used to define the stopping criteria in the process of training the MLPs as it is usually done in this kind of approaches.

Manually generated transcriptions were used to obtain frame-to-phone alignments needed for training the phonetic classifiers. In the case of TVBN data, the BNT system was used; while the TSR system was used to align telephone data, both fixed and mobile.

As in the case of the previous systems, a multistream system is built with PLP, log-RASTA and MSG feature parametrizations. Network characteristics are similar to those of the BNT system, but the size of the two hidden layers was fixed to 1500 due to the reduced amount of available data.

Phonetic networks are trained with 8 kHz sampling rate data, thus speech from the TVBN corpus was previously downsampled. In order to simulate more accurately telephone channel characteristics, we experimented to apply an additional pass-band filtering stage in the frequency range of telephone speech. However, not remarkable differences were found depending on wether telephone-like filtered or not filtered data was used for training the networks. Thus, the results shown in this work were obtained with downsampled data without filtering.

In next Table 2 the WER performance of the new proposed system referred to as the radio telephone transcriber (RTT) is compared to the two previous baseline systems. Two different test conditions are considered: planned and spontaneous speech speaking style. The WER average of the two conditions is also provided.

Table 2. WER results of the telephone speech recognizer (TSR), the broadcast news transcriber (BNT) and the radio telephone transcriber (RTT) in planned and spontaneous test conditions

	TSR	BNT	RTT
planned	66.2	51.7	43.7
spontan	85.2	69.3	64.3
average	72.0	58.4	51.5

In both planned and spontaneous speech, the new RTT system achieves a considerable improvement with respect to the best baseline system performance. The most noticeable improvement is obtained in planned speech condition. In this case, the impact of missmatched language model is less important and a 15.5% relative error reduction is obtained with respect to the BNT system thanks to better acoustic modeling. However, the relative improvement in spontaneous condition is quite lower due to main influence of inappropriate language modelling. For all conditions, the RTT system achieves a relative word error rate reduction of 28.5% with respect to the TSR system and 11.8% with respect to the BNT system.

5.3   Future Work and Challenges

According to the reported results of our ongoing research activities, it is clear that many challenges still need to be faced in order to achieve reasonable performance of transcription of telephone speech in radio broadcast applications.

On the one hand, there is a need for real conversational telephone speech data in Portuguese language to develop robust acoustic modelling. In this work, it has been shown that the use of other sources of data can help to alleviate this problem, but it is not a definite solution. With regards to other robustness issues, we are currently investigating the use of alternative feature parametrizations that might better match the telephone speech recognition problem, such as the standard advanced front-end of ETSI [14]. Additionally, the use of speaker normalization techniques like vocal tract length normalization (VTLN) [15] is being investigated.

On the other hand, adaptation of current broadcast news language model to better match the spontaneous speaking style of conversational speech appears as a necessary step for future improvements. Thus, the problem of corpora resources is present again, since there exists a limited amount of language model training data of the desired characteristics.

Finally, it must be noticed that both TV broadcast news transcription and telephone based information retrieval systems are usually limited by the need of real time functionality. However, there is not need for real time limitations in most applications of speech recognition to radio broadcast data. The typical application is to generate information (transcriptions) of already stored data. In this case, more computational demanding decoding strategies can be applied. For instance, multiple stage decoding steps based on adaptation and re-decoding of automatically transcribed data.

6   Conclusions

Everyday huge amounts of multimedia data are generated by broadcast media world-wide, which would be desirable to have automatically segmented and transcribed. Actually, there exist real systems capable of providing accurate transcriptions in some contexts, such as in TV broadcast news applications. In this work, we have started to investigate the possible re-usability of a broadcast news transcription system for the European Portuguese language to the similar radio broadcast transcription problem. The main challenges that one can find are the significant increase of both telephone speech and spontaneous speaking style. Thus, we have initially focused on the automatic detection of telephone speech and the improvement of phonetic acoustic modelling for particular conversational telephone speech. A relative WER reduction of 11.8% was achieved with respect to the broadcast news system, besides an high classification rate of the telephone channel detector proposed. Finally, some thoughts for future development have been provided.

Acknowledgements

This work was funded by PRIME National Project TECNOVOZ number 03/165.

References

1. Nguyen, L., Xiang, B., Afify, M., Abdou, S., Matsoukas, S., Schwartz, R., Makhoul, J.: The BBN RT04 English Broadcast News Transcription System. In: Proceedings of Interspeech 2005, Lisbon, Portugal (2005)

2. Gales, M.J.F., Kim, D.Y., Woodland, P.C., Chan, H.Y., Mrva, D., Sinha, R., Tranter, S.E.: Progress in the CU-HTK Broadcast News Transcription System. IEEE Transactions on Audio, Speech, and Language Processing 14(5), 1513–1525 (2006)
3. Galliano, S., Geoffrois, E., Mostefa, D., Choukri, K., Bonastre, J.-F., Gravier, G.: The ESTER Phase II Evaluation Campaign for the Rich Transcription of French Broadcast News. In: Proceedings of Interspeech 2005, Lisbon, Portugal (2005)
4. Meinedo, H., Caseiro, D., Neto, J., Trancoso, I.: AUDIMUS.media: A Broadcast News speech recognition system for the European Portuguese language. In: Proceedings of PROPOR- 2003, Portugal (2003)
5. Gauvain, J.-L., Lamel, L., Schwenk, H., Adda, G., Chen, L., Lefèvre, F.: Conversational telephone speech recognition. In: Proceedings of ICASSP-2003, pp. 212–215 (April 2003)
6. Matsoukas, S., Prasad, R., Laxminarayan, S., Xiang, B., Nguyen, L., Schwartz, R.: The 2004 BBN 1xRT Recognition Systems for English Broadcast News and Conversational Telephone Speech. In: Proceedings of Interspeech 2005, Lisbon, Portugal (2005)
7. Godfrey, J.J., Holliman, E.C., McDaniel, J.: Switchboard: Telephone speech corpus for research and development. In: Proceedings of ICASSP-1992, pp. 517–520 (March 1992)
8. Morgan, N., Bourlard, H.: An introduction to hybrid HMM/Connectionist continuous speech recognition. IEEE Signal Processing Magazine, 25–42 (1995)
9. Mohri, M., Pereira, F., Riley, M.: Weighted finite-state transducers in speech recognition. In: ISCA ITRW Automatic Speech Recognition, Paris, pp. 97–106 (2000)
10. Martins, C., Teixeira, A., Neto, J.: Dynamic language modeling for a daily broadcast news transcription system. In: Proceedings of ASRU-2007, Kyoto, pp. 165–170 (2007)
11. Hagen, A., Neto, J.: HMM/MLP Hybrid Speech Recognizer for the Portuguese Telephone SpeechDat Corpus. In: Proceedings of PROPOR-2003, Portugal (2003)
12. Lindberg, B., Johansen, F., Warakagoda, N., Lehtinen, G., Kacic, Z., Zgank, A., Elenius, K., Salvi, G.: A noise robust multilingual reference recogniser based on SpeechDat(II). In: Proceedings of ICSLP 2000, Beijing, pp. III, 370–373 (2000)
13. Junqua, J.-C., Haton, J.P.: Robustness in Automatic Speech Recognition: Fundamentals and Applications. Kluwer Academic Publishers, Dordrecht (1996)
14. ETSI standard doc.: Speech Processing, Transmission and Quality Aspects (STQ); Distributed speech recognition; Advanced feature extraction algorithm. ETSI ES 202 050 Ver. 1.1.5 (2002)
15. Kamm, T., Andreou, G., Cohen, J.: Vocal tract normalization in speech recognition: Compensating for systematic speaker variability. In: Proceedings of the 15th Annual Speech Research Symposium, Baltimore, USA (1995)

A Platform of Distributed Speech Recognition for the European Portuguese Language

João Miranda and João P. Neto

L²F - Spoken Language Systems Lab / INESC-ID,
Instituto Superior Técnico / Technical University of Lisbon
Rua Alves Redol, 9, 1000-029 Lisboa, Portugal
{jrsm,Joao.Neto}@l2f.inesc-id.pt

Abstract. In this paper we present a Distributed Speech Recognition system for the European Portuguese based on the *Audimus* system, that can be used in embedded systems such as PDAs. The obtained system has no significant degradation in what concerns word error rates or increased latency in processing, and only a small increase in word error rate when the audio is recorded using the microphone in the device.

Keywords: Automatic Speech Recognition, Distributed Speech Recognition, Embedded Systems.

1 Introduction

Audimus [1,2] is an Automatic Speech Recognition (ASR) system, tailored for the Portuguese Language. It is used for a variety of tasks, being parameterized by models that are adapted to each specific task. Since *Audimus* is prepared to work with tasks which involve large vocabularies, some of its models are very large and processing them requires large amounts of memory and processing power, so it has been designed to run on computers that can accomodate these requirements.

The existence and increasing popularity of a large number of embedded devices, such as PDAs and cell phones, combined with the difficulties of entering data using traditional input devices such as keyboards (due to their small size), makes it desirable to have a speech interface running on those devices. However, they have resources that are very limited when compared to desktop workstations. For example, their processors are usually RISC processors, which besides being much slower than workstation CPUs, often lack a floating point processing unit, so that arithmetic calculations must be rewritten as calculations involving only integers. Also, while it is not uncommon for modern workstations to have a few gigabytes of memory, the devices being targeted rarely have more than a few tens of megabytes available for processing.

One way to reduce the processing load on the embedded device would be to transfer all the processing to the server, in what is known as Network Speech Recognition (NSR), but for many networks the bandwidth is reduced and therefore it becomes necessary to employ a low bitrate coding, thereby reducing the

A. Teixeira et al. (Eds.): PROPOR 2008, LNAI 5190, pp. 182–191, 2008.

quality of the speech received at the server. These problems are overcome by Distributed Speech Recognition (DSR)[3,4] systems, which process the signal at the client device, in order to reduce the dimensionality of the set of features that must be transmitted - basically, the speech recognition system is divided into two parts:

- the front-end, which runs at the client (i.e., the embedded device) and performs the feature extraction and acoustic model calculation
- the back-end, which runs at the server and executes the most time and memory consuming module of the system - the decoder.

In this work, we intended to port the *Audimus* system to embedded devices. We have large vocabulary (over 100K words) models for several tasks in the European Portuguese Language, and wanted to be able to use these models in the resulting speech recognition system. To fulfill these requirements, we decided to adopt the DSR paradigm.

In the next section, the state-of-art in Speech Recognition techniques is briefly described, with a particular emphasis on the *Audimus* system. In the third section, the architecture of the implemented DSR system is described and, in the fourth section, some of the implementation options are discussed. Finally, in the fifth section, the results obtained with the system are presented.

2 Current Speech Recognition System

Figure 1 shows the architecture of a generic speech recognition system.

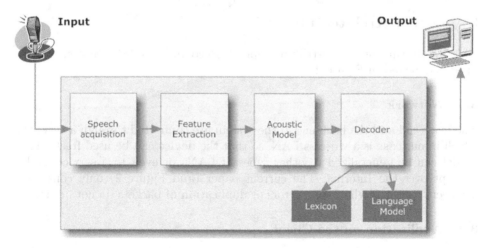

Fig. 1. A generic ASR system architecture as a cascade of processing blocks

State-of-art speech recognition systems generally employ some combination of these components. The speech acquisition module performs the capture of the speech that is to be recognized; it can be done in several ways, using a simple microphone, a microphone array, or a multimedia file.

The feature extraction component performs a form of transformation on the speech signal, producing a reduced set of features that are intended to better represent it. It is implemented using signal processing techniques, that essentially try to remove the signal information that does not help to identify the words being spoken. In particular, *Audimus* has a rich set of components, that compute ETSI, MFCC, PLP, and RASTA features [2], which are based on different algorithms.

The acoustic model estimates a set of *posteriori* probabilities: for each phone, the probability of it having been generated by the observations (that are the output of the feature extraction component). Currently the two main approaches to solve this problem are to use a Gaussian Mixture Model, which assumes that each feature vector is generated by a distribution which is a mixture of Gaussians, and a Multi-Layer Perceptron. *Audimus* uses the latter.

Finally, the decoder is the module that, based on a language model (which is a description of "how likely" is each sequence of words in the language) and on a lexicon, finds the sentence that is the most likely to have been uttered by the speaker. To approach the decoding problem, some systems use decoding strategies based on the Viterbi or stack decoder [5] (also known as the A* algorithm). *Audimus* uses the WFST-based approach to Large Vocabulary Speech Recognition [6].

In addition to the above structure, which is common to almost all state-of-art ASR systems, *Audimus* also has components to segment a speech signal into sentences, and to classify a frame of audio into speech / non-speech [2] (to reduce the load when there is no useful speech to process).

3   System Architecture

The architecture of the Distributed Speech Recognition (DSR) System we defined is depicted in Figure 2.

3.1   Network

The front-end and the back-end of the system are connected through a network, which in our case is a wireless LAN, so that the device can be used freely. This model can be generalized to other types of LANs or even to a service model concept over the Internet. The current paper (and Figure 2) only considers, however, networks where loss, errors or duplication of packets do not occur.

3.2   Application on the Client

In the client, speech acquisition can be done either using the device's embedded microphone or with a Bluetooth Microphone, which enables higher quality audio capture, since the speech source is closer to the microphone. We used the PortAudio Library [7] to capture the audio in a system-independent way, of which there is already a version for Windows Mobile, the target OS.

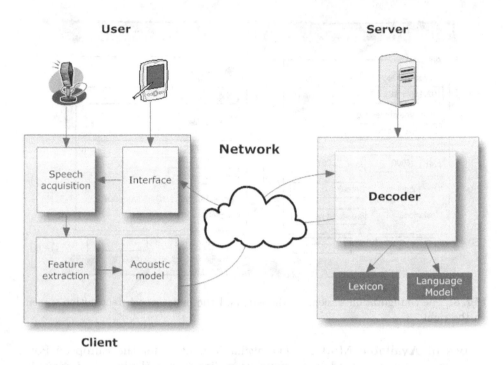

Fig. 2. Architecture of the DSR System

We also developed a simple interface in which the user can specify the parameters of the distributed recognition system (the IP of the server, models - all sizes and types of the *Audimus* system being supported) and the processing the user wants the server to do, speak to a dictation interface, and see the results, as can be seen in Figure 3.

3.3 Server Configuration

Initialization of Communication. Communication between client and server is initiated by the client, as it is usual in the client-server model. The server must therefore be running before a connection from the client is attempted. The client must provide the server's IP address and the port where it is listening to incoming requests, as can be observed in figure 3. The server accordingly creates a new *Audimus* process to handle the current request.

Model Synchronization. As seen in Figure 3, the models used for recognition can be specified before recognition using the interface in the client. The server application manages this change by using the *Audimus* API appropriately to change the models at the server.

Data Transfer. The data is transferred from the client to the server across a normal TCP/IP connection, as seen above. There is also a parallel control connection that enables asynchrounous control of the recognition process.

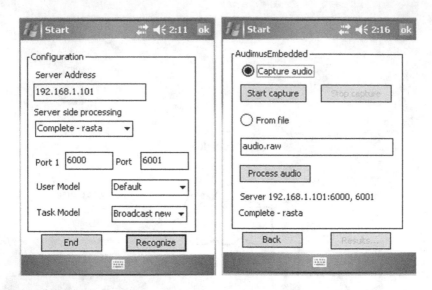

Fig. 3. The configuration screen, on the left, and the main speech recognition screen on the right

Types of Available Models. The available models, for the European Portuguese, are language models, pronunciation dictionaries (lexica), and acoustic models.

Acoustic models come in the form of a file containing the neural network weights (since *Audimus* uses MLPs for acoustic modelling), and they can be adapted to a user or set of users or be generic.

Lexica and langage models are, in *Audimus*, modelled by WFSTs. Language models can represent a contained task, where they can be used to obtain very good performance by restricting the possible sentences considerably, or larger tasks, with many different words, where, even compressed, they can grow to sizes of hundreds of megabytes.

4   Portability to an Embedded System

In this section we focus, at a lower level, on the portability issues that were solved in order to obtain a working system.

We chose to port the PLP component as the feature extraction module. This was motivated by the fact that we intend to recognize clean speech, not speech significantly distorted by noise. As the acoustic model component, we selected the *ForwardMLP* component, which performs forwarding on a MLP model. We decided to port the acoustic model also, since we want to progress into a totally embedded system, where all the processing (including the decoding) is performed at the client.

We also ported *Audimus' MacroComponent* system, to make the specification of different pipelines of components more flexible. In doing so, we had to port the ZThreads Library [8], over which the *MacroComponent* system is built, to the device.

The devices we were targeting with our system have limited processing capabilities, in particular, they do not usually possess floating-point processing units. Emulating the inexistent floating-point units through software incurs a slowdown of about 10x, which renders it unacceptable for applications that are processing intensive, such as the computation of PLP features, needed for speech recognition. The solution is the use of fixed point computations in all arithmetic operations. The memory limitations of the processor are also relevant in the acoustic processing stage, because some acoustic models occupy non-negligible amounts of space.

4.1   PLP Component

The main issues found while porting the PLP component to the embedded device are described below. Some of the issues considered in this section apply to other components as well.

Computation of the FFT. The FFT is a central component in the computation of PLP coefficients, and also represents about 50% of the execution time. The FFT is done in 32-bit fixed point in order to preserve the best possible accuracy. Also, a N-point FFT introduces a gain of N (the output vector has a magnitude which can be up to N times larger). Therefore, to avoid overflow, the input data is shifted right between each two *butterflies* (FFT subroutines). The FFT implementation in *Audimus* was further optimized by replacing calls to trigonometric functions, used to compute the twiddle factors, by table lookups that can be pre-computed.

Computation of the Power Spectrum. The power spectrum can be estimated from the complex output of the FFT by calculating the magnitude of each complex number. However, to avoid an expensive square root operation, most applications work with the squared magnitude instead. This increases the range of numbers that must be represented, which complicates a fixed point implementation. The adopted solution was to use a dual fixed point implementation [9], where a single bit selects one of two possible exponents. This ensures that the range is enough to cover the squared magnitude spectrum, while mantaining most of the speed gained by the fixed point approach. One alternative is to use a fast approximation of the magnitude of a vector, but that introduces errors that are up to 10%.

Approximation of the Cube Root Function. The auditory spectrum must be equal-loudness weighted and cube-root compressed to account for the characteristics of human audition. The cube-root function must therefore be implemented in fixed-point. One way to solve this problem is to tabulate some of the function's values and then to use linear interpolation to approximate it between those values. The property of the cube-root function f - $f(ax) = \sqrt[3]{ax} = \sqrt[3]{a}\sqrt[3]{x} = f(a)f(x)$ - and, in particular, the fact that it is an odd function (i.e.,

$f(-x) = -f(x))$, mean that we only need to consider the interval between 0 and 1, since any interval between 0 and 2^k can be reduced to the first using a multiplication. We chose to use a table with 256 equally-spaced entries (to avoid using expensive division operations), which leads to an average error of less than 10^{-5}. The resulting implementation was roughly $5 - 10$ times faster than the general-purpose function *pow* of the C library.

Further Optimisation of Operations. In addition to the use of fixed point arithmetic, most ARM processor's division operations are very slow or inexistent, being emulated in software. As a result, whenever possible, division operations (found, for example, in the computation of LPC coefficients from the autoregressive model) were replaced with multiplications by their inverse.

4.2   ForwardMLP Component

In the ForwardMLP component, it is necessary to compute the output of a MultiLayer Perceptron. To that effect, we need to calculate the output of a set of neurons, which is a linear combination of the outputs of the neurons of the layer immediately to the left. We also need to calculate the activation function, which is usually the sigmoid function, but that is easily done as a table lookup. This computation fits naturally within the framework of matrix multiplication, and the current implementation uses one of several highly optimized BLAS (Basic Linear Algebra System) libraries to perform this operation in reasonable time, since the matrices used are, for large networks, very large. Unfortunately, to the best of our knowledge, there is no BLAS system targeting ARM devices. It was, therefore, necessary to improve the baseline ($O(n^3)$) matrix multiplication algorithm directly with the following optimizations:

Locality of Reference Optimizations. The trivial matrix multiplication algorithm uses the processor cache sub-optimally, since to calculate a row of the output matrix, it will read all of the second matrix. Instead, we partition the first and second matrices into blocks, and multiply them "blockwise" (i.e. the elements of the higher level multiplication operation will be matrices themselves). The number of operations executed by the algorithm will be the same as before, but if we choose the size of the block so that it fits in the cache of the processor, the number of cache misses will be much lower, causing the algorithm to run much faster.

Reduction of the Network's Memory Footprint. By quantizing the matrices and all input values to 16 bits, it is possible to reduce the size of the neural network considerably, albeit at a small cost in precision. This not only saves memory but also improves the algorithm's locality (because more data can be made to fit in the cache) and speed (since our target processor can muliply four 16-bit values in one clock cycle). Also, as the input of our network consists not only of the current frame but also of the 3 that precede and follow it, each feature frame appears seven times in the first matrix of the first multiplication. This fact was explored to further reduce size by storing each feature frame only once.

5 Results

The PC used for our tests was a 2.4 Ghz Core Duo with 2 GB of RAM. The target device used in the tests was a 520 Mhz XScale, with 64 MB of RAM and 256 MB ROM running Windows Mobile 6.0. The network between the PC and device was a Wi-Fi 802.11b network.

We selected the radiology task to perform our tests, using an adapted language model existent in the laboratory, consisting of 13161 words. The test set was split into five tests, totalling 215 sentences and 2292 words. Each test contained one or more full radiology reports. The differences in recognition quality for each set were measured using two different acoustic models: the generic acoustic model, and the speaker dependent acoustic model. Also, we considered how different configurations - a) audio acqusition and processing in the PC, b) acquisition in the device and complete processing in the PC, c) distributed system with feature extraction in the device, d) distributed system that also includes MLP processing - would impact recognition speed, by measuring it against our test set. Finally, we also assessed the impact that using the PDA's internal, lower-quality microphone, would have in recognition quality. The results are summarized in the tables below.

Table 1. Duration of each test (column 2) and time taken to recognize the five tests when using only the PC (column 3), when transferring the audio from the PDA to the PC (column 4) and when executing the PLP component and the PLP and MLP components in the PDA (columns 5 and 6) respectively

Test	Duration	Time(PC)	Time(PDA to PC)	Time(PLP in PDA)	Time(MLP in PDA)
1	349s	125s	126s	129s	135s
2	484s	161s	163s	167s	184s
3	546s	204s	205s	207s	220s
4	567s	207s	208s	211s	226s
5	540s	188s	190s	193s	210s

There was no significant increase in delay incurred by the network transmission of the data from the device to the PC, as can be observed by comparing columns 3 and 4 of table 1. Additionally, the increase in latency noticed when considering the configurations of the fifth and sixth columns of the table was most likely due to overhead in the creation of the *MacroComponent* architecture in the PDA, which is more significant in the acoustic model case.

Table 2 shows that the use of speaker adapted models almost completely eliminated the errors in the radiology task. Furthermore, there was a small (0.51%) degradation in WER because of rewriting the PLP component in fixed point for the PDA. This is mainly due to the numerically unstable nature of delta calculation, used in the PLP component, which increases the relative error of the fixed point implementation by about one order of magnitude.

In the 2nd column of table 3, we show the results of trying to directly reuse the speaker adapted model, used in the discussion above, with audio recorded using

Table 2. Word error rates using the generic model, the adapted model, and the adapted model with PLP processing done in the PDA, respectively

Test	Words	WER (gen. model)	WER (ad. model)	WER (ad. model, in PDA)
1	317	7.89%	2.52%	3.47%
2	448	11.20%	2.23%	2.90%
3	520	8.85%	1.73%	2.31%
4	521	11.70%	2.50%	2.88%
5	486	9.05%	1.44%	2.06%
Average	458	9.86%	2.05%	2.66%
Total	2292			

Table 3. Word error rates using the PDA's internal microphone. The 2nd column refers to the acoustic model trained with the high quality microphone, while the 3 last columns refer to the acoustic model trained for the PDA's microphone. In the 4th and 5th columns, the PLP component / PLP + acoustic model, respectively, are executed in the PDA.

Test	Base SA model	Adapted model	Adapted model+PLP	Adapted model+PLP+MLP
1	14.83%	1.89%	2.84%	4.10%
2	19.20%	4.24%	4.24%	4.91%
3	17.12%	3.46%	3.46%	4.03%
4	25.34%	4.61%	4.22%	5.76%
5	13.40%	3.91%	3.91%	4.53%
Total	18.41%	3.75%	3.80%	4.71%

the PDA's microphone. This caused a large increase in word error rate, so we further adapted the above model using 100 additional sentences, recorded with the PDA's microphone. As a result, the WER dropped to 3.75% (3rd column of table 3), which represents a degradation of less than 2% when compared to the use of a high quality microphone. There was no significant degradation caused by the execution of the PLP component in the PDA (column 4); in one of the testcases, the WER actually decreased. However, also porting the MLP component (column 5) increased the WER by 0.91%, probably due to the reduced precision of the 16 bit matrix representation, discussed in section 4.

6   Conclusions

In this paper we have presented a Distributed Speech Recognition system for the European Portuguese. Our intent to progress, in future work, towards a fully embedded system, where all the processing resides on the device, has motivated the decision of investigating two configurations for the distributed system, with and without the acoustic model component. In both configurations it was possible to achieve word error rates below 5% , with a speaker and microphone adapted acoustic model, in a 13161 word radiology task.

Acknowledgements

This work was funded by PRIME National Project TECNOVOZ number 03/165.

References

1. Meinedo, H., Caseiro, D.A., Neto, J.P., Trancoso, I.: AUDIMUS.MEDIA: a Broadcast News speech recognition system for the European Portuguese language. In: Mamede, N.J., Baptista, J., Trancoso, I., Nunes, M.d.G.V. (eds.) PROPOR 2003. LNCS, vol. 2721. Springer, Heidelberg (2003)
2. Meinedo, H., Neto, J.P.: Automatic speech annotation and transcription in a broadcast news task. In: ISCA Workshop on Multilingual Spoken Document Retrieval, Macau, China (2003)
3. Ramabadran, T., Sorin, A., McLaughlin, M., Chazan, D., Pearce, D., Hoory, R.: The ETSI Distributed Speech Recognition (DSR) standard server-side speech reconstruction. IBM Research Report (2003)
4. Suk, S.Y., Jung, H.Y., Makino, S., Chung, H.Y.: Distributed Speech Recognition System for PDA in Wireless Network Environment. In: Proceedings of SPECOM 2004, St. Petersburg, Russia (2004)
5. Paul, D.B.: An Efficient A* Stack Decoder Algorithm for Continuous Speech Recognition with Stochastic Language Model. In: Proceedings of the International Conference on Acoustics, Speech and Signal Processing (ICASSP-1992), San Francisco, USA (1992)
6. Caseiro, D., Trancoso, I.: A Specialized On-the-Fly Algorithm for Lexicon and Language Model Composition. IEEE Transactions on Audio, Speech and Language Processing 14(4), 1281–1291 (2006)
7. PortAudio - portable cross-platform API, http://www.portaudio.com
8. ZThreads - A platform-independent, multi-threading and synchronization library for C++, http://zthread.sourceforge.net/
9. Ewe, C.T., Cheung, P.Y.K., Constantinides, G.A.: An Efficient Alternative to Floating Point Computation. In: FPL 2004. LNCS, vol. 3203, pp. 200–208. Springer, Heidelberg (2004)

Supporting e-Learning with Language Technology for Portuguese

Mariana Avelãs, António Branco, Rosa Del Gaudio, and Pedro Martins

University of Lisbon
Faculdade de Ciências, Departamento de Informática
NLX - Natural Language and Speech Group
Campo Grande, 1749-016 Lisbon, Portugal
{mariana,antonio.branco,rosa,pmartins}@di.fc.ul.pt

Abstract. In this paper we report on the development and adaptation of language technology tools for Portuguese aimed at supporting e-Learning via the extension of a Learning Management System with new functionalities. We also describe how these tools were integrated into this Learning Management System and present results of both their intrinsic and extrinsic evaluation.

Keywords: Automatic keyword extraction, ontology, definition extraction, e-Learning, LMS, annotated corpus.

1 Introduction

The immense potential of Language Technology to enhance e-Learning has been repeatedly pointed out, and to a very large extent such potential remains to be explored. In this paper, we report on some first steps in that direction, discussing the application of some tools and resources for the computational processing of Portuguese with the aim of supporting e-Learning. More specifically, we report on the development and application of three tools aimed at enhancing learning activities in the scope of a Learning Management System (LMS).

One of the tools is a Keyword Extractor, which supports a new functionality with which the LMS was extended: once a text-based Learning Object is selected, a list of candidate keywords for that object can be automatically generated. This list can be subsequently filtered out by the users so that only the more relevant are retained and are persistently associated with that Learning Object. This functionality can be used by tutors in their task of meta-data annotation and thus helps to alleviate the burden of hand writing them, speeding up that process. It can also be used by students, who can obtain on the fly a list of some core concepts for a Learning Object they may have just imported into the LMS, and rapidly have a first glimpse of their content or relevance.

A second tool which was developed is a Definition Extractor, which supports another new functionality of the LMS, the Glossary Candidate Detector (GCD): from a given Learning Object selected, it is possible to generate a list of tentative

A. Teixeira et al. (Eds.): PROPOR 2008, LNAI 5190, pp. 192–201, 2008.

definitions that form a draft glossary; this glossary in turn can also be subsequently filtered out and extended by the users. Again, this functionality can be used by tutors in their task of meta-data annotation and helps to speed up that process. It can be used by students as well, who can obtain a draft overview of the concepts being defined in a Learning Object imported into the LMS.

Finally, the third tool developed was a Semantic and Multilingual Search Tool. A key component of this tool is an ontology and the annotation of the Learning Objects with their concepts: in the Learning Objects, each natural language expression conveying one of those concepts is associated to such concept via metadata annotation. Accordingly, the search tool developed permits to retrieve Learning Objects given the concept entered and its occurrence in the retrieved objects. Since the ontology is common for Learning Objects from different idioms, the set of retrieved objects can include also those not written in the language of the user, thus supporting cross-language search.

The results reported in the present paper were obtained in the scope of the LT4eL project activities. This is an FP6 European project whose goals were pursued with the cooperative contribution of 12 partners, including our team, under the coordination of the University of Utrecht. In the present paper, we focus in the tools and results contributed by our team.

In Section 2, we describe the corpus collected and annotated in order to support the development and the intrinsic evaluation of the tools. In Section 3, a detailed presentation of the keyword extractor is offered, while the glossary candidate extractor is discussed in Section 4. In Section 5, we present the ontology and the semantic search mechanisms it supports. In Section 6, we briefly describe how these tools were integrated in the LMS and what was the outcome of their extrinsic evaluation. Finally, Section 7 is devoted to conclusions.

2   The Corpus

In order to support the development and the intrinsic evaluation of the tools a corpus was developed. Given that the corpus was to be used also for the extrinsic evaluation, viz. as a repository of learning material in the LMS, we selected documents that can be taken as Learning Objects.

A Learning Object (LO) is any small, reusable chunk of instructional media, digital or non-digital, which can be used, re-used or referenced during technology supported learning and should be enriched with metadata (the actual standard is the "Learning Object Metadata") [1]. Keeping this in mind, we selected 31 documents, mostly of a tutorial nature, apt to be used as LOs, covering three domain areas, namely Information Technology (IT) for non experts, e-Learning (eL) and Information Society (IS). Table 1 shows the composition of the corpus.

The XML-based format version of the corpus went through a process of linguistic annotation. The corpus was automatically annotated with morphosyntactic information using LX-Suite, a set of tools for the shallow processing of Portuguese with state of the art performance [2].

This annotation includes information about sentence and tokens boundaries, POS categories, and inflectional features and lemmas.

Finally, in the last step the output of the annotation tools was converted into a common, project internal, XML format, the LT4eLAna document format. The DTD of this format conforms to a DTD derived from the XCESAna DTD, a standard for linguistically annotated corpora [3]. This DTD structures the documents into paragraphs, sentences, chunks and tokens. The textual content of tokens is the actual text of the document while the attributes associated to the tokens encode linguistic and layout information. Markup for some other elements was yet added, namely for keywords, defined terms and defining text.

Over this version of the corpus in this final format, a phase of manual annotation of keywords and of definitions was carried out.

Concerning keywords, 29 documents were annotated (corresponding to 265 915 tokens) with 1 033 different types, which means a mean of 35.6 types per document.

Definitions were marked with the indication of the definiens and of the definiendum. Information regarding the type of definitions was also encoded, namely distinguishing four different kinds of definitions: definition introduced by the verb "to be", termed copula definitions; definitions introduced by other verbs; definitions introduced by a punctuation mark; and definitions of none of these previous three types. Table 2 displays the distribution of the different types of definitions in the corpus, and their breakdown by sub-corpora.

Table 1. Corpus domain composition

Domain	tokens
IS	92825
IT	90688
eL	91225
Total	274000

Table 2. The distribution of definitions

Type	IS	IT	eL	Total
Copula	80	62	24	166
OtherVerb	85	93	92	270
Punctuation	4	84	18	106
other	30	54	23	107
total	199	295	157	651

3   The Keyword Extractor

Keywords are (single or multi-word) terms that are presented to very briefly characterize a text and resume what it is about. In order to extract such terms automatically, a few algorithms, based on distributional statistics, were tested. In particular, project internal work provided an implementation of algorithms based on TF*IDF, RDIF and a term frequency adjusted version of IDF (ARDIF). Such tool, developed by Lemnitzer and Dergorski [4], took into account the linguistic information encoded in the corpora, in particular the base form of each word, the part of speech, and the morpho-syntactic features. These tools try to pay justice to the fact that good keywords have a typical, non random distribution in and across documents and that keywords tend to appear more often at certain places in texts (e.g. headings, etc.).[1]

---

[1] For full details, the reader is referred to [4].

These tools ran over the corpus described in the previous Section, and its outcome underwent a subsequent process of refinement. When looking at their results, it was apparent that some terms selected as candidate keywords were not apt to be considered keywords at all and could be very easily discarded. For instance, focusing on single-word keywords, this was the case of candidates made of punctuation marks or of a single preposition. Or, when taking multi-word expressions, for instance, that was the case of candidates starting with punctuation marks or prepositions.

In order to automatically refine such preliminary outcome, a system of pattern-based filters was developed. That filtering module is based on the use of four portmanteau tags that are in correspondence with the elements of the POS tagset used for the annotation of the corpus:

PLU - punctuation elements, that should be ignored completely.
FLU - lexical units that are not possible as single-word, though they can appear inside multi-word units but not at the initial or final position.
CMLU - lexical units which are admissible in multi-word lexical units, even at their beginning or end, but cannot form a single-word keyword.
MLU - admissible both as single-word keyword and as of a multi-word one.

Intrinsic evaluation was carried out at the output of this filtering of the first tentative results provided by the statistics-based tools. Scores for Precision, Recall and F-measure were obtained against the manually annotated documents reserved as test set. Table 3 displays the results for each base technique tried out, showing a slight advantage for the combination TFIDF-based algorithm followed by rule-based filtering.

Table 3. Keyword Extractor intrinsic results

	ADRIDF			RIDF			TFIDF		
	R	P	F	R	P	F	R	P	F
filtered	0.30	0.17	0.21	0.21	0.12	0.15	0.31	0.18	0.22

Given that the manual annotation of keywords was performed by a single annotator, in order to have a more reliable notion of the intrinsic performance of the tool an experiment was carried out to obtain a score for inter-annotator agreement on this specific task of keyword assignment.

Ten individual testers were given one LO from the corpus and were asked to extract the 10 keywords that should be assigned to that document. The agreement between testers was assessed by using the AC1 measure proposed in [5]. It scored 0.58 which indicates that the task is inherently quite difficult (even for humans).

Note that the scores displayed in the table above were obtained by comparison with the list proposed by a single annotator. Accordingly, a much more significant measure of the performance of the tool is to be collected with the AC1 score obtained for the comparison between the tool and the ten annotators.

The agreement between human testers and the tool scored 0.67. The list of keywords proposed by the "typical" tester (taken as the 10 most selected keywords by all the testers) is thus in agreement with the system more than the testers agree among each other. This is clearly an indicator of a very fgood performance of the system given the inherent difficulty of the task.

Finally, further pursuing the intrinsic evaluation of the keyword extractor, additional scores for the performance of this tool were obtained yet from another perspective. The first 20 keywords automatically extracted from a document were presented to 10 human testers. These testers were then asked to rate the keywords in a scale from 1 to 4 (very relevant, quite relevant, not relevant to the document, not a valid term). The average score was calculated over the entire set of 20 keywords, over the first 10 and the over the first 5. For the entire list of 20 keywords, a score of 2.34 was obtained; 2.08 was the score for the first 10 candidate keywords; and finally 1.94 was the score obtained for the first 5. These results are quite satisfactory: they indicate that the keywords automatically extracted are correctly ranked by the tool (with the more relevant being presented in the first positions) and that those higher ranked tend to be quite relevant.

4   The Glossary Candidate Detector

The Glossary Candidate Detector (GCD) was designed to automatically detect definitions, being able to tell apart the definiens from the definendum. A rule-based approach was adopted to develop this tool. The rules encode general patterns of candidate definitions whose basic components are some reserved words (e.g. verb "to be", etc.) and POS categories. The patterns were hand crafted on the basis of the analysis of the development data previously created, under the form of a corpus annotated with definitions.

To write down such rules, we resorted to lxtransduce. This is a tool that allows to build transducers specially suited to add or rewrite XML markup. It is a component of the LTXML2 tool set developed at the University of Edinburgh.[2]

In order to develop such transducer, three types of definition were identified and for each one a specific set of rules was written (for more details see [6]). Furthermore, the 274 000 token corpus was split in two parts, a development set, with 75% of the corpus, the remaining 25% for the test data.

Similarly to what was done for the keyword extractor, the GCD was evaluated both in a quantitative and in a qualitative manner. For the quantitative evaluation, the value of recall and precision was calculated at the sentence level. Recall here is the proportion of the sentences correctly classified by the system as containing a definition with respect to the sentences manually annotated as actually containing a definition. Precision is the proportion of the sentences correctly classified by the system with respect to the sentences automatically annotated. Furthermore, the F_2-measure[3] was also calculated. This score was

---

[2] http://www.ltg.ed.ac.uk/~richard/ltxml2/lxtransduce-manual.html

[3] $F_2 = \frac{(1+2)*Precision*Recall}{(2*Precisio)+Recall}$.

preferred to the simple F-measure in virtue of the type of task at stake. We are more interested in higher recall than in higher precision, given the application of the tool which is better to give more (possibly incorrect) definition candidates (with a higher recall, at the expense of a lower score in Precision) than to miss good definitions (in the opposite situation). We obtained a score of 0.14 for Precision, 0.86 for Recall and 0.33 for the F_2-measure.

On a par with this quantitative evaluation, a qualitative evaluation was carried out involving a group of users. We selected six MA students and presented them a LO with a list of definitions automatically generated using the tool—the LO was a 12 page introductory document on the use of Internet and the GCD had extracted 34 different definitions. Testers were instructed to read the document carefully and then score each definition using a rating scale from 1 to 4 (very good definition, good definition but not complete, acceptable definition, not a definition at all). The average score was 2.21, thus indicating that the candidate passages automatically extracted are on average considered good definitions according to human appreciation.

5   Semantic Search Tool

The Semantic and Multilingual Search Tool aims at allowing semantic search within a collection of documents; in more concrete terms in view of the application at stake, within a set of LOs. This means that it is possible to search for a term and retrieve, for example, all documents containing not only that term but also its synonyms and related concepts (such as super and sub-concepts). Since the tool is based on aligned ontologies developed for different languages,[4] it is possible to search for a term from a language A and retrieve documents in languages other than A, allowing for a multilingual retrieval.

The Semantic Search tool builds on the Lucene retrieval engine [7] embedded in the LMS and is based on three resources: a domain ontology, a lexicon, and an annotation grammar.

The ontology resulted from the merge between the DOLCE top-ontology, intermediate concepts from OntoWordnet, and a domain-specific ontology developed from scratch. This latter part was built in a bottom-up manner using as starting point the collection of keywords automatically generated for the corpora of every language in the project. This collection was translated into English in order to end up with a common collection. This final collection offered a first list of concepts to be covered by the domain ontology. Additionally, when entering these concepts in the ontology, concepts important to establish intermediate levels with the upper ontologies were added.[5] The domain covered by the final ontology is the realm of Computer Science for Non-Computer Scientists and includes concepts related to operating systems, applications, document preparation, computer networks, markup languages, world wide web, etc. The ontology

---

[4] The languages concerned are: Bulgarian, Czech, Dutch, English, German, Polish, Portuguese, Romenian and Maltese.

[5] For a fully detailed account of the core ontology building process see [8].

includes about 950 domain concepts, including 50 concepts from DOLCE and about 250 intermediate concepts from OntoWordNet.

The lexicon, in turn, was built by collecting every possible lexicalization for each one of the concepts in the ontology. By the end of the lexicon development process, we ended up with a list with 917 entries and 1019 lexicalizations, where each concept is associated with its possible lexicalizations.

Finally, an annotation grammar was developed which has the lexicon as its central component. A first common template of this grammar was put forward by Simov and Osenova [9] and subsequently worked out to develop different grammars for every language, and in particular for Portuguese by us. When applied to an input LO, this grammar detects possible (single or multi-word) lexical units and suggests all concepts of the ontology possibly expressed by that lexical unit. In the process of annotating the corpus with concepts, the output of this grammar is validated by human annotators who can select the right concept, or reject all and/or suggest a new one.

The lexicon constitutes the main relay resource between the query entered to start a search for documents, the ontology and, consequently the semantic-based search. The words entered are looked up in the lexicon, and the concepts that are associated to them are the items actually used in the search process, which will retrieve those documents containing some occurrences of those concepts in the markup semantic layer underlying the raw text.

Given the nature of the semantic search functionality, it was submitted only to an extrinsic evaluation, as described in the next Section.

6   Integration in the LMS and Extrinsic Evaluation

Besides the intrinsic evaluation of the tools developed when applicable, an extrinsic evaluation was also carried out after the integration of the new functionalities supported by them in the LMS. The LMS used was ILIAS, an open-source, fully edged web-based LMS that allows users to create, edit and publish learning material in an integrated system with normal web browsers.[6]

Fig. 1 displays the Graphical User Interface (GUI) by means of which it is possible to invoke the keyword extractor over a certain LO, and automatically obtain a list of candidate keywords for that document. The pane shows the candidate keywords list proposed after pressing the "Generate KeyWords" button. The user can accept the proposed keywords by checking the boxes and can also add new ones in the text field below them.

Fig. 2, in turn, presents a sample of the outcome of calling the GCD over a given LO.

Finally, Figure 3 shows the results of a semantic search triggered by the query made of the word "editor". It is worth noting that when a semantic search is performed, besides the relevant documents, the fragment of the ontology surrounding the concept used in the search is also displayed in the panel right

---

[6] http://www.ilias.de/

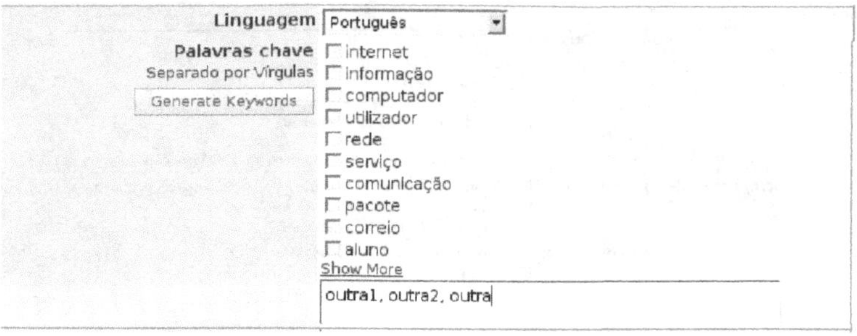

Fig. 1. User Interface ILIAS - Keyword Generator

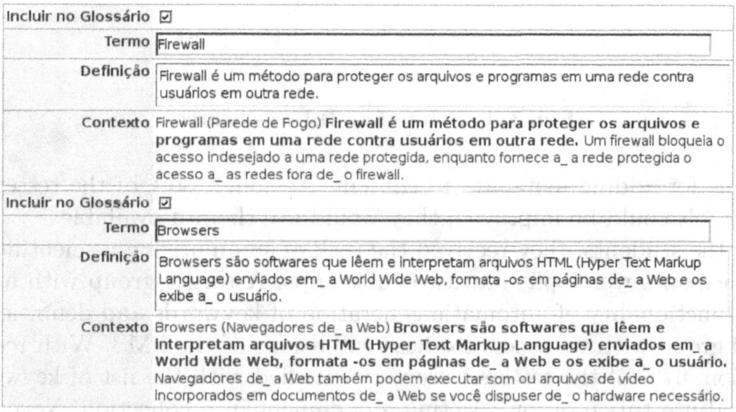

Fig. 2. User Interface ILIAS - Glossary Candidate Detector

inferior corner. The nodes of this fragment are clickable and allow the launching of a new search of LOs with occurrences of the concept clicked on.

The extrinsic evaluation was designed seeking to get some insight on the satisfaction of the potential end-users with respect to the new functionalities. This evaluation was based on the user scenario methodology [10]. Scenario here is meant to be "a story focused on a user, which provides information on the nature of the user, the goals he wishes to achieve and the context in which the activities will take place".

There were scenarios developed for two roles, i.e. for two kinds of users, students and tutors. For each role, two scenarios were created, one aimed at assessing the Keyword Extractor and GCD, and another aiming at assessing the semantic search. A group of at least 6 students participated in the student scenarios and the tutor scenarios were performed by 3 university teachers.

Regarding the extraction of keywords and the use of GCD by tutors, the participants were requested to generate a list of keywords and a glossary using the tools in order to make a certain LO available for a particular course. All testers (100% of score) agreed that both tools are useful, in particular for people

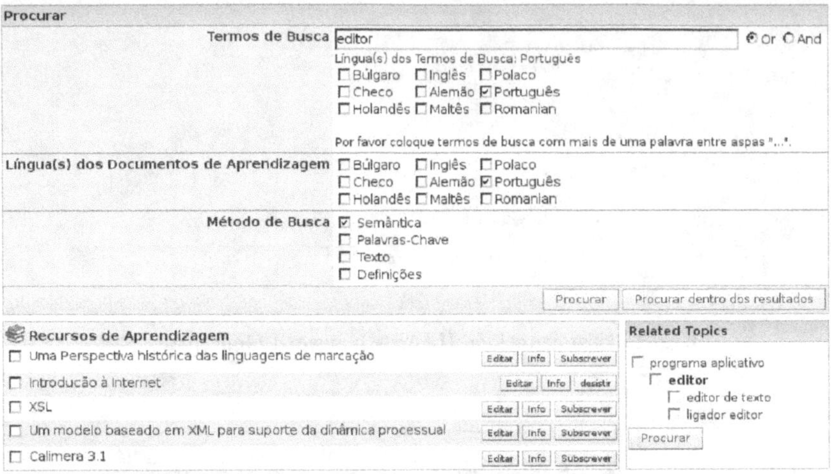

Fig. 3. User Interface ILIAS - Search

responsible for adding metadata to content. Although 30% of the testers think that the tools could be improved, they would use them if available.

As for the students, they received the task of summarizing a scientific paper. The participants were split into two subgroups. A target group with access to the new functionality of automatic generation of keywords and definitions, and a control group with no access to these extensions of the LMS. With respect to satisfaction, 67% of the testers were very satisfied with the list of keyword and 80% would use this tool for selecting a document in a collection. Nevertheless, 50% think that some important terms are missing. Regarding the glossary, all testers agreed that definitions were of a good quality, even if some definitions were missing. All testers agreed on the usefulness of this tool for this particular task and they would use the tool for extracting definitions from other papers. Besides checking satisfaction of users, the abstracts developed by the two groups were also evaluated using as metric the number of relevant concepts covered by abstracts. It turned out that the abstracts produced by the target group had a best coverage than the abstracts of the control group. On average, abstracts produced by the target group mentioned 5.5 relevant concepts while abstracts produced by the control group mentioned 4.2.

Regarding the semantic search functionality, tutors were given the task of refining a list of prerequisites for a given course, and to identify those LOs in the LMS repository which would help a student to learn about those prerequisites. Although for all testers it was easy to locate the relevant topics and identify relevant documents, 50% of them were not able to find some topics that they thought should be present. All testers agreed on the advantages of using such a tool in a virtual learning environment.

Students, in turn, were provided with a quiz with multiple choice questions, and were asked to try to find the documents containing the relevant answers. 83% of the testers found that their search terms returned mostly relevant content and 67% reported that the use of an ontology helped them in completing the task; 83% pointed out that ontology browsing and semantic search permit linking concepts in a way they were not aware of before.

7 Conclusions

In this paper we presented language technology resources and tools developed with the purpose of enhancing e-Learning by supporting new language processing-based functionalities embedded in an LMS. These tools were assessed under intrinsic and extrinsic evaluation.

Overall, the results coming out of the evaluation and reported above are positive and very encouraging. They provide an objective ground to the repeated claim that there is an important potential to be explored in what concerns the application of language technology to enhancing e-Learning.

References

1. LTSC: Learning technology standards committee website, http://ltsc.ieee.org/
2. Silva, J.R.: Shallow processing of Portuguese: From sentence chunking to nominal lemmatization. Master's thesis, Universidade de Lisboa, Faculdade de Ciências (2007)
3. N., I., K., S.: Xml, corpus encoding standard, document Xces 0.2. Technical report, Department of Computer Science, Vassar College and Equipe Langue et Dialogue, New York, USA and LORIA/CNRS, Vandouvre-les-Nancy, France (2002)
4. Lemnitzer, L., Lukasz, D.: Language technology for elearning: Implementing a keyword extractor. In: Fourth EDEN Research Workshop Research into online distance education and eLearning. Making the Difference, 2006 in Castelldefels, Spain (2006)
5. Gwet, K.: How to estimate the level of agreement between two or multiple raters. In: Handbook of Inter-Rater Reliability, Gaithersburg, Maryland (2001)
6. Gaudio, R.D., Branco, A.: Learning to identify definitions using syntactic feature. In: Progress in Artificial Intelligence, 13th Portuguese Conference on Aritficial Intelligence, Guimarēs, Portugal. Springer, Berlin (2007)
7. Gospodnetic, O., Hatcher, E.: Lucene in Action. Manning Publications (2004)
8. Osenova, P., Simov, K., Mossel, E.: Language resources for semantic document annotation and crosslingual retrieval. In: LREC 2008 (2008)
9. Simov, K., Osenova, P.: A system for a semi-automatic ontology annotation. In: Proceedings of Workshop on Computer-aided language processing CALP
10. Carrol, J.M.: Scenario-based design. John Wiley and Sons, Inc., Chichester (1995)

ParaMT: A Paraphraser for Machine Translation

Anabela Barreiro

Faculdade de Letras da Universidade do Porto & CLUP-Linguateca
New York University
`barreiro_anabela@hotmail.com`

Abstract. In this paper we present ParaMT, a bilingual/multilingual paraphraser
to be applied in machine translation. We select paraphrases of support verb constructions and use the NooJ linguistic environment to formalize and generate
translation equivalences through the use of dictionary and local grammars with
syntactic and semantic content. Our research shows that linguistic paraphrasal
knowledge constitutes a key element in conversion of source language into controlled language text that presents more successful translation results.

Keywords: ParaMT, paraphraser, paraphrase, support verb construction, multiword expression, machine translation, controlled language, NooJ, Lexicon
Grammar Theory.

1 Introduction

The benefits of paraphrasal knowledge to *Natural Language Processing* have been
quantified in areas such as summarization [1]-[3], question-answering [4]-[5], information extraction [6], and machine translation [7]-[8], among others. Recent ACL
workshops dedicated exclusively to paraphrasing reveal the growth in this field of
knowledge. However, most published works describe statistics-based approaches to
gather paraphrases. Statistical methods to acquire paraphrases are based on word co-
occurrences and word combinations and have little or no linguistic knowledge. They
also apply algorithms to corpora that may be inadequate or insufficient.

In this paper, we claim that effective results from linguistically based research on
paraphrases can save substantial effort and resources employed by statistically based
machine translation systems, by providing the opportunity to improve linguistic precision as a means to drive machine translation, rather than statistics. We argue that
science in general is founded on direct analytical observation and believe that good
quality machine translation relies on intimate language knowledge, not on probabilistic calculations. We have taken one important linguistic phenomenon, a particular
type of phrase, a support verb construction, and built a body of lexical, syntactic and
semantic knowledge around this phrasal type. Then, we applied this knowledge to a
bilingual/multilingual paraphraser, which we intend to integrate in machine translation systems. Our hypothesis was that linguistic knowledge applied to a machine
translation system would improve its output quality. We verified that support verb
constructions is an area where statistics tend to "trap" systems. If statistical systems
are not sensitive to these constructions, the consequence may be misleading

A. Teixeira et al. (Eds.): PROPOR 2008, LNAI 5190, pp. 202–211, 2008.

translations. We argue that our linguistic system provides a statistical system with special training data that could correct this problem.

2 Support Verb Constructions

Support verb constructions are predicate noun constructions (noun + arguments) where the main verb has a weak semantic value, such as in *make a promise*. Semantically weak verbs are called support verbs in Lexicon-Grammar theory, but they are also known as light verbs [9]-[10]. Support verb constructions are multiword expressions that, within the area of corpus linguistics, have been subcategorized as collocation phenomena [11]-[14]. The term *"collocation"* is generally used to define words or terms that 'go together' with a precise meaning. Most works on collocations consist mainly in identifying collocations within a corpus, with the goal of including them in extended dictionaries. Even though it is widely known and used, collocation is a 'sort of' statistical related term (co-location means positioning side by side or close together) that is too broad for linguistic analysis. We look at collocations as multi-layered linguistic phenomena which, in our opinion, must be identified and studied individually, as proposed in [15]. We consider that the more we know about multiword expressions, the more sophisticated their descriptions are in the electronic dictionaries or the more accurately they are formalized in computer grammars, the better the quality of machine translation output and of natural languages applications in general.

Identifying source language multiword expressions such as support verb constructions is not a trivial task, but it is the starting point for paraphrasal knowledge, as it is for translation. As early as 1988, as demonstrated inter alia by [16]-[18], the suggestion of conceptually separating monolingual paraphrasing from translation in machine translation has been put forward by the insertion of a "style transfer" module which selects the "best or chosen translation" from multiple "possible" translations. The idea of dynamically invoking monolingual grammars to perform translation of multiword expressions was raised by developers on the working prototype built by the IBM-INESC Scientific Group back in the late eighties [ibidem]. Our approach uses monolingual grammars for the identification of support verb constructions and bilingual/multilingual grammars for translation and bilingual/multilingual paraphrasing. We can use paraphrasing in a monolingual text as a pre-editing procedure for controlled language writing and generate and translate paraphrases allowing their insertion directly in machine translation. We will load the paraphraser with Portuguese to English data and use the NooJ linguistic environment [19] to formalize and translate support verb constructions through finite-state transducers (local grammars) for bilingual/multilingual purposes. The theoretical framework behind this study is the Lexicon-Grammar [20]-[21], which stands on the principles of the transformational grammar of Harris, [22]-[23]. According to the Lexicon-Grammar, simple sentences (predicate and its arguments), also known as elementary sentences, and not the individual words, represent basic syntactic-semantic units. Natural language processing systems, particularly machine translation systems that take into account these linguistic units yield more opportunities for success.

3 Machine Translation Problem Evidence

Our experience with machine translation confirms that currently the results are far from perfect [24]. Translation results extracted from METRA [25], and described in [26] prove that machine translation engines are unsuccessful particularly at handling the translation of support verb constructions. A literal and unnatural translation is provided by most machines. For example, the English support verb construction *make a decision* is translated into Portuguese as *fazer uma decisão* instead of *tomar uma decisão* or even as the strong verb *decidir*, which represent its optimal paraphrase. This inaccuracy means that the English support verb *make* is directly translated into the Portuguese support verb *fazer* (default translation), instead of being recognized as part of the support verb construction which embeds semantic meaning as a whole.

We have tried to replace some support verb constructions with lexical verbs and verified that overall machine translation engines showed significantly better results. For example, machine translation engines are unanimous in choosing the Portuguese verb *decidir* as the correct translation for the English verb *decide*. This pre-editing, or more precisely controlled language writing by paraphrasing, improves translation results and makes output sentences more comprehensible overall. This proves that, if we consider pre-editing of the input sentences where support verb constructions occur, changing each instance into a lexical verb, we are not changing the meaning of the source sentence and we are giving the machine translation engine a distinctly better chance of improving the output result, by filtering out some noise, i.e., the weak verb. The support verb construction *make a decision* is a stylistic alternative to the verb *decide*, where neither the support verb *make* nor the determiner *a* add any meaning to the expression. In fact, in support verb constructions, the support verb is often void of meaning. Trying to translate them brings additional difficulties to machine translation systems, which is unnecessary until/unless they become more sophisticated. Our idea is to have several possibilities and not limit the system to only one possibility, as long as the system translates with precision. However, we believe that it is pointless to challenge one limited system with structures that we know *a priori* this system cannot translate well. For equivalent paraphrasing, the support verb must be recognized as part of a support verb construction which must be considered as a single semantic unit. The default assumption of all machine translation systems which cannot discern whether a word, in this case a support verb, adds semantic meaning to a phrase, is to assign equal semantic value to each word individually, unless, otherwise instructed. The system fails by incorrectly assigning semantic value to a support verb, resulting in a loss in equivalence of the output sentence. This is the problem of direct translation.

In sum, empirical evidence shows that application of linguistic knowledge to proper handling of support verb constructions by machine translation systems or NLP applications is effective. We believe that our methodology leads to attainable paraphrasing translation solutions. This paper demonstrates that we can create an instrument of some utility to the research community. We chose support verb constructions because they have been extensively studied from both theoretical and practical perspectives, in several different languages, by many authors, over a considerable period. They are fairly systematic, and therefore quite suitable ·for formalization and integration with machine approaches. Support verb constructions are abundant in language and their formalization is generally essential for machine translation. Lastly, they often represent paraphrases. For example, the English support verb construction [*pay a visit to NP*] is a phrasal alternative to the transitive construction [*visit NP*]. Both expressions have equivalent meaning and can be translated in the same way into Portuguese, [*visitar NP*].

4 ParaMT Resources and Methodology

In any language processing application, the linguistic resources represent the foundation. High-quality linguistic descriptions lead to sophisticated resources that help improve systems. In machine translation especially, the linguistic resources are the driving force that boosts the translation process. Our paraphrasing system is based on *Port4NooJ*, the open source NooJ Portuguese linguistic module, which integrates a bilingual extension for Portuguese-English machine translation. Port4NooJ is developed on two original sources: NooJ linguistic environment and OpenLogos lexical resources. The module and the linguistic resources are described thoroughly in [27] and available online in [28] and [29]. The elements that we want to emphasize here are the ones directly concerned with processing of support verb constructions. Accordingly, each dictionary entry includes, beyond the commonly used part-of-speech and inflectional paradigm, a description of the·syntactic and semantic attributes (*Syn-Sem*), as well as the associated distributional and transformational properties, such as predicate arguments, information about which determiners and prepositions occur with predicate nouns in "less variable" expressions, and derivational descriptions. Derivation is a very important issue, because it has implications not only at the lexical level, but also at the syntactic level. Derivational suffixes often apply to words of one syntactic category and change them into words of another syntactic category, while semantically they maintain their integrity. For example, the affix *–ção* changes the verb *adaptar* (*to adapt*) into the noun *adaptação* (*adaptation*) and the affix *-mente* changes the adjective *rápido* (*quick*) into the adverb *rapidamente* (*quickly*). This is extremely important for support verb constructions because it permits the establishment of equivalence grammars that map (i) support verb constructions such as *fazer uma adaptação (de)* (*to make an adaptation (of)*) to the verb *adaptar* (*to adapt*), where the predicate noun *adaptação* (*adaptation*) has a semantic and morpho-syntactic relationship with the verb *adaptar* (*to adapt*) or (ii) support verb constructions such as *ter um final rápido* (*to have a quick ending*) to the verbal expression *terminar rapidamente* (*to end quickly*), where the autonomous predicate noun *final* (*ending*) has a semantic relationship with the verb *terminar* (*to end*), and the adverb *rapidamente* (*quickly*) has a semantic and morpho-syntactic relationship with the adjective *rápido* (*quick*). Thus, our verb entries contain the identification of derivational paradigms for nominalizations (annotation *NDRV*) and a link to the derived noun's support verbs (annotation *NVSUP*), as in Fig. 1 below. Nominalizations are followed by their inflectional paradigm properties. Any other lexical constraints, such as prepositions, determiners, specific arguments, etc., will be added. Autonomous predicate nouns (non-nominalizations), such as *favor* (*favor*) are lexicalized and classified with the annotation *Npred* and have associated with them support verb and other lexical constraints, such as a preposition (*NPrep*), and a lexical verb (*VRB*) with the same semantics. We have also classified predicate adjectives and established the link between them and the corresponding verbs (*ADRV*), such as between the verb *adoçar* (*to sweeten*) and the adjective *doce* (*sweet*). We have started the assignment of corresponding copula verbs to these adjectives.

adaptar,V+FLX=FALAR+Aux=1+INOP57+Subset132+EN=adapt+*VSUP=fazer* +*DRV*=NDRV00:CANÇÃO +*NPrep*=de
favor,N+FLX=MAR+Npred+AB+state+EN=favor+*VSUP*=fazer+*NPrep*=a+*VRB*=ajudar
rápido,A+FLX=RÁPIDO+PV+eagerType+EN=quick+*DRV*=AVDRV06:RAPIDAMENTE
adoçar,V+FLX=COMEÇAR+Aux=1+OBJTRundif75+Subset604+EN=sweeten+*DRV*=ADRV11:VERDE+*VSUP*=tornar

Fig. 1. Sample of the dictionary

According to these linguistic constraints, we have created relationship properties at the dictionary level and then apply those properties in local grammars in order to recognize support verb constructions in corpora and generate them for applications such as controlled language writing and machine translation. In section 5, we describe how we use these resources to recognize and generate paraphrases automatically.

Our strategy to formalize idiomatic expressions and distinguish them from expressions with a more complex syntactic behavior is to lexicalize them. Therefore, semi-frozen support verb constructions, where the support verb is the only variable word in the whole expression, are lexicalized in the dictionary of multiword expressions. For example, in *dar a mão à palmatória* (*to acknowledge being wrong*) or *fazer vista grossa* (*to ignore*), the support verbs *dar* (*to give*) and *fazer* (*to make*) are assigned an inflectional paradigm and the rest of the words in the expression remain invariable. As our electronic dictionaries provide enhanced meaning of single words, including contextual significance and increasingly more valuable tagging data, we also intend to enlarge and refine the role of a bilingual dictionary to include entries for multiword expressions that consider the understanding and analysis of each type of multiword expression, by beginning with support verb constructions and their paraphrases. The ability to give the machine translation user multilingual paraphrasing ability constitutes an important step towards achieving better quality machine translation.

5 Paraphrases for Machine Translation

As we have mentioned above, in order to obtain monolingual paraphrases or to translate support verb constructions from Portuguese to English using NooJ, we combine the properties formalized in the Portuguese dictionary with local grammars. Local grammars are ways of formalizing language constructs using input and output symbols, i.e., they are language descriptions in the form of graphs containing an input entry (with linguistic information) and an output entry (with linguistic constraints to the output, or simply the binary information of the recognized or not recognized sequence). In NooJ, these local grammars are represented by finite-state transducers, and are widely applied to texts/corpora, for identification and analysis of local linguistic phenomena of a natural language, extraction of named entities from texts, recognition and tagging of words, or multiword expressions, identification of syntactic constituents such as noun phrases and completives, extraction of semantic relations, and disambiguation. Among these possible applications, we extended local grammars to recognize, paraphrase and translate support verb constructions, creating ParaMT, a bilingual/multilingual automatic paraphraser. In order to establish relations of equivalent morpho-syntactic predicates in the same language (Portuguese) or between two languages (particularly between Portuguese and English), we use the dictionary properties. Since we have classified all predicate nouns in the dictionary as [*NPred*], we can now use this lexical information in a syntactic grammar to identify the predicate in a support verb construction and apply this grammar in corpora. Fig. 2 represents a simple local grammar used to recognize and generate support verb constructions and transform them into their verbal paraphrases.

Fig. 2. Grammar to recognize and paraphrase support verb constructions

This grammar matches verbs, which are marked in the dictionary as support verbs that are followed by a left modifier (determiner, adjective or adverb or other quantifiers), a predicate noun and optionally a preposition. The elements in parentheses ( ) are stored in variables V, N or PREP. If a dictionary entry has a lexical constraint, such as NPrep=*a* in the phrase [*dar um grande abraço a*] (*to give a big hug to*), the support verb construction will be recognized by the grammar and mapped to the verb *abraçar* (*to hug*), the lemma of the noun specified in the variable $N_. The elements in bold <$V_=$N$VSUP>, and $PREP_=$N$NPrep> represent lexical constraints that are displayed in the output, such as specification of the support verb or the preposition that belongs to a specific support verb construction. The predicate noun is identified, mapped to its deriver and displayed as a verb, the other elements of the phrase are eliminated. Fig. 3 shows a concordance where Portuguese support verb constructions are recognized and paraphrased as lexical strong verbs.

gosto de ver o comboio a	fazer corridas /correr	à velocidade máxima ao lon
o de cheque especial para	fazer doações /doar	às entidades que escolher. A
pres e, quando é preciso ir	fazer filmagens/filmar	fora do estúdio, às vezes fic
je queria trocar de pares e	fazer um jogo /jogar	ao melhor de três sets , mas
dra deu-me um papel para	fazer uma lista de/listar	todas as coisas boas que ex
res foram à caracterização	fazer uns retoques/retocar	, outros estão a descansar n

Fig. 3. Recognition and monolingual paraphrasing of support verb constructions (Support verb construction / corresponding verb)

ParaMT makes possible the recognition of Portuguese support verb constructions in a text and their automatic conversion into an English verb (bilingual paraphrasing), as in Fig. 4.

a fazer um estágio para	dar aulas de/teach	religião, mas não se impor
m -- os filhos -- juntos e	fizeram a mudança para/change	Johannesburg, e ensinaram
. Necessitava apenas de	ter a certeza de/know	que não escapara à sua
ente hipotética. -- Deves	ter alguma ideia/know	. Dorothy andava a fazer u
não podemos deixar de	ter cautela/beware	. Pobre Caro, pensou Lynd
ra dos chinelos, antes de	ter chance de/can	mudar de idéia. Como pos
ope a Jean, esta pareceu	ter dificuldade em/avoid	olhá-lo nos olhos. Deixou
ao Kiss dela. Apesar de	ter falta de/lack	amor-próprio, isso não sigr
igos e imprensa estava a	ter lugar /occur	numa longa galeria com car
uiu ter filhos. -- Tens de	ter mão /control	nessa confusão toda. Sam
spondi, minha mãe deve	ter medo de/fear	cobras. Eu disse no Gabin
da loja antes de ele	ter tempo de/could	chamar a brigada de narcó
a triste aventura havia de	ter um fim/finish	
Ela ouvira a tia Velma	ter uma discussão com/argue	Jack acerca de mostarda r
de olhos fechados para	ter uma ideia de/know	como seria ser cego e
ter paciência.» «Voltei a	ter uma imensa vontade de/want	viver. A conversa parecia

Fig. 4. Recognition and bilingual paraphrasing of support verb constructions (Portuguese support verb construction / corresponding English verb)

6 Preliminary Quantitative Evaluation

Currently, our bilingual Portuguese-English general dictionary comprises about 60,000 entries distributed by 30,000 nouns, 11,000 verbs, 2,800 adjectives, 4,700 adverbs, and 11,500 other part of speech entries. The dictionary of proper names comprehends about 6,000 entries. Our multiword expression dictionary comprehends about 40,000 entries, 20,000 nominal; 10,000 verbal; 5,000 adjectival and 5,000 adverbial multiword expressions. We have over 8,000 derivational links between verbs and nominalizations and about 1,000 derivational links between verbs and predicate adjectives. A few general multiword expression grammars cover over 5,000 expressions of several other types. We have not yet evaluated the coverage of the multiword expressions dictionary and grammars in corpora, but we have some preliminary results for the evaluation of support verb constructions. In order to obtain these results, we selected from COMPARA [30], a parallel corpus of English-Portuguese fiction, all sentences where the infinitive form of the Portuguese verbs *fazer* (*to do*), *dar* (*to give*), *pôr* (*to put*), *tomar* (*to take*) and *ter* (*to have*) occurred with a noun or with a left modifier and a noun. First, we manually classified these combinations as to whether they corresponded to support verb constructions or not. We confirmed that these verbs occur very frequently in a support verb construction. 89% of the occurrences of *dar*, 88% of *tomar*, 77% of *pôr*, 47% of *fazer* and 20% of *ter* were in a support verb construction. This means that globally in 64.2% of the times, these verbs are used as support verbs, that corresponds to nearly 2/3 of the occurrences.

Subsequently we selected randomly a sub-corpus with 500 sentences (100 for each selected verb), containing instances of only support verb constructions. We classified them manually and compared these results with the results obtained automatically. We tried to have constraining recognition rules so that paraphrasing would be more precise. Currently, we can recognize 62.6% of the support verb constructions with high scores in precision. Furthermore, we not only recognize the support verb

constructions, as we also paraphrase them with high degree of success. Fig. 5 shows
the results of the support verb construction recognition (precision and recall) and the
results (precision) of our automatic paraphraser.

	SVC Recognition Precision	SVC Recognition Recall	SVC Paraphrasing Precision
Pôr	73/73 - 100%	73/100 – 73%	72/73 - 98.6%
Tomar	75/75 - 100%	75/100 – 75%	68/73 - 93.1%
Ter	65/65 - 100%	65/100 – 65%	59/65 - 90.7%
Dar	57/60 - 95%	57/100 – 57%	46/51 - 90.1%
Fazer	43/45 – 95.5%	43/100 – 43%	40/45 - 88.8%
Average	62.6/63.6 - **98.4%**	62.6/100 - **62.6%**	57/61 - **93.4%**

Fig. 5. Evaluation of simultaneous recognition and paraphrasing of support verb constructions

7 Conclusions

In this paper we have tried to answer the question of whether paraphrase information
can improve machine translation output and how the analysis and formalization of
paraphrases can contribute to the larger task of machine translation. We have
addressed linguistic analysis and computational formalization of bilingual short
paraphrases for support verb constructions using NooJ linguistic environment. We
have demonstrated the scope of the phenomenon as a basis for a machine translation
multiword expression dictionary, which can be used both in machine translation
development or machine translation evaluation and in the extension of the scope in
current dictionary functionality.

The discovery process has provided results in two areas. First, it has led to the
creation of a primitive multiword expression electronic dictionary that addresses
monolingual Portuguese and bilingual Portuguese-English paraphrases of equivalent
meaning between support verb constructions and their noun counterparts. Second, it
has helped to further specify the definition of multiword expressions. The interface
between user and software that is presented is not finished yet, but once the sub-task
is well-understood this interface can be simplified and, hopefully, will be usable and
as easily integrated into the larger task of machine translation, as the single word
electronic dictionary that has already been integrated.

Our work based on support verb constructions illustrates what can be done with
ParaMT for any kind of multiword expression. The method is repeatable.
Furthermore, the tool is extensible to cover larger and more complex linguistic
phenomena, including sentence level paraphrases that can be used for controlled
language writing or translation. While this research is intended to find a place in ideal
machine translation, it can be used as an electronic multiword dictionary. From a
monolingual point of view, it is useful to simplify pre-translated source text
(rendering the text less complex, less flowery, etc.). Converting support/weak verbs
into lexical strong verbs helps to simplify and reduce the number of words in a text
which has a positive impact on translation cost, in circumstances where word count or
"white space" is sensitive. From a bilingual point of view, it helps reduce ambiguity
and verbosity. It can be used as an on-line linguistic aid for translators so they can

determine the best translation (evaluation purposes), and for automated machine translation evaluation. This knowledge is useful to machine translation development, because it permits deeper understanding of source text, and it provides a successful methodology to analyze paraphrasing, given that paraphrasal intelligence is crucial in both machine translation development and machine translation evaluation.

Acknowledgements

We would like to thank Max Silbeztein and Slim Mesfar for providing support on technical aspects of NooJ, and Sérgio Matos and Diana Santos for helpful comments on this paper.

This work was partly supported by grant SFRH/BD/14076/2003 from *Fundação para a Ciência e a Tecnologia* (Portugal), co-financed by POSI.

References

1. Barzilay, R., McKeown, K.: Extracting Paraphrases from a Parallel Corpus. In: Proceedings of the ACL/EACL, Toulouse, pp. 50–57 (2001)
2. Barzilay, R.: Information Fusion for Multidocument Summarization. Ph.D. Thesis, Columbia University (2003)
3. Hirao, T., Suzuki, J., Isozaki, H., Maeda, E.: Dependency-based Sentence Alignment for Multiple Document Summarization. In: Proceedings of the COLING, pp. 446–452 (2004)
4. Ibrahim, A., Katz, B., Lin, J.: Extracting structural paraphrases from aligned monolingual corpora. In: Proceedings of the Second International Workshop on Paraphrasing (ACL 2003), pp. 10–17 (2003)
5. Duboué, P.A., Chu-Carroll, J.: Answering the question you wish they had asked: The impact of paraphrasing for Question Answering. In: HLT-NAACL 2006 (2006)
6. Shinyama, Y., Sekine, S.: Paraphrase Acquisition for Information Extraction. In: The Second International Workshop on Paraphrasing: Paraphrase Acquisition and Applications (IWP2003), Sapporo, Japan (2003)
7. Callison-Burch, C., Koehn, P., Osborne, M.: Improved Statistical Machine Translation Using Paraphrases. In: Proceedings NAACL 2006 (2006)
8. Callison-Burch, C.: Paraphrasing and Translation. PhD Thesis, University of Edinburgh (2007)
9. Kearns, K.: Light verbs in English (manuscript, 2002)
10. Butt, M.: The light verb jungle. Harvard Working Papers in Linguistics 9, 1–44 (2003)
11. Quirk, R., Greenbaum, S., Leech, G., Svartvik, J.: A comprehensive grammar of the English language. Longman, London (1985)
12. Biber, D.: Variation Across Speech and Writing, pp. 3–27. Cambridge University Press, Cambridge (1988)
13. Crystal, D.: A dictionary of linguistics and phonetics, 3rd edn. Blackwell Publishers, Oxford (1991)
14. Sinclair, J.: Corpus Concordances Collocations. Oup, Oxford (1991)
15. Meyers, A., Reeves, R., Macleod, C.: NP-External Arguments: A Study of Argument Sharing in English. In: Proceedings of the ACL 2004 Workshop on Multiword Expressions: Integrating Processing, Barcelona, Spain, July 26, 2004, pp. 96–103 (2004)

16. Santos, D.: A fase de transferência de um sistema de tradução automática do inglês para o português, Tese de Mestrado, IST, UTL (1988)
17. Santos, D.: Lexical gaps and idioms in Machine Translation. In: Karlgren, H. (ed.) Proceedings of COLING 1990, Helsinki, August 1990, vol. 2, pp. 330–335 (1990)
18. Santos, D.: Broad-coverage machine translation. INESC Journal of Research and Development 3(1), 43–59 (1992)
19. Silberztein, M.: NooJ: A Cooperative, Object-Oriented Architecture for NLP. In: INTEX pour la Linguistique et le traitement automatique des langues. Cahiers de la MSH Ledoux, Presses Universitaires de Franche-Comté (2004)
20. Gross, M.: Méthodes en syntaxe. Hermann (1975)
21. Gross, M.: Les bases empiriques de la notion de prédicat sémantique. In: Guillet, A., Leclère, C. (eds.) Formes Syntaxiques et Prédicat Sémantiques, Langages, Larousse, Paris, vol. 63, pp. 7–52 (1981)
22. Harris, Z.: Co-occurrence and transformation in linguistic structure. Language 33, 293–340 (1957)
23. Harris, Z.: Mathematical Structures of Language, p. 230. Wiley, New York (1968)
24. Barreiro, A., Ranchhod, E.: Machine Translation Challenges for Portuguese. In: Linguisticæ Investigationes 28.1. John Benjamins Publishing Company, Amsterdam, pp. 3–18 (2005)
25. Sarmento, L.: Ferramentas para experimentação, recolha e avaliação de exemplos de tradução automática. In: Santos, D. (ed.) Avaliação conjunta: um novo paradigma no processamento computacional da língua portuguesa, pp. 193–203. IST Press, Lisboa (2007)
26. Barreiro, A. Formalization of Support Verb Constructions and their Paraphrases: Applications in Machine Translation (provisory title). PhD dissertation (forthcoming, 2008)
27. Barreiro, A.: Port4NooJ: Portuguese Linguistic Module and Bilingual Resources for Machine Translation. In: Blanco, X., Silberztein, M. (eds.) Proceedings of the 2007 International NooJ Conference, Univ. Autonoma de Barcelona, June 7-9, 2007. Cambridge Scholars Publishing (forthcoming, 2008)
28. NooJ, http://www.nooj4nlp.net/
29. Linguateca, http://www.linguateca.pt/Repositorio/Port4Nooj/
30. Frankenberg-Garcia, A., Santos, D.: Introducing COMPARA, the Portuguese-English parallel translation corpus. In: Zanettin, F., Bernardini, S., Stewart, D. (eds.) Corpora in Translation Education, pp. 71–87. St. Jerome Publishing, Manchester (2003), http://www.linguateca.pt/COMPARA/

Second HAREM: New Challenges and Old Wisdom

Diana Santos[1], Cláudia Freitas[2], Hugo Gonçalo Oliveira[2], and Paula Carvalho[3]

[1]SINTEF ICT, Norway
[2]CISUC, DEI-FCTUC, Portugal
[3]DI-FCUL, Portugal
Diana.Santos@sintef.no, maclaudia.freitas@gmail.com, hroliv@dei.uc.pt,
pqfcarvalho@gmail.com

Abstract. Discussion of the Second HAREM: changes to the guidelines, introduction of new tracks, improvement of evaluation measures and description of the new evaluation resources.

1 Introduction

In this paper we present the second evaluation contest of named entity recognition in Portuguese, the Second HAREM[1] which started September 2007 and whose submission period took place 14-28 April 2008. We are mainly concerned with presenting the options while designing this new event, describing the process followed and the differences compared to the First HAREM.

After a successful first event, the First HAREM, in whose workshop most participants stated their definite interest in a new edition, and which culminated with the production of a book [1], call for participation in a second edition was issued in September 2007. A record number of 22 different prospective participants or interested parties enrolled. In early 2008, 16 systems registered, although only 10 would eventually participate.

Discussion took place until November 2007, and two new tracks or tasks were proposed: (i) temporal recognition and normalization, according to extensive guidelines proposed by a group of participants, Hagège et al. [2]; and (ii) relation detection between named entities (dubbed ReRelEM), including, but not limited to, coreference identification.

As for training material, a fully annotated example collection – six fully annotated texts, ca. 1,500 words – were made available to the participants at the HAREM site in January 2008, while the golden collections from First HAREM were converted to the new format: ca. 140,000 words, corresponding to 9,000 NEs from 257 different documents. Later, a fully annotated subsection of previous material – ca. 3,500 words and 279 NEs – was also created by the authors of the time guidelines.

---

[1] http://www.linguateca.pt/HAREM/

A. Teixeira et al. (Eds.): PROPOR 2008, LNAI 5190, pp. 212–215, 2008.

2 "Old", Persistent, Features of HAREM

The most important features of the First HAREM remained: (i) its semantic model (that describes the use of a NE in context, and not its dictionary meaning) [1, chapter 4] and (ii) the flexibility of its evaluation setup (in particular the existence of selective scenarios) [3].

To introduce the semantic model to newcomers, let us take the example of continents: These are, basically, lexically covered by the five words *Oceânia*, *Ásia*, *América*, *Europa* and *África*. In addition to requiring that systems decide whether these continents are actually being mentioned and thus the words do not refer to a ship (COISA), a deity (PESSOA) or a book (OBRA), as is usual in NER contests, HAREM asks systems to decide also whether the particular mention in context refers to an entity of the physical domain (LOCAL FISICO), an entity belonging to human geography (LOCAL HUMANO), the people inhabiting that continent (PESSOA POVO), supra-national political organizations (ORGANIZACAO ADMINISTRACAO), or even just the abstract concept (ABSTRACCAO), whatever that may be. This shows that HAREM tasks are considerably more difficult, and fine-grained, than the classical NER task as represented for example by MUC [4].

As to the selective scenario property, it allows HAREM to encompass different systems with different goals and different applications in mind, enabling comparsion of every system also relative to its preferred view (its selective scenario).

3 Improvements

The problems of artificially separating identification from classification, noted and discussed in [1, chapter 7], were solved by assuming a more consistent (although more complex) identification strategy, as well as removing from the NE task the identification of objects which only accidentally include names (as is the case of *pastéis de Belém* and *bolas de Berlim*).

A number of finer distinctions in the geographic domain were also added, mirroring, in a way, the finer grained classification of time expressions.

This will allow for an empirical investigation of the possibility or need of a more detailed subcategorization in NER.

We also made explicit the difference between not knowing (a negative statement) and knowing that **none** of the explicit choices was right (a positive one). Now we have OUTRO for the latter case, while not knowing is conveyed by no value at all, and the two cases are differently scored.

There were also several changes on the technical side. The format of collections and submissions was changed to XML, to achieve easier processing and validation of the material. A validator was deployed and made public for systems to test their output previous to submission.

Instead of allowing submissions doing either identification only or classification plus identification, both tasks were merged in one syntax only; and a more consistent use of OUTRO label across CATEG, TIPO and SUBTIPO was implemented.

Substantial changes to the evaluation machinery, while preserving backward compatibility, became necessary. In fact, one the most important contributions of Fisrt HAREM was to define a set of measures and metrics for NER, at the same time making available a set of open source programs to compute them.

Those measures, however, were based on a fixed depth of categories and types: each category had a number of types, while now we have a four level hierarchy, with everything optional. We have therefore extended and made more robust the evaluation measure, in order to account, in the same fell swoop, for everything covered by the previous measures (except for types-only).

We have also taken measures to give a more adequate treatment to vagueness in the evaluation, accepting and expecting submissions to also produce more than one classification (|) or delimitation (ALT).

In parallel, we removed the partial identification feature of First HAREM (which was blind) by consistently annotating, with the ALT feature, all possible meaningful parts. Only those should be rewarded.

The new measure, an extension of the combined measure of First HAREM, accounts for the existence of subtypes and for the optionality of all values, as well as dealing more adequately with vague NEs (with N categories): $1 +$

$$\sum_{n=1}^{N} (1 - 1/num_{cat}) * cat_{certa} * \alpha + (1 - 1/num_{tipos}) * tipo_{certo} * \beta + (1 - 1/num_{sub}) * sub_{certo} * \gamma -$$
$$\sum_{n=0}^{M} (1/num_{cat}) * cat_{espuria} * \delta + (1/num_{tipos}) * tipo_{espurio} * \epsilon + (1/num_{sub}) * sub_{espurio} * \phi$$

4   The New HAREM Collection and Its Annotation

As to the constitution of the golden collection for the Second HAREM, i.e., the subset of the collection that will be used as comparison stock, this time – accompanying the flow of time – we included, and made heavy use of, new genres such as blogs, wikis, encyclopedia (Wikipedia) entries, and questions (such as used in question answering), in addition to the more traditional kinds of newspaper text and standard Web pages. Oral transcriptions and literary text, due to the difficulty of obtaining this kind of text, were far more scarcely used.

To create the full collection that includes the golden collection and is provided for the systems to analyse, we used a different strategy from the previous HAREM, in which we had tried to mirror the genre distribution of the golden collection. This time we just included in the full collection all training materials already available, while all remaining material came from the CHAVE collection [5], newspapers from 1994-1995. The texts themselves were chosen from the recall base of last GeoCLEF: for each topic, all the relevant documents and ten irrelevant documents were taken from the Portuguese pool.

Human annotation of the golden collection was performed in several stages: (i) initial independent annotation of each text by two annotators, using a specially developed tool, Etiquet(H)AREM[2] [6] (ii) automatic comparison using another tool, comp(H)AREM; (iii) discussion and revision of these differences; (iv) full sequential revision; (v) revision per category.

---

[2] Available from http://linguateca.dei.uc.pt/index.php?sep=recursos

When disagreement showed that several different interpretations of the same text are possible, use of vagueness was encouraged. A detailed set of guidelines, as well as specific difficulties in interpreting Portuguese text, was gathered.

5   New Challenges

One of the hallmarks of Linguateca's activities, fully reflected in HAREM, is not to repeat merely what has been done for other languages. On the contrary, we are keen on innovation and on doing first class research, which means that we try to do both something original **and** conceived with Portuguese in mind. We firmly believe that progress in NLP can be done in any language. So, when considering the possibility of offering co-reference as a new track in HAREM – and resources and time available would prohibit us to do something similar to ARE[3] for Portuguese, we would not do a simple identification of NEs referring to the same entity as in MUC or ACE [7]. Rather, we proposed the task of identifying the most frequent (and less controversial) relations among NEs.

Acknowledgments. This work was done in the scope of the Linguateca project, jointly funded by the Portuguese Government and the European Union (FEDER and FSE) under contract ref. POSC/339/1.3/C/NAC. We are grateful to David Cruz and Luís Miguel Cabral for participating in the organization of HAREM and to Luís Costa for relevant comments.

References

1. Santos, D., Cardoso, N. (eds.): Reconhecimento de entidades mencionadas em português: Documentação e actas do HAREM, a primeira avaliação conjunta na área. Linguateca (2007)
2. Hagège, C., Baptista, J., Mamede, N.: Proposta de anotação e normalização de expressões temporais da categoria TEMPO para o HAREM II (2008)
3. Santos, D., Seco, N., Cardoso, N., Vilela, R.: HAREM: An Advanced NER Evaluation Contest for Portuguese. In: Calzolari, N., et al. (eds.) Proceedings of LREC 2006, 22-28 May 2006, pp. 1986–1991 (2006)
4. Grishman, R., Sundheim, B.: Message understanding conference-6: a brief history. In: Proceedings of the 16th conference on Computational linguistics, pp. 466–471. Association for Computational Linguistics, Morristown (1996)
5. Santos, D., Rocha, P.: The key to the first clef with portuguese: Topics, questions and answers in chave. In: Peters, C., Clough, P., Gonzalo, J., Jones, G. (eds.) Multilingual Information Access for Text, Speech and Images: Results of the Fifth CLEF Evaluation Campaign, pp. 821–832. Springer, Heidelberg (2005)
6. Carvalho, P., Oliveira, H.G.: Manual de utilização do Etiquet(H)AREM (2008)
7. Doddington, G., Mitchell, A., Przybocki, M., Ramshaw, L., Strassel, S., Weischedel, R.: The Automatic Content Extraction (ACE) Program: Tasks, Data and Evaluation. In: Lino, M.T., et al. (eds.) Proceedings of LREC 2004, Lisbon, Portugal, ELRA, 26-28 May 2004, pp. 837–840 (2004)

---

[3] See http://clg.wlv.ac.uk/events/ARE/

Floresta Sintá(c)tica: Bigger, Thicker and Easier

Cláudia Freitas[1], Paulo Rocha[2], and Eckhard Bick[3]

[1,2]Linguateca, DEI, Universidade de Coimbra, Portugal
{freitas,parocha}@dei.uc.pt
[3]Syddansk Universitet, Odense, Denmark
eckhard.bick@mail.dk

Abstract. In this paper, we describe the resumption of activities of Floresta Sintá(c)tica, a treebank for Portuguese. We present some underlying guidelines around the project and how they influence our linguistic choices. We then describe the new texts added to the treebank, proceed to mention the new syntactic information added to the old texts, and finally describe the new user-friendly search system and the plans for its expansion.

Keywords: Treebank, corpus, syntax, Portuguese language.

1 Introduction

The Floresta Sintá(c)tica[1] is a publicly available treebank for Portuguese. It was created in 2000 as a collaboration between Linguateca[2] and VISL Project[3], and consists of European and Brazilian Portuguese-language texts automatically annotated by the parser PALAVRAS (Bick, 2000). As the project resumed in 2007, the goal of this paper is to present Floresta's new features, namely, (i) additional texts; (ii) linguistic information; and (iii) search interface. A detailed description of the project, as well as its main motivations, objects, building process and usefulness were described elsewhere (see Afonso et al, 2001 and the Floresta documentation page, at the website).

Floresta has a subset corpus, Bosque, manually revised. Since 2007, Bosque has been undergoing a re-revising process, which guarantees more consistent material, regarding not only annotation aspects, but also the documentation of the underlying linguistic choices. In addition, in this new phase we created Selva, an intermediate corpus between Floresta and Bosque, in both size and degree of revision. Finally, we're developing a new search interface, Milhafre.

Although the usefulness of a treebank like Floresta has already been documented (Afonso et al. 2001), we would like to reinforce here the underlying ideas that guide Floresta's choices: to reflect a consensus among the possible syntactic analysis of a given phenomenon, or, at least, to offer an informed choice. As a result, we expect to be able to (i) offer material to the widest possible range of users; (ii) serve as a research space, and not as a one-theory demonstration space (though of course we are

---

[1] http://www.linguateca.pt/Floresta/
[2] http://www.linguateca.pt/
[3] http://visl.sdu.dk

A. Teixeira et al. (Eds.): PROPOR 2008, LNAI 5190, pp. 216–219, 2008.
© Springer-Verlag Berlin Heidelberg 2008

aware that we can not escape from an underlying theory to the syntactic annotation). So, we have to balance (a) the need for a grammar that is rich and complex enough in order to process real language (our corpora); (b) the absence of a consensual syntactic model; and (c) the linguistic background of the users. In other words, one of our challenges is to make the material useful, regardless of the "quantity and quality" of the users' linguistic background.

The remainder of the paper is organized as follows: in section 2 we describe Selva; in section 3, we describe some of the new linguistic information that is available; section 4 presents Milhafre, a new search system and its interface for queries; finally, section 5 shows our conclusions and directions for future work.

2 Bigger: The "Selva"

We are aware that Bosque is limited, from both the linguistic and the computational-statistical point of view, by its small size. Additionally, both Bosque and Floresta are composed only of newspaper texts from two single sources. Therefore, we decided to build Selva, a corpus that contains around 300.000 words and 30.000 sentences, divided into three roughly equal shares of scientific, literary and transcribed spoken texts, further subdivided in approximately equal shares of Portuguese and Brazilian texts. These texts were mainly selected for their free availability, which means that the literary texts are mainly late 19th century and early 20th century works (around 10.000 words by each of five Portuguese and five Brazilian authors), while the spoken texts are composed of interviews previously included in the AC/DC project (Santos & Bick 2000) and parliamentary transcripts. Scientific texts were mainly taken from Wikipedia articles on scientific subjects and a small set of academic theses. Selva is intended to be a partially reviewed corpus, where some characteristics of the corpus are reviewed one by one, instead of the complete annotation being revised tree by tree as in Bosque.

3 Thicker

One of our tasks was to map the new tags from the parser into the previously reviewed files of Bosque, and then review them manually; Selva had those tags from the start.

First, we reviewed some new function tags. The tags N<ARGS and N<ARGO were introduced to mark arguments of the head noun related to subjects and objects, respectively, when the head noun is a deverbal noun. We used N<ARG to those that are not related to deverbal nouns. Noun modifiers continue to be marked as N<, as in the examples below:

1. nenhuma delas tem *medo* de não encontrar — N<ARG
2. A *poluição* das águas — N<ARGO (= poluir águas)
3. A *participação* de ONGs — N<ARGS (= ONGs participam)
4. A *poluição* de origem humana — N<

Another novelty of Bosque is the "searchable" tags, added to either terminal or non-terminal nodes or both, and introduced to simplify the search for some complex structures, which can now be found looking for a single tag. At clause level, "searchable" tags were implemented marking the presence of elliptic subjects and types of subclauses (relative clauses, comparative clauses, consecutive clauses, etc.). Other topics included complex verbal tenses (marked on the main verb), passives, and partitives. Focusing on non-verbal structures, we revised "searchables" related to relative-clauses, substantive clauses and partitive constructions.

4   Easier: Milhafre

Since its inception, the usefulness of Floresta has been somewhat limited by the absence of an effective search interface/tool. There are several interfaces available, mainly for Bosque, such as CorpusEye (Bick, 2005) and the in-house developed Águia (*eagle*). Besides, several different formats of Bosque can be obtained from the website (Vilela et al., 2005) for use with other tools - including the TigerXML format, for use with TigerSearch (Lezius 2002), or the PennTreebank, which can be used e.g., with TGrep2 (Rohde, 2005). However, we didn't consider these tools ideal, considering the richness of Floresta and its typical user.

As a first stage, we updated Águia to deal with the changes in format. Águia uses the CQP toolkit (Christ el al., 1999); this toolkit is however not appropriate for searches in tree structures, and doesn't handle well the nested structures which are usual in syntactic trees. Therefore, we chose to use Tgrep2, a tool appropriate to that kind of search, and developed an interface, Milhafre (*goshawk*), which allows the user to bypass both Tgrep2's complex syntax and the need to learn the extensive list of tags used in Floresta. This new JavaScript-based interface handles the users' requests and transforms them into a query to be answered by TGrep2.

Currently, the system handles not only searches for words, structures, PoS, and their functions, but for also lemma, morphology, and "searchables" mentioned above. Milhafre may return also aggregate results (like the distribution by function of NPs, or the distribution by lemma of prepositions following NPs). All results are made available in text format as well.

5   Concluding Remarks

In this paper, we presented some of the new features of Floresta Sintá(c)tica – its size, interface and linguistic information. We know that size is a crucial factor in a Treebank, as is a friendly search interface. That is the reason Selva continues to undergo revision, and the Milhafre search tool is still improving. Rather than subscribing to one specific school of syntax, our linguistic options try to suit the widest range of linguistic users, reinforcing our main role as resource providers for research on Portuguese PLN and corpus studies.

Acknowledgement

This work was done in the scope of the Linguateca, contract n°339/1.3/C/NAC, project jointly funded by the Portuguese Government and the European Union.

References

Afonso, S., Bick, E., Haber, R., Santos, D.: Floresta sintá(c)tica: um treebank para o português. Actas do XVII Encontro da Associação Portuguesa de Linguística (APL) (2000)

Bick, E.: The Parsing System Palavras, Automatic Grammatical Analysis of Portuguese in a Constraint Grammar Framework. Aarhus University Press (2000)

Bick, E.: CorpusEye: Et brugervenligt web-interface for grammatisk opmærkede korpora. In: Widell, P., Kunøe, M. (eds.) 10. Møde om Udforskningen af Dansk Sprog, Proceedings. Århus University (2005)

Christ, O., Schulze, B.M., Hofmann, A., Koenig, E.: The IMS Corpus Workbench: Corpus Query Processor (CQP): User'sManual. Institute for Natural Language Processing, University of Stutgart (CQP v2.2) (1999)

Lezius, W.: TIGERSearch - Ein Suchwerkzeug für Baumbanken. In: Busemann, S. (ed.) Proceedings der 6. Konferenz zur Verarbeitung natürlicher Sprache (KONVENS 2002). Saarbrücken (2002)

Rohde, D.: TGrep2 User Manual, version 1.15, May 10 (2005)

Santos, D., Bick, E.: Providing Internet access to Portuguese corpora: the AC/DC project. In: Gavrilidou, M., Carayannis, G., Markantonatou, S., Piperidis, S., Stainhauer, G. (eds.) Proceedings of LREC (2000)

Vilela, R., Simões, A., Bick, E., Almeida, J.J.: Representação em XML da Floresta Sintáctica. In: Ramalho, J.C., Simões, A., Correia Lopes, J. (eds.) XATA 2005 (2005)

The Identification and Description of Frozen Prepositional Phrases through a Corpus-Oriented Study

Milena Garrão[1], Violeta Quental[1], Nuno Caminada[2], and Eckhard Bick[3]

[1]Pontifícia Universidade Católica do Rio de Janeiro
migarrão@terra.com.br, violetaq@puc-rio.br
[2]Instituto Militar de Engenharia
nuno.caminada@gmail.com
[3]Institute of Language and Communication, University of Southern Denmark
eckhard.bick@mail.dk

Abstract. This research is a corpus-based analysis of Brazilian Portuguese prepositional phrases that have a frozen status. Based on a previous list, elaborated for parser PALAVRAS (Bick, 2000), we examine the PPs frozen syntactic behavior and propose three different PPs syntactic-semantic sets, which are true assets for lexicographic purpose and NLP lexical resource.

1 Introduction

By focusing on the lexicon as a core resource in any NLP system, the inspiration for this work is twofold. First, we take a pure lexical path and concentrate on the prepositional phrases (PPs) previously listed by Bick (2000), revising their frozen status through corpora evidences. We start this part of research from a list of PPs, classified as frozen PPs by the parser PALAVRAS (Bick, 2000). The following step is searching for the concordance of these expressions in three tagged corpora - Natura/Público and Nilc São Carlos (http://www.linguateca.pt); Corpus do Português (Davies & Ferreira: http://www.corpusdoportugues.org) -, and in the WEB, using Google.

Then, we describe 3 syntactic-semantic sets for classifying frozen PPs which function as an adverbial phrase. We are also extending the initial limit of PPs to a bigger unit, which includes the verbs with which they occur, the recognition of other parts-of-speech (POS) between the verb and the PP, and relevant semantic annotation. The corpora returned results were discriminated as Brazilian (BP) or European Portuguese (EP). Here, we consider only BP since we are relying on our judgment as Brazilian native speakers to identify a syntactic freeze from an occasional combination.

A PP may function as an adjective or an adverb. For instance, em flor ("in bloom") has an adjectival function when this expression appears adjuncted to nouns related to flora: cerejeiras em flor ("cherry trees in bloom"). On the other hand, a esmo ("at random"), has an adverbial function, modifying verbs like

A. Teixeira et al. (Eds.): PROPOR 2008, LNAI 5190, pp. 220–223, 2008.

atirar ("to shoot") as in *atirar a esmo* ("to shoot at random"). Here, we'll focus on this second type of PP. In other words, we are interested in describing PPs that are accompanying a verb. Therefore, we also want to ascertain whether the structure V+PP functions as a frozen structure as a whole, such as *suar em bicas* ("to sweat in a copious manner") or whether it has an autonomous syntactic status, such as *na moda* ("in fashion").

For that matter, we are considering a consistent list of 948 adverbial PPs collected from PALAVRAS parsing lexicon, which are already tagged as syntactic freezes attached to a verb, such as *nas=últimas VPP <sc> <estar+>*. Our present aim is 1) to check these PPs' frozen status in corpora; 2) to redefine these tagging whenever necessary; 3) to include some new adverbial PPs that could have been left out. In the following section, we give a brief critical review of traditional accounts on syntactic freezes and in section 3 we present what we consider to be a suitable description of distinctive syntactic-semantic patterns of frozen adverbial PPs.

2   Traditional Criteria for the Description of Frozen PPs Structural Patterns

The traditional way of defining syntactic freezes is by combining these criteria: a) non-compositionality: the compound overall meaning does not correspond to the sum of its parts; b) non-substitution or arbitrariness: it is not possible to change a word in a syntactic freeze, even for a synonym; c) non-modification or inflexibility: syntactic freezes cannot be modified by addition of lexical items or by syntactic transformation.

We believe that all these criteria have theoretical implications and counter-evidence. If we take, for instance, non-compositionality, we would rule out combinations such as "estar na moda" and "casar no cartório", because their overall meaning would be thought to correspond to the sum of its parts. If we take the criterion non-substitution or arbitrariness, we would also rule out these very same combinations since they could be respectively changed into "estar dentro da moda" and "casar no civil". And finally, if we take the inflexibility criterion, we would as well rule out compounds such as "estar super na moda" and "casar novamente no civil".

For that reason, we consider feasible to describe patterns in which these frozen PPs behave, but we don't agree that there is a solid theory of collocations. There is always an exception that contradicts the pattern (even prototypical cases such as "bater as botas", could be substituted for "bater a caçoleta"). Therefore, what we did in this research was to group combinations which reveal a similar syntactic-semantic pattern).

3   Sets of Frozen Adverbial PPs

In this section, we identify 3 syntactic-semantic sets of frozen PPs, as follows:

i) Frozen PPs that follow support verbs: The most clear-cut pattern re-garding adverbial PPs is the case in which the PP usually follows a distinctive set of interchangeable support verbs, but also occur with other POS. By sup-port verb, we mean those verbs normally described as having a linking syntac-tic role (e.g.: *ser, estar, ficar, continuar* etc.) or verbs displaying an inchoa-tive semantic aspect in a particular structure (such as "entrar" in "*entrou em vigor*"). Examples of this pattern are: Na=moita (estar, continuar, ficar, NP, Ø); De=vento=em=popa (ir, seguir, continuar, NP, Ø); Em=pânico (entrar, estar, continuar, NP, Ø). PPs displaying this pattern also have a higher level of inde-pendence if compared to other patterns. Therefore, they should not be described as an inseparable part of a V+PP, as it was presented in the prior list.

ii) Frozen PPs that belong to a wider frozen structure: Cases in which the PP occurs with other POS but are frequently headed by one verb, or a limited set of verbs usually semantically related. Again, as well as in i), these PPs should not be tagged as part of a V+PP structure, since they were also detected with other POS. Examples found in corpora are: na=mesma=tecla VPP (bater, insistir, tocar, Ø); em=ovos (pisar, andar, Ø). If we think about frequency, however, we could consider these PPs as "borderline cases", since they were seldom found without a special verb or a limited set of verbs.

iii) Built-in PPs: In this set, the PPs do not occur by themselves, and therefore they could be tagged as belonging to a V+PP structure. In some cases, they could be linked to more than one verb, such as: à=tona VPP <advs> <vir+> <trazer+> <voltar+>; do=sério VPP <advs> <sair+> <tirar+> ; do=riscado VPP <advs> <entender+>. Therefore, we consider these PPs as incorporated to a wider frozen structure. In this set, thus, the PPs should not be tagged as separated from the V+PP compound. We also identified that all sets could display three different semantic markers breaking the V+PP: V (frequency adv.) PP; V (intensity adv.) PP; V (poss) PP.

4   Quantitative Results

From an initial list of 948 combinations of V+PPs, 124 could not be supported by corpora evidence. That means zero occurrences in all corpora, including the Web. Apart from these null cases, other 127 were found only in EP. We also excluded 12 cases of mistagged combinations, (such as sozinho VPP <sc> <estar+>), which were not V+PPs. On the other hand, 92 combinations were added to the list, based on intuition double-checked in corpora. Thus, our final list number of PPs under evaluation is 777.

From the 777 examples under evaluation, 90 (11.6%) could not be grouped in any of the 3 sets outlined, confirming freezeness phenomenon as very unsafe to frame in static criteria.

Table 1. An overall look at Adverbial PPs distinctive sets

Distinctive Sets	Comparative frequency	Some examples
i) free PPs (Support verbs + other POS)	353 examples (45.5%)	de=papo=para=o=ar; de=ressaca; em=boas=mãos; na=pindaíba; no=bem-bom; no=prelo; de=olho; na=berlinda;
ii) free PPs / borderline cases (Special verb(s)+ infrequent other POS)	254 examples (32.6%)	na=telha (dar); no=calo=de (pisar); no=gosto=do=povo (cair); pelas=tabelas (cair); de=brisa (viver)
iii) Not free PPs (Special verb(s))	80 examples (10.3%)	na=veneta VPP <piv> <dar+> <vir+>; pelos=cotovelos VPP <advs> <falar+>; com=os=burros=n'água VPP <sc> <dar+>

5 Concluding Remarks and Future Work

We chose to establish different syntactic-semantic sets for adverbial PPs since we claim that their behavior is slippery to formalization. We identified that set i) should be tagged separately from the verbs; set iii) should be tagged as part of a V+PP structure and set ii), a borderline group, could take a different path depending on the NLP task.

The next step (Caminada, forthcoming), is to analyze V+PP pattern in two corpora: PLN-BR corpus (Aluisio, 2007) and the web, using a true statistics framework and compare the results. Through a semantically blind statistics tool, using mathematical measures to spot real collocations, we may obtain more clearcut answers, mainly for borderline cases (set ii), and for the identification of most prototypical insertion markers. For now, we could say that Adverbial PPs may also act as Adjectival PPs; therefore, they have a rather free status and cannot be tagged as part of a V+PP. We aim to carry out the same analysis with the 270 adjectival PPs provided by Bick (2000).

References

1. Aires, R.V.X., Aluísio, S.M.: Criação de um corpus com 1.000.000 de palavras etiquetado morfossintaticamente. Série de Relatórios do NILC, NILC-TR-01-8 (2001)
2. Bick, E.: The Parsing System Palavras - Automatic Grammatical Analysis of Portuguese in a Constraint Grammar Framework, Århus (2000)
3. Caminada, N.: Sistema para Identificação de Expressões Cristalizadas da Língua Portuguesa. Master dissertation, forthcoming. IME - Instituto Militar de Engenharia (2008)
4. Davies, M., Ferreira, M.J.: Corpus do Português,
   http://www.corpusdoportugues.org
5. Guenthner, F., Blanco, X.: Multi-lexemic expressions: an overview. In Lingüística Investigaciones Supplementa, pp. 201–218. Benjamins, Amsterdam (2004)
6. Jackendoff, R.: The Architecture of lhe Language Faculty. MIT Press, Cambridge (1997)
7. Kilgarriff, A., Rychly, P., Smrz, P., Tugwell, D.: The Sketch Engine. In: Proc. Euralex, Lorient, France, pp. 105–116 (July 2004)
8. Manning, C., Schutze, H.: Foundations of Statistical Natural Language Processing. MIT Press, Cambridge (1999)

CorrefSum: Referencial Cohesion Recovery in Extractive Summaries

Patrícia Nunes Gonçalves[1], Renata Vieira[1], and Lucia Helena Machado Rino[2]

[1] Pontifícia Universidade Católica do Rio Grande do Sul – Porto Alegre – Brasil
[2] Universidade Federal de São Carlos – São Carlos – Brasil

Abstract. A common problem in extractive summaries is the occurrence of referential expressions which are of difficult interpretation. In this paper we propose and evaluate a system for summary post-edition, which aims at replacing referential expressions, trying to avoid problems of broken referential linkage. To propose expressions that best represent the evoked entity, the system uses knowledge about coreference chains.

1 Introduction

Automatic document summarization is a field that has received increasing attention in recent years. The main goals in this area are the selection of the most relevant information in a text and their representation in a new reduced text [7], preserving its quality.

There are two main approaches for automatic summarization. The shallow approach makes use of experimental and statistical methods for selection of the most relevant sentences to compose the summary also called as extractive sumaries. The deep approach is based on formal and linguistic theories.

A common problem in extractive summaries is the occurrence of referential expressions which are of difficult interpretation. In this paper we propose and evaluate a system for summary post-edition, which aims at replacing referential expressions, trying to avoid problems of broken referential linkage. The system is based on a hybrid approach: it employs methods from deep approaches to verify and recover quality of summaries which were first generated by a superficial approach. CorrefSum is thus a system developed to verify the referential cohesion of the extractive summaries using knowledge about coreference chains. This paper presents some experiments on the basis of this tool.

This paper has the following structure: in Section 2 describes the CorrefSum tool. Section 3 presents experiments using CorrefSum. Conclusions are presented in Section 4.

2 The CorrefSum System

The goal of the developed system is to treat problems of referential cohesion found in extractive summaries, using knowledge about coreference chains from the source text to replace noun phrases whose referential interpretation may be

A. Teixeira et al. (Eds.): PROPOR 2008, LNAI 5190, pp. 224–227, 2008.

difficult. The development and evaluation of this tool was based on the Summ-it corpus [4] composed by 50 newspaper texts from Folha de São Paulo, science section, written in Brazilian Portuguese.

The corpus was first processed by the PALAVRAS parser [1]. Then, the corpus has been manually annotated with coreference information using the MMAX tool [8]. Each text was annotated and reviewed by 2 annotators using the same annotation reference manual [3]. Experiments were undertaken with summaries produced by the Portuguese summarizers GistSumm [9] and SuPor-2 [5].

The CorrefSum system has the following modules: file reading, reference score processing and summaries reviewing. The goal of the reading file module is to read and store information about the summaries and the source text. The reference score processing module is designed to search in the source text for the sentences included in the summary, select all coreference chains related to noun phrases which are present in the summary and apply a scoring scheme for the coreference chain elements. The scores are based on the following criteria, the presence of the criteria adds one point to the score of the noun phrase:

Proper Name: if the noun phrase contains any proper name.

Size: if the noun phrase is the longest one in the chain, considering character length.

First: if the noun phrase is the first element in its chain.

Apposition: if the noun phrase contains commas (generally used as an apposition mark).

All chain elements are scored on the basis of the above features. The points are cumulative. They are the selection criteria for replacement of referring expressions in the original summary. The summary recovery module uses the scoring scheme to select the most complete chain element for replacing the original summary term. This module is also responsible for maintaining the compression rate as configured by the user. If the recovered summary exceeds the rate, and there is an apposition in the noun phrase to be replaced, then only the first part of the appositon is selected. Also parenthesis may be disconsidered when the compression rate is exceeded.

3   Experiments and Evaluation

The recovered summaries generated by CorrefSum were evaluated automatically using Rouge [6]. As Rouge makes use of reference summaries for comparison, we used summaries manually generated by professional summarizers [2]. We have adopted Rouge-1, which makes use of unigram for comparison.

Table 1 shows total number of coreference chains in the source texts of the whole corpus and average per text, total number of chains in the resulting summaries, total number of replacements performed by CorrefSum and compression rates before and after replacements.

On average, the source-texts convey 11,72 coreference chains whereas summaries contain 6,60. CorrefSum has performed a total of 89 replacements, with 1,78 on average per text. For SuPor-2 summaries we had 75 replacements (1,5

Table 1. Summaries processed by CorrefSum

	# Source Chains	# Summary Chains	# Replacements	Compress. Rate Original (%)	Compress. Rate Recovered (%)
GistSumm Total	586	330	89	-	-
GistSumm Average	11,72	6,60	1,78	25,30	28,36
SuPor-2 Total	586	338	75	-	-
SuPor-2 Average	11,72	6,76	1,5	23,14	25,52

average per text). The largest number of replacements that occurred in a summary alone was 4, there were also cases in which no replacements were necessary, however, there were replacements in most summaries.

The average compression rate for the original summaries were 25,30%/23,14% and the summaries recovered after the application of CorrefSum system got an average compression rate of 28,36%/25,52%, an increase of about 3% when compared to their original size, due to replacements of less informative referential expressions by more complete ones. Although there is an increase in the compression rate, it is still below the aimed 30%.

Rouge Evaluation of Recovered Summaries
Table 2 shows the Rouge measures for the original summaries generated by GistSumm and SuPor-2 agaist the recovered summaries by CorrefSum.

Table 2. Rouge Evaluation

	ORIGINAL			RECOVERED		
	Recall	Precision	F-measure	Recall	Precision	F-measure
GistSumm	45,59%	54,90%	49,26%	50,85%	54,74%	52,28%
SuPor-2	48,37%	63,07%	54,33%	53,15%	64,08%	57,36%

We noticed that by applying CorrefSum, F-measure which shows a balanced average between precision and recall increased from 49,26% to 52,28% and 54,33% to 57,36%.

The replacements performed with the goal of recovering the referential cohesion indicate, according to this measure, improvements in the informativity of the summaries.

4   Conclusions

Some of the most common problems in extractive summaries is the occurrence of referential expression which are of difficult interpretation. This informational gap, can often cause reading misunderstandings. In this paper we have proposed and evaluated a system for automatic summary post-edition, which aims at rewriting referential expressions in the most coherent possible way, trying to avoid problems of referential linkage. In order to achieve this, the noun phrases in the summaries are analyzed according to their coreference chains, with the goal of identifying expressions which best represent the evoked entity and performing the corresponding substitution when appropriate.

The experiments conducted in this paper considered two summarizers previously evaluated for Portuguese: GistSumm and SuPor-2. The results show an increase in F-measure for both.

For the moment, the linguistic knowledge is given by a corpus of manually annotated coreference chains. We are currently integrating our system with a coreference resolution [10]. Also, in order to improve the performance of the system, we intend to consider the classification of anaphoric expressions to verify the need for substitution, to solve the referential cohesion problems in internal noun phrase, and also to generate alternative referential expressions based on the coreference chains, instead of just replacing them. A further step in this research is to build and evaluate automatic summarizers which take into consideration the coreference chains in the choice of relevant sentences.

Acknowledgments

This work has been partially supported by CAPES and CNPq.

References

1. Bick, E.: The Parsing System "PALAVRAS" - Automatic Grammatical Analysis of Portuguese in a Constraint Grammar Framework. PhD thesis, Department of Linguistics, University of Århus, DK (2000)
2. Barbosa Coelho, J.C.: Uso de informação de correferência e anáfora para verificação da coesão e coerência textual na sumarização automática. Trabalho de Conclusão de Curso de Letras. Unisinos - São Leopoldo (Junho 2007)
3. Barbosa Coelho, J.C., Collovini, S., Vieira, R.: Instruções para anotação de relações anafóricas e referência dêitica. Universidade do Vale do Rio dos Sinos, São Leopoldo, RS, versão 2.6 edn. (November 2006)
4. Collovini, S., Carbonel, T., Fuchs, J.T., Coelho, J.C., Rino, L., Vieira, R.: Summit: Um corpus anotado com informações discursivas visando á sumarização automática. In: Proceedings of the SBC, 5° Workshop em Tecnologia da Informação e da Linguagem Humana (TIL 2007), Rio de Janeiro, RJ (2007)
5. Leite, D., Rino, L.: Supor: extensões e acoplamento a um ambiente para mineração de dados. Technical report, Departamento de Computação, Universidade Federal de São Carlos. São Carlos-SP. NILC-TR-06-07 (2006)
6. Lin, C.-Y.: Rouge: A package for automatic evaluation of summaries. In: Proceedings of ACL 2002 Workshop on Automatic Summarization, Philadelphia, USA (2000)
7. Mani, I.: Automatic Summarization. John Benjamins Publishing Co., Amsterdam (2001)
8. Müller, C., Strube, M.: Mmax: A tool for the annotation of multi-modal corpora. In: Proceedings of the 2nd IJCAI Workshop on Knowledge and Reasoning in Practical Dialogue Systems, Seattle, Washington, pp. 45–50 (2001)
9. Pardo, T.A.S.: Gistsumm - gist summarizer: Extens oes e novas funcionalidades. Technical report, NILC-TR-05-05. São Carlos-SP (2005)
10. Souza, J.G., Gonçalves, P.N., Vieir, R.: Automatic coreference resolution applied to portuguese. In: 8th Workshop on Computational Processing of Written and Spoken Language (PROPOR 2008), Aveiro, Portugal. Springer, Heidelberg (2008)

Answering Portuguese Questions

Luís Fernando Costa and Luís Miguel Cabral

Linguateca, Oslo Node, SINTEF ICT, Norway
{Luis.Costa,Luis.M.Cabral}@sintef.no

Abstract. Esfinge is a general domain Portuguese question answering system that participated in the last four editions of CLEF. This system uses the Web as a fundamental resource in its architecture, using information redundancy rather than sophisticated annotations of the document collections to retrieve answers. In this paper we describe experiments that took as starting point the version of Esfinge that participated at the evaluation contest CLEF 2007. These experiments consisted in using different types of search patterns to retrieve relevant documents for questions, as this issue (document retrieval) was responsible for most of the errors occurred at CLEF 2007.

Keywords: Question answering, Portuguese, question reformulation.

1 Architecture of Esfinge

In this paper we will give a short description of the Portuguese question answering system Esfinge [1], as well as of a set of experiments performed with this system using different types of search patterns to retrieve relevant documents to answer questions.

The architecture of Esfinge is composed by a pipeline of modules that handles each question in order to provide one answer.

The questions are initially fed to an *Anaphor Resolution* module which caters for the resolution of anaphors. This module adds, to the original question, a list of alternative questions where the anaphors are (hopefully) resolved.

Then, Esfinge iterates over the set of alternative questions created in the previous module:

- *The Question Reformulation* module transforms the question into patterns of plausible answers. This is done using two different approaches: a) using a set of predefined pattern pairs that associate patterns of questions with patterns of plausible answers, producing a set of pairs (answer pattern, score) or b) using PALAVRAS [2] analysis to identify the main verb, its arguments and adjuncts and some entities from previous topic questions which are used to create search patterns.
- The *Search Document Collections* module then uses these patterns to search in document collections. If no documents are retrieved, execution stops and NIL is returned meaning that the system is not able to answer the question.
- Otherwise it is possible to proceed by searching the same patterns in the Web using Google's and Yahoo's search APIs (this is optional).

A. Teixeira et al. (Eds.): PROPOR 2008, LNAI 5190, pp. 228–231, 2008.
© Springer-Verlag Berlin Heidelberg 2008

- Then, all text passages retrieved by the previous modules are analysed by the named entity recognition system SIEMES [3] and an n-grams module in order to obtain candidate answers, ranking then according to their frequency, length and the score of the passage from where they were retrieved (these parameters are multiplied in order to define the score of each candidate answer).
- This ranking is in turn adjusted using the BACO database of co-occurrences [4].
- Then, the candidate answers (by ranking order) are analysed to check if they pass a set of filters (these filters are used to exclude answers that are contained in the questions, very frequent words and answers where the constituent words have an unlikely sequence of PoS[1]). Answers are also checked where it regards to the existence of documents in the collections supporting them.
- From the moment that Esfinge finds a possible answer, it will only check candidates that include that answer in order to find more complete answers.

After iterating over all alternative questions, Esfinge has a set of possible answers. That is when the module *Answer Selection* comes to play. This module aims to select the best answer to the given question, which will be the final answer to be returned.

2   Experimental Setup

The error analysis in [1] pointed out several causes for the wrong answers provided by Esfinge. These included among others: wrong or incomplete search patterns, document retrieval failure, missing patterns to identify the type of answer (type of named entities) and problems with the search in Wikipedia.

Our initial work evolved around adding more patterns to identify answers which are named-entities and updating the existing ones based on the results of the aforementioned error analysis. Additionally the Wikipedia collection was re-indexed for not allowing searches on words shorter than 3 characters and for lacking the last sentence in some cases.

In the baseline results in this paper, Esfinge uses therefore an updated answer type identification functionality and a new Wikipedia index. Additionally we did the following experiments:

More complete search patterns. According to the error analysis in [1], wrong or incomplete search patterns were the main cause for wrong answers (63 of the 165 wrong answers). We found that out of this 63, in 41 of them the problem was that the interrogative noun phrase had not been catered for in the created search patterns. This meant that important words were being left out, which frequently led to the retrieval of not relevant text passages. For example for the question *Que país declarou a independência em 1291?, the* word *país* was not included in the search patterns.

We adapted the part of the *Question Reformulation* module that uses PALAVRAS analysis to create search patterns in order to include the complete noun phrases.

Search patterns without verbs. Document retrieval failure was the second more frequent cause for wrong answers (33 out of the 165 wrong answers).

The search patterns created by Esfinge use the words as they appear in the questions, but observing the solutions of the questions in QA@CLEF 2007 one can realize

---

[1] Jspell (http://search.cpan.org/dist/Lingua-Jspell/) was used for this purpose.

that sometimes in the supporting snippets some of the words in the question do not appear (synonyms appear instead of them, verbs appear in different tenses, etc.).

Since Esfinge does not use annotated document collections or dictionaries, we decided to experiment what could be achieved by not including verbs in the search patterns used to retrieve relevant passages. For that purpose we created an option in the *Question Reformulation* module to create search patterns without verbs. These patterns were used when no answer could be retrieved with the complete patterns.

Combining two types of search patterns. As described in section 1, Esfinge uses two different techniques to create search patterns to retrieve relevant passages: a) Using a pattern file that associates patterns of questions with patterns of plausible answers; b) Creating patterns using PALAVRAS analysis to identify the main verb, its arguments and adjuncts.

[1] reports experiments where sets of answers obtained using different information sources were combined/merged. The results of these experiments were worse than some of the original sets of answers. In this paper we decided to test a different answer combination approach, namely what could be achieved combining the two types of search patterns.

3 Evaluation and Discussion of the Results

The questions used to test the system were the 200 questions used at QA@CLEF 2007 for the PT-PT track (questions and answers in Portuguese) [5]. We are aware that it is questionable to use the same set of questions in the error analysis and in a subsequent evaluation, but creating a new set of questions is a very time-consuming task. However our experiments are not tailored to this particular set of questions, instead they try to address general problems detected in the error analysis.

Table 1 presents the results obtained in the experiments described in the previous section. The line "CLEF 2007" refers to the results of the best run described in [1].

Table 1. Results of the experiments (F: Factoid questions; D: Definition questions)

Description	Right Answers				Unsupported Answers	Inexact Answers (missing words)	Inexact Answers (too many words)	Good supporting snippets
	All	NIL	F	D				
CLEF 2007	35	5	28	7	1	6	1	59
Baseline	34	5	29	5	4	7	1	58
More Complete Search Patterns	35	7	31	4	4	7	1	60
Without verbs	41	4	37	4	7	7	1	71
Combination	44	3	39	5	8	6	1	76

Table 2 gives an overview of the main causes for errors in the experiment with the best results.

Table 2. Causes for wrong answers in the best run

Cause	CLEF 2007	Combination
Co-reference resolution	25	23
Wrong or incomplete search patterns	63	15
Document retrieval failure	33	12
Answer scoring algorithm	24	60
Answer support testing	7	27
Other	6	19
Total	165	156

The best results appeared in the run which combined two different types of search patterns which not surprisingly also had a lower number of correct NIL answers. It is also worth to note that the improvements were obtained only in the factoid questions.

Nevertheless, the most significant result of our evaluation was obtained in the error analysis performed for the best run: even though the final results were not strikingly better, Table 2 shows that we managed to move the errors to a later stage in the system execution. Whereas at CLEF 2007, most of the errors were due to wrong and incomplete search patterns and document retrieval failure, in the combination experiment described in this paper most of the errors occurred in the answer scoring algorithm and in testing whether an answer is supported by a text snippet.

Acknowledgments. This work was done in the scope of the Linguateca project, jointly funded by the Portuguese Government and the European Union (FEDER and FSE) under contract ref. POSC/339/1.3/C/NAC. We would also like to thank Diana Santos for the work related to the use of PALAVRAS.

References

1. Cabral, L.M., Costa, L.F., Santos, D.: What happened to Esfinge in 2007? In: Peters, C., Jijkoun, V., Mandl, T., Müller, H., Oard, D.W., Peñas, A. (eds.) CLEF 2007. LNCS, vol. 5152, pp. 261–268. Springer, Heidelberg (2008)
2. Bick, E.: The Parsing System "Palavras": Automatic Grammatical Analysis of Portuguese in a Constraint Grammar Framework. Aarhus University Press, Aarhus (2000)
3. Sarmento, L.: SIEMÊS - a named entity recognizer for Portuguese relying on similarity rules. In: Vieira, R., Quaresma, P., Nunes, M.d.G.V., Mamede, N.J., Oliveira, C., Dias, M.C. (eds.) PROPOR 2006. LNCS (LNAI), vol. 3960, pp. 90–99. Springer, Heidelberg (2006)
4. Sarmento, L.: BACO - A large database of text and co-occurrences. In: Proceedings of LREC 2006, Genoa, Italy, May 22-28, 2006, pp. 1787–1790 (2006)
5. Giampiccolo, D., Forner, P., Peñas, A., Ayache, C., Cristea, D., Jijkoun, V., Osenova, P., Rocha, P., Sacaleanu, B., Sutcliffe, R.: Overview of the CLEF 2007 Multilingual Question Answering Track. In: Peters, C., Jijkoun, V., Mandl, T., Müller, H., Oard, D.W., Peñas, A. (eds.) CLEF 2007. LNCS, vol. 5152, pp. 200–236. Springer, Heidelberg (2008)

XisQuê: An Online QA Service for Portuguese

António Branco, Lino Rodrigues, João Silva, and Sara Silveira

University of Lisbon, Portugal
{antonio.branco,lino.rodrigues,jsilva,sara.silveira}@di.fc.ul.pt

Abstract. This paper describes XisQuê (http://xisque.di.fc.ul.pt)
an online service for real-time, open-domain question answering (QA) on
the Portuguese Web.

Keywords: QA, question answering, real-time QA, open-domain QA,
web-based QA, factoids.

1 Introduction

In this paper we present XisQuê a real-time, on-line service for open-domain
Question Answering (QA) over the Portuguese Web.

Paper structure. Section 2 presents the architecture adopted for the QA system
and in Section 3, the performance of the system is described in terms of its speed
and ability to deliver appropriate answers.

2 The Underlying QA System

XisQuê is supported by a QA system developed to comply with the following
major design features:

Portuguese input: the admissible input are well-formed questions from Portuguese (e.g. *Quem assassinou John Kennedy?*).

Real-time: the system provides the output in real-time.

Web-based: the answers are searched in documents retrieved on the fly from
the Web.

Portuguese Web: the documents are obtained in the Portuguese web, that is
the collection of documents written in Portuguese and available on-line.

Open-domain: the questions may address issues from any subject domain.

Extraction-based: the answers returned are excerts of the retrieved documents
without additional processing.

At the system's heart lies the QA infrastructure described in [1], which is
responsible for handling the basic non-linguistic functionality. Its architecture

A. Teixeira et al. (Eds.): PROPOR 2008, LNAI 5190, pp. 232–235, 2008.

follows what has become a quite standard configuration that has been explored and perfected in similar QA systems for other natural languages [2]:

Question Processing. This phase involves three tasks: (i) detection of the expected semantic type of the admissible answers; (ii) gathering of relevant keywords; (iii) extraction of the main verb and major supporting NP of the input question.

Document Retrieval. In this phase, the system acts as a client of search engines (viz. Ask, Google, MSN Live and Yahoo!), submitting the list of keywords obtained in previous phase and retrieving relevant documents.

Answer Extraction. The last phase includes two tasks performed over the retrieved documents: (i) the sentences most likely containing an admissible answer are selected; (ii) candidate answers are extracted from the selected sentences. XisQuê delivers up to 5 candidate answers (termed "short answers" below) together with the sentences from which they were extracted ("long answers"). It may happen that for some answers only "long answers" are provided. See the example of an outcome in the Annex.

On top of this infrastructure, the natural language driven modules were implemented by using state-of-the-art shallow processing tools developed at our group. They include tools for sentence and token segmentation, POS annotation, morphological analysis, lemmatization and named entity recognition, specifically designed to cope with the Portuguese language [3,4,5].

3   Performance

The online service was evaluated along two simensions: (i) timeliness, or the speed at which answers are returned; and (ii) appropriateness, or the ability of the system to answer appropriately. A total of 60 test question were randomly picked from Trivial Pursuit® cards, by selecting 15 questions for each of the four interrogative pronouns the system handles (viz. Quem, Quando, Onde and Que) This test set is at http://xisque.di.fc.ul.pt/features.html

Table 1. Timeliness and Appropriateness scores obtained March 3-5, 2008

	Question type				Overall
	Quem	Quando	Onde	Que	(average)
Total time (msec.)	18896	20026	22706	25093	21680
"Outside" time	11569	12058	12421	17488	13839
Core QA system time	7327	7968	8465	7605	7841
Answers returned (short)	60.00%	66.67%	46.67%	53.33%	56.67%
Answers returned (all)	100.00%	100.00%	100.00%	100.00%	100.00%
Accuracy (short)	60.00%	66.67%	40.00%	53.33%	55.00%
Accuracy (all)	93.00%	100.00%	100.00%	100.00%	98.33%
MRR (short)	0.5167	0.4778	0.4333	0.5000	0.4819
MRR (all)	0.6489	0.6667	0.7444	0.8889	0.7372

3.1 Timeliness

The service was assessed with respect to time it takes on average to return answers to the input questions. From a development point of view, it is instructive to also determine how much of that time is spent searching for and downloading documents, since those tasks are contingent on third-party search engines that lie outside the QA system proper.

Table 1 shows the average running time in miliseconds. There are some variations when we consider different questions types, but it is mostly caused by fluctuations in the retrieval time (2 696 std. dev.) since the variations for system time are much smaller (492 std. dev.). Overall, the system takes an average of 22 seconds to display the page with the results, with 14 (ca. 64%) of those being spent "outside" the system.

3.2 Appropriateness

Evaluating the appropriateness of a QA system that runs over the Web poses specific problems since the Web is mutable and the results that are obtained for the same set of test questions under different evaluation runs may vary due to external factors, such as website availability of the relevant documents. As a consequence, there is no fixed gold standard against which the output of the system can be automatically compared. Nevertheless, it is possible to obtain an indicative measure of the system's performance through sampling, by manually evaluating the answers to the set of questions.

Table 1 summarizes the scores for a few evaluation metrics: **Answers returned** is the proportion of questions for which the system provided at least a candidate answer — regardless of its rank in the five answer list or even regardless its being a correct answer. Overall, the system provides candidate answers (short- or long-) to 98.33% of the questions in the test set. In turn, it provides short candidate answers to 58.33% of the test set questions. **Accuracy** is the proportion of questions for which a correct answer was provided — regardless its rank in the five returned answer list. In the "all" line, a long-answer is counted in the lot of the correct ones in case it is correct and no short-answer (correct or not) was extracted from it. The system provides a correct short-answer to 45.00% of the test set questions and a correct answer (short- or long-) to 98.33% of that same set. **MRR** stands for mean reciprocal rank: it is a measure commonly adopted in QA evaluation of how highly, on average, the first correct answer is ranked in the answer list [6]. For instance, if all questions have a correct answer and these all appear in position 1, the MRR scores 1; in case they would all appear in position 2, the MRR would score 0.5. The overall value obtained for the QA system is 0.7539 when short- and long-answers are considered, and is 0.4819 when only short-answers are taken into account (a value of 0 was assigned for questions without any short-answer).

4 Conclusion

In this paper we presented the first QA service that complies with all of the following desgin features: it is a real-time, open-domain, freely accessible on-line factoid QA service for the Portuguese Web.

References

1. Rodrigues, L.: Infra-estrutura de um serviço online de resposta a perguntas com base na web portuguesa. Master's thesis, Universidade de Lisboa, Portugal (2007)
2. Paşca, M.: Open-Domain Question Answering from Large Text Collections. CSLI Studies in Computational Linguistics. CSLI Publications (2003)
3. Silva, J.R.: Shallow processing of Portuguese: From sentence chunking to nominal lemmatization. Master's thesis, Universidade de Lisboa, Portugal (2007)
4. Nunes, F.: Verbal lemmatization and featurization of Portuguese with ambiguity resolution in context. Master's thesis, Universidade de Lisboa, Portugal (2007)
5. Ferreira, E., Balsa, J., Branco, A.: Combining rule-based and statistical methods for named entity recognition in Portuguese. In: Actas da 5ª Workshop em Tecnologias da Informação e da Linguagem Humana (2007)
6. Voorhees, E.: The TREC8 question answering track report. In: Proceedings of the 8th Text REtrieval Conference (TREC) (1999)

Annex – System Output Example

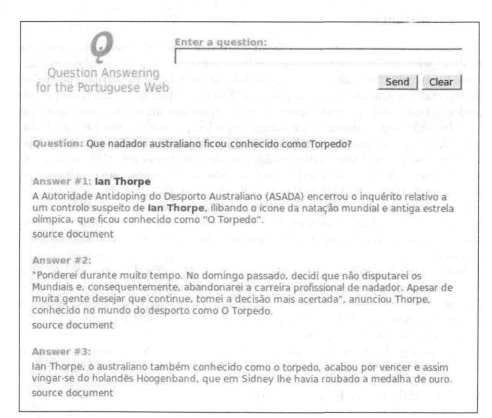

Using Semantic Prototypes for Discourse Status Classification

Sandra Collovini[1], Luiz Carlos Ribeiro Jr.[1], Patricia Nunes Gonçalves[1], Vinicius Muller[2], and Renata Vieira[1]

[1] Pontifícia Universidade do Rio Grande do Sul - Porto Alegre - Brasil
[2] Universidade do Vale do Rio dos Sinos - São Leopoldo - Brasil

Abstract. Discourse status is related to different aspects of entity mention in the discourse, such as whether they are first or subsequently mentioned and on what grounds. This paper presents the evaluation of semantic prototype as input feature for discourse status classification considering Decision Trees as machine learning algorithm. We show that the semantic prototypes improves classification of two specially difficult and scarce classes.

1 Introduction

Anaphora Resolution (AR) is a difficult discourse processing task that needs to be dealt with, for its importance and usefulness to the development of several natural language processing systems, especially those related to textual knowledge interpretation, generation and acquisition. The development of such tools (which perform tasks such as text summarization, question answering, and machine translation) needs, among other things, an effective way to resolve anaphora. Indeed, to find out anaphoric relations for processing information in natural language texts, we need also to distinguish discourse status of referring expressions appropriately: finding out whether they are anaphoric or not, and of which type of anaphora they are. This is due to the fact that the most frequent type of referring expression, definite descriptions - nouns phrases with a definite article (o, a, os and as in Portuguese) are highly ambiguous in regard to their discourse status. Also, different types of referring expressions need different computational treatment.

Related work usually consider a binary classification for discourse status, new or old (or in other words, anaphoric or not). When refining the problem, we can distinguish four classes, two classes of new and two classes for old (as explained in detail in Section 2 below). However the classification of four different classes is a much more challenging task.

The main goal of the current work is thus to verify if such semantic information can improve classification considering four distinct classes of discourse status.

2 Classification Experiments

Based on previous studies ([5], [3]), four classes of discourse status are considered in this work. Below we explain each of them with examples (antecedents are underlined).

A. Teixeira et al. (Eds.): PROPOR 2008, LNAI 5190, pp. 236–239, 2008.

Discourse-new (brand-new): they introduce entities which are new in the discourse, that is, they are not mentioned in the previous text. For example: ***The 430 Km from Assis Chateau Briand road***.

Associative (anchored-new): they introduce new entities in the discourse; however, their interpretation is anchored in an antecedent expression. For example: *the computer* – ***the HD***.

Direct (plain-old): they have an antecedent in the text, the semantic relation with the antecedent is identity and both expressions have the same head-noun: *students* – ***the students***.

Indirect (related-old): they also have an identity relation with their antecedents; however, the expressions have different head-nouns. For example: *the employees* – ***the workers***.

We can see that semantic relation plays an important role in the distinction of these classes; however, no previous work concerning Portuguese has made use of semantic information to classify discourse status. On the availability of semantic prototype information provided by the parser PALAVRAS [1] we can test whether new semantic features can improve classification scores.

The experiments were carried out on 24 newspaper articles from Folha de São Paulo (FSP), written in Brazilian Portuguese. They were automatically annotated with linguistic information using the parser PALAVRAS, and manually annotated for anaphoricity using the MMAX tool [4].

Table 1. Manual Annotation – FSP

Classes	# (%)	Sub-classes	# (%)
New	644 (62%)	*Discourse-new*	550 (53%)
		Associative	94 (9%)
Old	401 (38%)	*Direct*	285 (27%)
		Indirect	116 (11%)
Total			**1045** (100%)

The distribution into the four previously presented classes is rather unbalanced, as shown in Table 1. The tendency for a learning algorithm is to generalize towards the most frequent classes.

The corpus was divided in two parts, we used one part to learn classification models on the basis of 10-fold classification experiments (some of the 700 original examples were replicated in order to balance the data set, resulting in a data set of 1440 examples).

The learned model was then evaluated on the second part, which is unbalanced and consisting of previously unseen data (containing about 350 examples).

The 16 initial features used in [3] were defined in terms boolean values, almost all relating to the syntactic structure of the noun phrase, such as, presence of prepositional phrases; appositions; relative clauses; adjectival phrases; size of the noun phrase, the presence of other determinant besides the definite article, and others. Another important feature considers whether the head of the noun phrase is a word that does not occur previously in the text.

In our work, in addition to these 16 previously considere features, we have used two semantic ones. The new semantic features are based on the semantic information provided by the parser PALAVRAS.

SEM_NOT_DIR is a boolean feature. It is true for a definite description that has an antecedent in the previous text with at least one identical semantic tag, but with a different head noun.

The *SEMANTIC_WINDOW* feature is a numerical value representing the total number of nouns that satisfy the same condition of *SEM_NOT_DIR*, considering a limited number of previous sentences to be examined (8 in our experiments, based on empirical testing).

The learning technique we consider is Decision Trees, we use the J48 algorithm as implemented in Weka [6].

Table 2. Results

Experiments	Classes	P	R	F	C
10-fold cross validation **Baseline**	*Discourse-new*	45%	50%	47%	51%
	Associative	44%	78%	56%	
	Direct	72%	74%	73%	
	Indirect	64%	5%	9%	
new unbalanced data set **Baseline**	*Discourse-new*	77%	48%	59%	53%
	Associative	19%	69%	30%	
	Direct	76%	75%	75%	
	Indirect	0%	0%	0%	
10-fold cross validation **SemInf**	*Discourse-new*	62%	37%	47%	61%
	Associative	58%	80%	67%	
	Direct	73%	77%	75%	
	Indirect	51%	49%	50%	
new unbalanced data set **SemInf**	*Discourse-new*	76%	37%	49%	47%
	Associative	18%	42%	25%	
	Direct	75%	80%	78%	
	Indirect	**12%**	**28%**	**16%**	

We now proceed to present the classification results. First we present results obtained on the basis of 10-fold cross validation, and then results regarding a new set of examples (previously unseen), which is another way to test the robustness of the new features that we are investigating. We show the results in terms of precision (P), recall (R), F-measure (F) and accuracy, the number of correctly classified instances (C).

First, we take as *Baseline* the classification based on the set of 16 features, as previously used in [2]. Results are shown in Table 2. For the new unbalanced validation set no examples of indirect cases were classified as such.

By including the two new semantic feature we see that, F-measure is generally maintained. However, we can see significant improvements for the *Indirect* class, although, there is some loss in the global accuracy for the new validation data set.

We believe that we can achieve better global figures on the basis of more elaborated semantic features. So far we have compared all semantic types independently and directly, but other heuristics may consider groups of semantic prototypes. We also plan to use the semantic features for the more challenging task of learning coreference resolution.

Acknowledgments

This work has been partially supported by CAPES and CNPq.

References

1. Bick, E.: The Parsing System PALAVRAS: Automatic Grammatical Analysis of Portuguese in a Constraint Grammar Framework. PhD thesis, Arhus University, Arhus (2000)
2. Collovini, S., Vieira, R.: Anáforas nominais definidas: balanceamento de corpus e classificação. In: IV Workshop de Tecnologia da Informação e Linguagem Humana TIL, Ribeirão Preto, SP, 2006. Proceeding of the Brazilian Symposium on Artificial Intelligence (2006)
3. Collovini, S., Vieira, R.:Learning discourse new references in portuguese texts. In: IFIP Conference on Artificial Intelligence - IFIP AI 2006. IFIP World Computer Congress (WCC2006), Santiago, Chile (2006)
4. Muller, C., Strube, M.: Mmax: A tool for the annotation of multi-modal corpora. In: Proceedings of the 2nd IJCAI Workshop on Knowledge and Reasoning in Practical Dialogue Systems, Seattle, Washington, pp. 45–50 (2001)
5. Vieira, R., Poesio, M.: An empirically-based system for processing definite descriptions. Computational Linguistics 26(4), 525–579 (2000)
6. Witten, I.H., Frank, E.: Data Mining: Practical Machine Learning Tools and Techniques with Java Implementation. Morgan Kaufmann, San Francisco (2000)

Using System Expectations
to Manage User Interactions

Filipe M. Martins, Ana Mendes, Joana Paulo Pardal,
Nuno J. Mamede, and João P. Neto

Spoken Language Systems Laboratory, L²F – INESC-ID
IST / Technical University of Lisbon
R. Alves Redol, 9 - 2° – 1000-029 Lisboa, Portugal
{fmfm,acbm,joana,njm,jpn}@l2f.inesc-id.pt
http://www.l2f.inesc-id.pt

Abstract. This paper presents a new approach to parse multiple data types in Dialogue Systems. In its initial version, our spoken dialogue systems platform had a single and generic parser. However, when developing two new systems, the parser's complexity increased and data types, like numbers, dates and free text messages, were not correctly interpreted. The solution we present to cope with these problems allows the system to rely on expectations about the flow of the dialogue based on the dialogue history and context. Because these expectations guide the parsing process, a positive impact is achieved in the recognition of objects in the user's utterance. However, if the user fails to match the system's expectations, for instance by changing the focus of the conversation, the system is still capable of understanding the input and recognizing the referred objects.

1 Introduction

DIGA (DIaloG Assistant) is a domain-independent framework for spoken dialogue systems [1] that was the basis of two distinct applications: a butler that controls an home intelligent environment; and an interface to remotely access information databases (like bus timetables). STAR, the Dialogue Manager of DIGA is frame-based: every domain is described by a frame, composed by domain slots that are filled with user requests until a service can be executed [2]. In the first working version of DIGA, the language understanding module of STAR grabbed every domain keywords in users utterances and matched them against the domain roles specified in the domain frame. Slots were filled with tokens collected solely by a generic parser, which is still being used. The unique functionality of the parser was to split the utterance into tokens. From the resulting set of tokens the relevant keywords were selected and used to fill the corresponding domain slots. Tokens not matching any slot were discarded. However, when creating two new telephone-based services (home banking and a personal assistant) we faced several challenges [3].

This paper addresses the challenges that arise at the parser level, to deal with ambiguity in user utterances during spoken interactions. Similar approaches can

A. Teixeira et al. (Eds.): PROPOR 2008, LNAI 5190, pp. 240–243, 2008.

be found on TRIPS architecture [4] where a parsing module with a linguistically motivated grammar is used [5]. Alternative parses are scored with hand-
tuned factors coded into lexical descriptions and grammar rules. The VerbMobil
project [6] uses three different parsers based on different approaches, which are
allowed to run in parallel. The idea is to take the benefits each approach can
deliver while overcoming their related problems. The RavenClaw framework [7]
takes into account the dialogue flow to ease the interpretation of users utterances, by embedding this information into a statistical model. Grammar-rules
are manually generated and domain-specific.

Next, we present our problem and solution (Sect. 2), then the evaluation
(Sect. 3), and finally, conclusions and future directions (Sect. 4).

2 Using Expectations in Parser Selection

The problem with our parsing technique came to our attention when developing
two new telephone-based dialogue systems. In the configuration of the parser for
the home-banking domain, the main problem was to cope with account numbers
as they usually are big and users prefer to spell them instead of reading them.
When creating Lisa, a digital personal assistant, we faced serious difficulties
when trying to write the domain objects' recognition rules.

To answer to these problems, the existing unique parser was replaced by a
module that manages the execution of a set of parsers: the Parsing Manager
(PM). This module allows the definition of parallel and independent sets of
parsers through a divide-and-conquer approach. It is configured with an XML
file that declares the data type of each parser and the sequence of parsers.

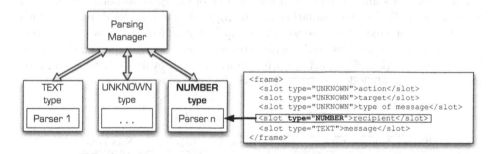

Fig. 1. Association between parsers and slot data types

Three parsers were built: NUMBER, which normalizes successive digits and numbers and corrects predictable recognition errors; DATE, which normalizes temporal expressions; and TEXT, which treats the input as a single chunk. With the
definition of this set of parsers the dialogue manager only needs to select the adequate passer at each turn. In order to help the system with this decision, both
the parser and the frame slot need to state its data types. Having the frame slots
tagged with its data types, the system can inform the PM of the expected data

type for the next utterance. Having the parsers also tagged with their data type the PM knows which parser to use by selecting the matching data type. The association between parsers' and frame slots' type can be seen in Fig. 1. This information is used to select the parser expected to be most accurate. When the system takes the initiative and asks something to the user in order to fill an empty slot in the frame, the Interpretation Manager (IM) and the PM are informed about the expected data type to select the best expected parser.

As an example, let us consider the service that helps to send a short text message. Firstly, the system needs to request the recipient's phone number: After receiving the user's response to the question, the interpretation manager uses the data type of the slot being asked to select the parser to be used. In the example, it is being asked a NUMBER and the adequate parser returns the intended result: '918765131'. This approach allows the system to focus on the expected data type which improves object recognition scores and performance, provided that the user keeps up with system's initiatives. If the user decides not to answer the system's question, the selected parser may fail to interpret the utterance. In this case, the IM requests a generic interpretation to the PM.

Moreover, and since this is a frame-based system, the user can state a set of parameters of the request in the same utterance: *I want to send a short text message to nine eighteen seven six five thirteen one*[1]. When this occurs, it is necessary to execute the parsers sequentially to maximize the object chunking process and the extracted information. The sequential execution of parsers is possible by the definition of parsers composed by a sequence of other parsers.

3   Evaluation and Results

To evaluate our solution, we built a test corpus of interactions between Lisa and a novice user. While the evaluation was being performed, system's expectations about the user next utterance were automatically added to the corpus. Afterwards, the corpus was manually annotated with the correct expected data type for each interaction. Comparing both annotations we evaluated the system for two data types: NUMBER and TEXT. The results[2] are shown on Table 1.

Table 1. Evaluation results

System's Expectation	Interactions	Hits	Non Hits
NUMBER	382	258	124
TEXT	49	44	5

Data type expectations were met 90% for TEXT, and 67.5% for NUMBER. The overall treatment of the user input improved by 28%, meaning that from the

---

[1] Although our system only processes the Portuguese language, we used English in order to allow a broader understanding of this paper.

[2] A "hit" happens when the system's expectations match the user utterance; when the user utters something unexpected by the system, a "non-hit" occurs.

total number of interactions with the user (1085), in 302 interactions the most adequate parser was used, because the system had the correct expectation about what would be the next user utterance.

4   Conclusions and Future Work

The technique of using the system's expectations about the user's next utterance improved the domain objects recognition accuracy. Nevertheless, only 40% of the interactions with the user benefited as only those provided the system with expectations. In the other 60%, the system did not create an expectation, and the generic parser was used. More parsers and grammars for new data types will be needed as new dialogue systems are built with this framework. A future enhancement is the inclusion of morphological, syntactic and even semantic linguistic-based interpretation. A more sophisticated parser is needed to identify the objects in the utterances and to explore the relations and dependencies between them. We plan to include another generic parser for the Portuguese language that we currently use for text analysis. The used grammar will need to be tailored to allow common spoken language phenomena and ungrammaticalities that usually do not occur in written language.

Acknowledgments. This work was funded by PRIME National Project TEC-NOVOZ number 03/165. Joana Paulo Pardal is supported by a PhD fellowship from Fundação para a Ciência e Tecnologia (SFRH/BD/30791/2006).

References

1. Neto, J.P., Mamede, N., Cassaca, R., de Oliveira, L.C.: The development of a multi-purpose spoken dialogue system. In: EUROSPEECH (2003)
2. Mourão, M., Cassaca, R., Mamede, N.: An independent domain dialogue system through a service manager. In: Vicedo, J.L., Martínez-Barco, P., Muñoz, R., Saiz Noeda, M. (eds.) EsTAL 2004. LNCS (LNAI), vol. 3230. Springer, Heidelberg (2004)
3. Martins, F., Mendes, A., Viveiros, M., Paulo Pardal, J., Arez, P., Mamede, N., Neto, J.P.: Reengineering a domain-independent framework for spoken dialogue systems. In: Proc. Software engineering, testing, and quality assurance for natural language processing, Workshop of ACL (to appear, 2008)
4. Allen, J., Ferguson, G., Swift, M., Stent, A., Stoness, S., Galescu, L., Chambers, N., Campana, E., Aist, G.: Two diverse systems built using generic components for spoken dialogue (recent progress on TRIPS). In: ACL Demo Sessions (2005)
5. Swift, M., Allen, J., Gildea, D.: Skeletons in the parser: using a shallow parser to improve deep parsing. In: COLING, ACL (2004)
6. Rupp, C., Spilker, J., Klarner, M., Worm, K.: Verbmobil: Foundations of Speech-to-Speech Translation. In: Verbmobil: Foundations of Speech-to-Speech Translation (2000)
7. Bohus, D., Raux, A., Harris, T., Eskenazi, M., Rudnicky, A.: Olympus: an open-source framework for conversational spoken language interface research. In: Workshop on Bridging the Gap: Academic and Industrial Research in Dialog Technology. HLT-NAACL (2007)

Adaptive Modeling and High Quality Spectral Estimation for Speech Enhancement

Luís Coelho[1] and Daniela Braga[2]

[1] Instituto Politécnico do Porto, Portugal
[2] Microsoft Language Development Center, Portugal

Abstract. In this work an adaptive modeling and spectral estimation scheme based on a dual Discrete Kalman Filtering (DKF) is proposed for speech enhancement. Both speech and noise signals are modeled by an autoregressive structure which provides an underlying time frame dependency and improves time-frequency resolution. The model parameters are arranged to obtain a combined state-space model and are also used to calculate instantaneous power spectral density estimates. The speech enhancement is performed by a dual discrete Kalman filter that simultaneously gives estimates for the models and the signals. This approach is particularly useful as a pre-processing module for parametric based speech recognition systems that rely on spectral time dependent models. The system performance has been evaluated by a set of human listeners and by spectral distances. In both cases the use of this pre-processing module has led to improved results.

1 Introduction

The problem of accurately recovering an underlying signal from a noisy channel was explored by several authors. Traditional proposed solutions are based on spectral-subtraction [1], signal-subspace embedding [2], spectral [3] and time-domain analysis [4]. In this approach one of the objectives is to accurately estimate the power spectral density (PSD) of the signal in order to improve the quality of noise treatment. The proposed methodology is based on an underlying autoregressive (AR) structure.

2 Signal and Noise Modeling and PSD Estimation

The discrete Kalman filter based tracking approach requires a discrete state-space model with the form:

$$\mathbf{x}(k) = \mathbf{F}\mathbf{x}(k-1) + \mathbf{G}\mathbf{w}(k) \tag{1}$$

$$\mathbf{y}(k) = \mathbf{H}\mathbf{x}(k) + \mathbf{v}(k) \tag{2}$$

where $\mathbf{x}(k)$ is the state vector, $\mathbf{y}(k)$ is the output or measurement vector, $\mathbf{F}$ is the (state-space) process matrix that relates previous and present states, $\mathbf{G}$ is the input weight vector, $\mathbf{w}(k)$ is the input vector, $\mathbf{H}$ is the output matrix and $\mathbf{v}(k)$ is a possible output disturbance.

A. Teixeira et al. (Eds.): PROPOR 2008, LNAI 5190, pp. 244–247, 2008.

An AR recursion has been used for signal representation with $s(k)$ as the speech signal under analysis at instant k, $\{a_i\}_{i=1}^{M}$ are the model parameters, $\{s(k-i)\}_{i=1}^{M}$ are delayed samples of the signal and $w(k)$ is assumed to be the noise component at time instant k. This can be written using a compatible state vector $\mathbf{x}(k)$ and a process matrix in controllable canonical form. The input weight matrix $\mathbf{G}$ that interfaces the driving noise $\mathbf{w}(k)$ and the process output matrix is:

$$\mathbf{G}^T = \mathbf{H} = (1\overbrace{0\ldots0}^{M-1}) \tag{3}$$

Coherently with equation 3 the output $y(k)$ and $v(k)$ are values and interface with the measurement error, with the last having variance σ_s^2.

The noise signal $\mathbf{n}(k)$ is also modeled by an AR process with order N with the combined signal-noise state-space model as:

$$\mathbf{x}(k) = \begin{pmatrix} \mathbf{S}(k) \\ \mathbf{N}(k) \end{pmatrix} \tag{4}$$

$$\mathbf{F}(k) = \begin{pmatrix} \mathbf{F}_s(k) & 0 \\ \hline 0 & \mathbf{F}_n(k) \end{pmatrix} \tag{5}$$

The remaining matrices are arranged in a comparable way.

In the described model the AR coefficients in the process matrix must be updated using the previous estimated values that result from the recursion. For improving time-frequency resolution, this work proposes a role inversion between the speech signal part of state-vector and the related part in the transition matrix. The new model comes as:

$$\begin{pmatrix} \mathbf{a}(k) \\ \mathbf{n}(k) \end{pmatrix} = \begin{pmatrix} \mathbf{I} & 0 \\ \hline 0 & \mathbf{F}_n(k) \end{pmatrix} \begin{pmatrix} \mathbf{a}(k) \\ \mathbf{n}(k) \end{pmatrix} + \begin{pmatrix} 0 & 0 \\ \hline 0 & 1\underbrace{0\ldots0}_{N-1} \end{pmatrix} \begin{pmatrix} 0 \\ w(k) \end{pmatrix} \tag{6}$$

$$y(k) = \begin{pmatrix} s(k) \ldots s(k-M-1) & 1\underbrace{0\ldots0}_{N-1} \end{pmatrix} \mathbf{x}(k) + v(k) \tag{7}$$

The speech model (whose coefficients are now included into the state vector) is updated by statistical behavior analysis of the previous samples.

The Kalman filter can recursively estimate the state of a linear stochastic process such that the mean squared error is minimized. With the given model the AR parameters can be estimated in parallel with the state vector by two discrete Kalman filter. The best linear estimate $\hat{\mathbf{x}}(k|k-1)$ at instant k using the knowledge up to instant $k-1$ is calculated as:

$$\hat{\mathbf{x}}(k|k-1) = \mathbf{F}(k-1)\mathbf{x}(k-1|k-1) + \mathbf{G}\mathbf{w}(k-1) \tag{8}$$

and the related prediction error covariance matrix, which is time dependent, is

$$\mathbf{P}(k|k-1) = \mathbf{F}(k-1)\mathbf{P}(k-1|k-1)\mathbf{F}(k-1)^T + \mathbf{G}\mathbf{D}\mathbf{G}^T \tag{9}$$

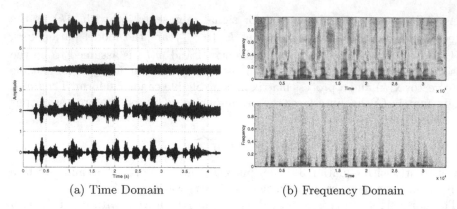

(a) Time Domain   (b) Frequency Domain

Fig. 1. Performance evaluation in the presence of non-stationary noise. (a) Time domain. From top to bottom we have the original signal, the artificially generated noise, the noisy signal (max SNR=2) and the estimated signal. Signals represented with an added bias for a clear picture. The noise signal is white in the beginning ($\mu = 0$) and the variance is increased ($\sigma^2 = 0.1$ to $\sigma^2 = 0.5$). After a small pause the noise signal has an AR(2) structure (freq. peaks at 2-KHz and 3.6-KHz). The original signal is the acoustical representation of "No próximo mês de Fevereiro já se saberão quais as vontades dos nossos irmãos." pronounced by a male speaker. (b) Frequency domain. Above the original signal spectrum and below the recovered speech spectrum. The XX axis is in sample units and the YY axis is in normalized frequency ($f_s = 16$-KHz).

where

$$\mathbf{D} = \begin{pmatrix} \sigma_s^2 & 0 \\ 0 & \sigma_n^2 \end{pmatrix} \tag{10}$$

The Kalman recursion can be performed by:

$$\mathbf{S}(k) = \mathbf{H}(k)\mathbf{P}(k|k-1)\mathbf{H}^T(k) \tag{11}$$

$$\mathbf{K}(k) = \mathbf{P}(k|k-1)\mathbf{H}^T(k)\mathbf{S}(k)^{-1} \tag{12}$$

$$y(k) = x(n) \tag{13}$$

$$\hat{y}(k) = \mathbf{H}(k)\hat{\mathbf{x}}(k|k-1) \tag{14}$$

$$\hat{\mathbf{x}}(k|k) = \hat{\mathbf{x}}(k|k-1) + \mathbf{K}(k)\left[y(k) - \hat{y}(k)\right] \tag{15}$$

$$\mathbf{P}(k|k) = [\mathbf{I} - \mathbf{K}(k)\mathbf{H}(k)]\mathbf{P}(k|k-1) \tag{16}$$

with $\mathbf{S}(k)$ as the state vector prediction and $\mathbf{K}(k)$ as the Kalman gain.

Instantaneous estimates for the PSD can be given by:

$$\hat{P}_x(e^{jw}, k) = \frac{\left|\hat{b}(0, k)\right|^2}{\left|1 + \sum_{i=1}^{M} \hat{a}_i(k)e^{-jwi}\right|^2} \tag{17}$$

For achieving more accurate results it is possible to average the model coefficients across several consecutive time frames.

3   Results

The described algorithm was tested with a set of noise corrupted speech signals. The original speech was in European Portuguese language and the noise recordings were made inside a car, inside a train and on industrial environment. Both signal were then mixed using several signal to noise ratios. An example of one of the experiments is presented in Fig. 1. Using a relative Itakura-Saito based metric the system achieved a 83% similarity with the clean speech. In a perceptive test, 9 volunteers, within the age range 19-23, were asked to classify 20 sentences according to intelligibility using a 1 to 5 points scale (1 for the worst result and 5 for the best result). The recordings included 10 male and 10 female speakers, speaking at their normal speaking rates (each sentence with around 18 words). The set of 20 sentences used for testing was composed by 10 noise corrupted sentences and 10 sentences which were noise filtered. The last obtained an average classification of 4.2 points while the former had only 3.3 points.

4   Conclusion

In this work an adaptive modeling and spectral estimation scheme based on Kalman Filtering is proposed for speech enhancement. It was shown how the speech signal, the noise signal and the speech model can be simultaneously integrated in a single state-space model. A dual Kalman filter algorithm is applied to this model in order to obtain instantaneous high quality PSD estimates. The system's modeling ability can be controled and adapted on the run. The obtained results are very encouraging but some improvements are still foreseen.

References

1. Martin, R.: Spectral subtraction based on minimum statistics. In: Proceedings of the Seventh European Signal Processing Conference (EUSIPCO 1994), Edinburgh, pp. 1182–1185 (1994)
2. Wan, E.A., Merwe, R.: Noise-Regulated Adaptive Filtering for Speech Enhancement. In: Proceedings of Eurospeech 1999, Budapest (1999)
3. Ephrain, Y., Malah, D.: Speech enhancement using a minimum mean-square error log-spectral amplitude estimator. IEEE Trans. on Acoustics, Speech, and Signal Processing 33(2), 443–445 (1985)
4. Zavarehei, E., Vaseghi, S.: Speech Enhancement in temporal DFT trajectories using Kalman Filters. In: Proceedings of Interspeech 2005, Lisboa, pp. 2077–2080 (2005)

On the Voiceless Aspirated Stops in
Brazilian Portuguese

Mariane Antero Alves, Izabel Christine Seara, Fernando Santana Pacheco,
Simone Klein, and Rui Seara[*]

LINSE – Circuits and Signal Processing Laboratory
Department of Electrical Engineering
Federal University of Santa Catarina, Brazil
{mariane,izabels,fernando,klein,seara}@linse.ufsc.br

Abstract. This research work presents a study on voiceless stop variants for
Brazilian Portuguese (BP). The analysis of these variants is based on voice
onset time (VOT) measurements. By considering a semi-spontaneous speech
corpus, the presence of aspiration in BP voiceless stops is verified. Concerning
velar and alveolar stops, the aspirated variant is more frequent than the unaspirated one. Through VOT measurements, the presence of slightly aspirated stops
is pointed out in such an analysis. The distribution of the variants with respect
to the stress position in the syllable is also assessed.

Keywords: Voiceless stops, Brazilian Portuguese, voice onset time, long lag.

1 Introduction

In several acoustic phonetics studies, experiments are based on speech data recorded
in controlled laboratory conditions. However, some results found in such conditions
may not be confirmed in a spontaneous speech context. Spontaneous speech analyses
have verified the presence of some phenomena so far unknown or not sufficiently
studied in the literature, due to their rare presence in controlled speech data.

Brazilian Portuguese (BP) voiceless stop consonants are one of these cases. The
major part of the studies point out to palatalized alveolars in front of high non-back
vowels as the only variants for this class of consonants (i.e., allophones [t^s, $t\int$])
[1]. In BP, aspirated variants are not described as voiceless stop allophones. However,
we have observed in speech signal analyses that aspiration occurs for any voiceless
stop with a high occurrence rate, especially in alveolar and velar contexts.

One of the measures used for the characterization of stop consonants is the voice
onset time (VOT) [2]. Acoustic phonetics literature distinguishes three different categories for VOT: short lag, long lag, and voicing lead. In BP, such consonants are
characterized only by short lag (voiceless) and voicing lead (voiced). The aim of the
present research is to show that aspirated voiceless stops occur in BP, which are determined by VOT values and defined by the long lag area.

This paper is organized as follows. In Section 2, a classification of stop consonants
as well as a brief review of VOT and place of articulation are provided. In Section 3,

---

[*] This work was partially supported by the Brazilian National Council for Scientific and Technological Development (CNPq), Studies and Projects Funding Body (FINEP), and Dígitro
Tecnologia Ltda.

A. Teixeira et al. (Eds.): PROPOR 2008, LNAI 5190, pp. 248–251, 2008.
© Springer-Verlag Berlin Heidelberg 2008

the speech data and the analysis method are presented. The obtained results are discussed in Section 4. Finally, conclusions and directions for future research are presented in Section 5.

2 Stop Consonants

The standard classification states that a stop (occlusive) is formed by a closure at any point of the vocal tract (leading to a period of silence) and then by the fast release of the air stream. Brazilian Portuguese stops can be divided into bilabial, alveolar and velar, according to the place where the air closure occurs. Such phonemes can also be classified as voiced, when vocal folds vibrate, or voiceless otherwise.

Besides vibration, another distinctive feature can be perceived in the production of stops. It is termed aspiration, acoustically perceived as a long delay before the following vowel, in which the air rushes out. Contrast among aspirated voiceless, unaspirated voiceless and voiced stops can be measured through VOT. Thereby, three distinct categories for VOT are established: (i) long lag, when voicing starts at approximately 35 ms after the release of the closure; (ii) short lag, when voicing occurs either simultaneously (VOT = 0) or slightly after the release; (iii) voicing lead, when voicing starts before the release of the occlusion [3].

The open literature concerning BP stops recognizes only two categories, namely: voicing lead (voiced stop) for /b, d, g/ and short lag (voiceless unaspirated stop) for /p, t, k/ [4], [5]. According to [2, p. 120-121], "in Romance Languages like French and Spanish, the voiceless stops have virtually no aspiration, and the contrast is between fully voiced stops and voiceless unaspirated stops". Although [4] concludes that BP voiceless stops are included in the short lag category, such an author also suggests that there might be a slight presence of aspiration in this language. However, the analysis conducted by [4] considered no distinction between aspirated and unaspirated stops. Thus, in [4], the maximum value observed for velar stop VOT (54.90 ms) can be an evidence of aspiration, since data from other languages mark the boundary of long lag at a minimum between 35 ms and 40 ms [3], [6].

3 Analysis Procedure[1]

Data here assessed is from a semi-spontaneous speech corpus, designed for speech recognition training, named BDVOX [7]. Such a corpus was recorded by 35 volunteer subjects. All participants are native speakers of BP, mainly of the South and South-east of Brazil. VOT values are measured manually by an expert, considering the simultaneous observation of waveform and spectrogram by using the software PRAAT[2].

4 Results of VOT Analysis and Discussion

The first aspect to be observed is the presence of aspiration in BP voiceless stops. For velar and alveolar stops, the aspirated variant (allophone) is more frequent than the

---

[1] Supplementary information regarding the analysis can be found at http://www.linse.ufsc.br.
[2] PRAAT: *doing phonetics by computer* (www.praat.org).

Table 1. Percentage distribution of position with respect to stress for each stop

Consonants	Variants	Percentage (%)	Position with respect to stress (%)		
			Pre-stressed	Stressed	Post-stressed
Bilabial	[pʰ]	49	17.15	22.86	8.57
	[p]	51	31.42	20.00	–
Alveolar	[tʰ]	57	11.65	12.62	32.00
	[t]	26	0.97	15.53	9.71
Velar	[kʰ]	18	25.93	46.30	9.26
	[k]	82	9.26	3.70	5.55

unaspirated one. For the bilabial stop, aspiration occurs in approximately 50 % of the samples (see Table 1).

Moreover, we can verify that the values found for the aspirated consonants (25-82 ms) are within the established range for long lag (35–135 ms according to [3] and [6]). This fact evidences the presence of this third area of VOT (long lag) as a variant of the voiceless stop phoneme in BP. However, VOT values seem to indicate that BP aspirated stops are classified in an intermediate region (slightly aspirated stops) [6], while English, for example, clearly presents aspirated stops at the long lag category. The considered data also shows differences which are statistically significant between the values of VOT that characterize the long lag and short lag areas.

Table 2. Mean VOT with respect to stress

Consonants	Variants	Mean (ms)	Mean VOT(ms)		
			Pre-stressed	Stressed	Post-stressed
Bilabial	[pʰ]	37.49	35.90	32.10	49.60
	[p]	14.96	15.50	14.10	–
Alveolar	[tʰ]	40.67	33.33	41.30	42.39
	[t]	18.28	–	19.01	17.18
Velar	[kʰ]	47.24	41.72	45.84	41.18
	[k]	17.18	17.34	16.36	17.44

VOT measurements have shown a distribution which depends on the stop consonants as well as the stress position in the syllable. For alveolar stops, the prevalence is for the aspirated variant in post-stressed position (32.00 %). In the case of the velar stops, the predominance is for the aspirated variant in stressed position (46.30 %). For bilabials, the prevalence is for the unaspirated variants in pre-stressed position (31.42 %) (see Table 1). Alveolars followed by [i] only present as variants: affricate (72 %) and aspirated stops (28 %).

A comparative analysis of our results with those discussed in [4] shows similar VOT values for both unaspirated bilabial and alveolar stops; however, it points out some differences for velar stops. At the same time as we have found a mean VOT of 17.18 ms for velar stops (see Table 2), [4] has obtained 33.90 ms. We believe that the disagreement is due to the incorporation of aspirated samples in the data analyzed by [4].

5 Conclusions

This work examined BP unvoiced stops concerning the presence of aspiration and VOT measurements. Although unaccounted for in the literature, we found a strong occurrence of aspirated stops (characterized by long lag VOT). Therefore, we can assume that these BP aspirated stops are phonetic variants (allophones) of the unvoiced stops. Thereby, the phonetic sequences [tʰ] and [tʃ] are valid variants for [t] in front of [i], while [tʰ] and [t] are the variants before other vowels. For bilabials and velars, we can account the phonetic variants [pʰ], [p], and [kʰ], [k], respectively. The variant distribution is dependent on the syllable stress position. The language dynamics allows that changes occur. However, in this study, we have not searched for the root of such innovations. Traditional theories do not seem to consider some aspects discussed here. Nevertheless, some theories, such as Use Phonology [8] and Exemplar Theory [9], allow incorporating the observed variants as a new process in the language system. Such current theories establish that phonetic details are essential for a correct phonological representation. This novel approach becomes very interesting since the information concerning phonetic variations have relevance for the mental representation. Such representations are realized from the mapping of the speech signal. These theories reveal that the variants with a larger frequency of occurrence are strengthened in detriment of those with a smaller rate. Thus, we can conclude that probably the perception of such consonants, in several contexts, must take into account long lag area (with aspiration). For this verification, we are elaborating perceptual tests by using these results, which will be subject to a future publication.

References

1. Cristófaro-Silva, T.: Discarding Phonemes: A Mental Representation in Use Phonology. In: Hora, D., Collischonn, G. (eds.) Linguistic Theory: Phonology and Other Subjects (in Portuguese), Ed. Universitária, João Pessoa, Brazil, pp. 200–231 (2003)
2. Ladefoged, P.: Vowels and Consonants: An Introduction to the Sounds of Languages. Blackwell Publishers, Massachusetts (2001)
3. Smith, B.L.: Temporal Aspects of English Speech Production: A Developmental Perspective. Journal of Phonetics 6, 37–67 (1978)
4. Klein, S.: Study of VOT in Brazilian Portuguese (in Portuguese). M.Sc. Dissertation. Federal University of Santa Catarina, Florianópolis, Brazil (1999)
5. Istre, G.L.: A Study of VOT on Brazilian Speakers (in Portuguese) (unpublished work)
6. Cho, T., Ladefoged, P.: Variation and Universals in VOT: Evidence from 18 Languages. Journal of Phonetics 27, 207–229 (1999)
7. Seara, I.C., Pacheco, F.S., Seara, J.R., Kafka, S.G., Klein, S., Seara, R.: BDVOX: Data Base for Automatic Speech Recognition of the Speech Multi Speakers (in French). In: 3ème Journées Linguistique de Corpus et Linguistique Apliquée, pp. 197–206. Actes des Troisièmes Journées de la Linguistique de Corpus, Lorient, France (2003)
8. Bybee, J.: Phonology and Language Use. Cambridge Studies in Linguistics, vol. 94. Cambridge University Press, Cambridge (2001)
9. Pierrehumbert, J.: Exemplar Dynamics: Word Frequency, Lenition and Contrast. In: Bybee, J., Hopper, P. (eds.) Frequency and the Emergence of Linguistic Structure, pp. 137–157. John Benjamins, Amsterdam (2001)

Comparison of Phonetic Segmentation Tools for European Portuguese

Luís Figueira and Luís C. Oliveira

L²F Spoken Language Systems Lab.
INESC-ID/IST,
Rua Alves Redol 9, 1000-029 Lisbon, Portugal
{luisf,lco}@l2f.inesc-id.pt
http://www.l2f.inesc-id.pt

Abstract. Currently, the majority of the text-to-speech synthesis systems that provide the most natural output are based on the selection and concatenation of variable size speech units chosen from an inventory of recordings. There are many different approaches to perform automatic speech segmentation. The most used are based on (Hidden Markov Models) HMM [1,2,3] or Artificial Neural Networks (ANN) [4], though Dynamic Time Warping (DTW) [3,4,5] based algorithms are also popular. Techniques involving speaker adaptation of acoustic models are usually more precise, but demand larger amounts of training data, which is not always available.

In this work we compare several phonetic segmentation tools, based in different technologies, and study the transition types where each segmentation tool achieves better results. To evaluate the segmentation tools we chose the criterion of the number of phonetic transitions (phone borders) with an error below 20ms when compared to the manual segmentation. This value is of common use in the literature [6] as a majorant of a phone error. Afterwards, we combine the individual segmentation tools, taking advantage of their differentiate behavior accordingly to the phonetic transition type. This approach improves the results obtained with any standalone tool used by itself. Since the goal of this work is the evaluation of fully automatic tools, we did not use any manual segmentation data to train models. The only manual information used during this study was the phonetic sequence.

The speech data was recorded by a professional male native European Portuguese speaker. The corpus contains 724 utterances, corresponding to 87 minutes of speech (including silences). It was manually segmented at the phonetic level by two expert phoneticians. It has a total of 45282 phones, with the following distribution by phonetic classes: vowels (45%), plosives (19.2%), fricatives (14.6%), liquids (9.9%), nasal consonants (5.7%) and silences (5.5%). The data was split in 5 training/test sets — with a ratio of 4/1 of the available data, without superposition. For this work we selected the following phonetic segmentation tools:

Multiple Acoustic Features–Dynamic Time Warping (MAF–DTW): tool that improves the performance of the traditional DTW alignment algorithm by using a combination of multiple acoustic

A. Teixeira et al. (Eds.): PROPOR 2008, LNAI 5190, pp. 252–255, 2008.

features depending on the phonetic class of the segments being aligned [5]. The implementation of the MAF–DTW used in this experiment uses a synthetic European Portuguese male voice from a different speaker than the recorded in the corpus;

Audimus: is a speech recognition engine that uses a hybrid HMM/Multi-Layer Perceptron (MLP) acoustic model combining posterior phone probabilities generated by several MLP's trained on distinct input features [7,8]. The MLP network weights were re–estimated to adapt the models to the speaker;

Hidden Markov Model Toolkit: (HTK) [9], using unsupervised speaker-adapted, context-independent Hidden Markov Models (HMM). The models were adapted based on initial segmentations generated by the MAF–DTW tool. The models have ergodical left–right topology, with 5 states each (3 emitting states);

eHMM: phonetic alignment tool oriented for speech synthesis tasks [10], developed in Carnegie Mellon University and distributed together with a set tools for building voices for Festival, called Festvox 2.1 [11]. The adopted model topology is the same as described for HTK; eHMM was also used doing acoustic model adaptation to the speaker.

In Table 1 we present the overall performance of each segmentation tool. From this table, it can be seen that the MAF–DTW is the tool with the worst performance in terms of Absolute Mean Error (AME): 41ms. This value is almost twice as much as the second worst result (eHMM). This was already expected, as DTW algorithms are usually very accurate, but simultaneously prone to gross labelling errors, when compared to speaker adapted algorithms [3]. Audimus has the best AME results, and also the smaller standard deviation results, showing that its errors are not widely spread (unlike DTW's). Both HMM based segmentation tools (eHMM and HTK) have a similar behavior.

Each tool's perfomance was evaluated for all the transition types. This study allowed the creation of a new segmentation tool by choosing the best tool for each transition type — using the highest number of borders inside the 20ms tolerance to the manual segmentations as the criterion. Table 2 shows the configuration of this segmenter (S1). Its overall results show that though its AME (16.60ms) is worst than Audimus' or eHMM's, there is an improvement in the number borders placed inside the 20ms error threshold (82.5%). This is due to the fact that the criterion used to choose the best segmenter for each transition is the 20ms error threshold performance, and not the AME. The S1 segmenter's composition shows

Table 1. Absolute Mean Error (AME), Root Mean Square Error (RMSE), Standard Deviation (σ) and borders with error below the 20ms tolerance ($< 20ms$)

	AME(ms)	**RMSE**(ms)	σ(ms)	$< 20ms$(%)
DTW	41.18	117.23	109.76	64.1
eHMM	20.54	33.07	25.92	68.1
HTK	15.44	24.00	48.9	76.9
Audimus	15.23	22.48	16.54	75.9

Table 2. S1 configuration: best combination of segmentation tools

	Nasal	Fricative	Liquid	Plosive	Vowel	Silence
Nasal	HTK	eHMM	eHMM	HTK	HTK	HTK
Fricative	Aud	Aud	eHMM	DTW	HTK	DTW
Liquid	Aud	eHMM	eHMM	Aud	HTK	Aud
Plosive	eHMM	HTK	HTK	HTK	HTK	HTK
Vowel	Aud	eHMM	Aud	Aud	Aud	DTW
Silence	eHMM	eHMM	eHMM	Aud	DTW	—

Table 3. SoM2 configuration: best combination of simple/pairs of segmentation tools

	Nasal	Fricative	Liquid	Plosive	Vowel	Silence
Nas	HTK	eHMM	eHMM	eHMM, HTK	Aud, HTK	eHMM,HTK
Fri	Aud	Aud	eHMM, Aud	DTW, Aud	HTK	DTW
Liq	Aud, HTK	eHMM, Aud	eHMM	Aud, HTK	Aud, HTK	Aud
Plo	eHMM	eHMM, Aud	Aud, HTK	eHMM, Aud	Aud, HTK	HTK
Vow	Aud, HTK	eHMM, Aud	Audi , HTK	Aud, HTK	Aud, HTK	DTW, HTK
Sil	eHMM	eHMM	eHMM	DTW, Aud	DTW, HTK	eHMM

that, as expected, the tools that involve acoustic model training have a better performance, though the DTW based algorithm performed better in some phonetic transitions — namely Fricative–Plosive, Silence–Vowel, Vowel–Silence and Fricative–Silence. The most important conclusion was that no segmentation tool obtained far superior results than the others: every tool had some transitions in which it performed better than any of the others, and transitions in which it performed worse.

Another configuration we studied was which pairs of segmenters obtained better results when its borders were combined linearly—*i.e.* for each transition the border was placed in the the average value of the two segmenters which yielded better results — again the criterion being the number of border inside the 20ms threshold. This new segmenter (M2) obtains better results than any of the individual segmenters, and even better than S1's, with an AME of 13.95ms, and 84.3% of the phonetic transitions with an error below 20ms.

The final configuration studied was the best combination of a single tool or the average of a pair of tools (SoM2). This presented the best results on the number of borders placed correctly: 84.6%. Its AME is 14.3ms, which is only worse when compared to the M2 configuration; Tab. 3 shows the configuration of SoM2.

In the future we plan to expand this work to more databases, to ensure its validity for different speakers of both genders. We also plan to use this method in larger speech inventories, so that we are able to measure its effect on the output speech quality. Another research topic will be using a combination of multiple individual segmentation tools to evaluate the confidence of third–party segmentations of speech databases.

Keywords: Automatic Phonetic Segmentation, Speech Synthesis, Hidden Markov Models, Dynamic Time Warping.

Acknowledgments. The authors would like to thank M. Céu Viana and Helena Moniz for providing the manually aligned reference corpus. The authors would also like to thank Hugo Meinedo and Sérgio Paulo for providing some of the tools used in this study. This work was funded by PRIME National Project TECNOVOZ number 03/165.

References

1. Toledano, D.T., Gómez, L.A., Grande, L.V.: Automatic phonetic segmentation. IEEE Transactions on Speech and Audio Processing 11 (November 2003)
2. Huggins-Daines, D., Rudnicky, A.I.: A Constrained Baum-Welch Algorithm for Improved and Efficient Training. In: Proc. Interspeech 2006s-9th International Conference on Spoken Language Processing, Pittsburgh, USA (2006)
3. Black, A.W., Kominek, J., Bennett, C.: Evaluating and Correcting Phoneme Segmentation for Unit Selection Synthesis. In: Proc. Eurospeech, Geneva, Switzerland, pp. 313–316 (2003)
4. Malfrre, F., Deroo, O., Dutoit, T.: Phonetic alignment: speech synthesis based vs. hybrid HMM/ANN. In: Proc. 5th International Conference on Spoken Language Processing (1998)
5. Paulo, S., Oliveira, L.C.: DTW-based Phonetic Alignment Using Multiple Acoustic Features. In: Proc. Eurospeech, Geneva, Switzerland, pp. 309–312 (2003)
6. Adell, J., Bonafonte, A.: Toward Phone Segmentation for Concatenative Speech Synthesis. In: Proc. 5th ISCA Workshop on Speech Synthesis (2004)
7. Neto, J.P., Martins, C., Meinedo, H., Almeida, L.B.: AUDIMUS — Sistema de Reconhecimento de Fala Contínua para o Português Europeu. In: PROPOR 1999 - IV Encontro para o Processamento Computacional da Língua Portuguesa Escrita e Falada, Évora (1999)
8. Meinedo, H., Caseiro, D., Neto, J.P., Trancoso, I.: AUDIMUS.Media: A Broadcast News Speech Recognition System for the European Portuguese Language. In: Mamede, N.J., Baptista, J., Trancoso, I., Nunes, das Graças Volpe Nunes, M. (eds.) PROPOR 2003. LNCS, vol. 2721, pp. 9–17. Springer, Heidelberg (2003)
9. Young, S., Ollason, D., Valtchev, V., Woodland, P.: The HTK Book (for HTK Version 3.2). Cambridge University Engineering Department (2002)
10. Prahallad, K., Black, A.W., Ravishankar, M.: Sub-phonetic Modeling for Capturing Pronunciation Variations for Conversational Speech Synthesis. In: Proc. ICASSP (2006)
11. Black, A.W., Lenzo, K.A.: Building Synthetic Voices, For FestVox, 2.1 edn. Language Technologies Institute, Carnegie Mellon University and Cepstral, LLC (2006), http://www.festvox.org

Spoltech and OGI-22 Baseline Systems for Speech Recognition in Brazilian Portuguese

Nelson Neto[1], Patrick Silva[1], Aldebaro Klautau[1], and Andre Adami[2]

[1] Universidade Federal do Pará, Signal Processing Laboratory,
Rua Augusto Correa. 1, 660750110 Belém, PA, Brazil and
[2] Universidade de Caxias do Sul,
Rua Francisco Getúlio Vargas. 1180, 95070-560 Caxias do Sul, RS, Brazil
{neto,krlospatrick,aldebaro}@ufpa.br, agadami@ucs.br
http://www.laps.ufpa.br
http://www.ucs.br

Abstract. Speech processing is a data-driven technology that relies on public corpora and associated resources. In contrast to languages such as English, there are few resources for Brazilian Portuguese (BP). This work describes efforts toward decreasing such gap and presents systems for speech recognition in BP using two public corpora: Spoltech and OGI-22. The following resources are made available: HTK scripts, pronunciation dictionary, language and acoustic models. The work discusses the *baseline* results obtained with these resources.

Keywords: Speech recognition,Brazilian Portuguese, HMMs, pronunciation dictionary.

1 Introduction

This work discusses current efforts within the *FalaBrasil* initiative [1]. The overall goal is to develop and deploy automatic speech recognition (ASR) resources and software for BP, aiming to establish baseline systems and allow for reproducing results across different sites. More specifically, the work presents resources and results for two baseline systems using the Spoltech and OGI-22 corpora. All corrected transcriptions and resources can be found in [1].

2 UFPAdic: A Pronunciation Dictionary for BP

In [2], a hand-labeled pronunciation dictionary *UFPAdic version 1* with 11,827 words in BP was released within the *FalaBrasil* initiative. The phonetic transcriptions adopted the SAMPA alphabet and were validated by comparing results with other publicly available pronunciation dictionaries for other languages. All the UFPAdic 1 was used for training a decision tree and adopting the procedure described in [2], a new dictionary was built by selecting the most frequent words in the CETENFolha corpus [3]. The new dictionary, called UFPAdic 2, has approximately 60 thousand words.

A. Teixeira et al. (Eds.): PROPOR 2008, LNAI 5190, pp. 256–259, 2008.

3   Building Language Models from CETENFolha

Several bigram language models were trained and tested using the HTK tools [4]. The models were trained using 32,100 sentences selected from the CETENFolha and OGI-22 corpora. Vocabularies with different sizes were created by choosing the most frequent words in the training set, which were also present in UFPAdic 2. The bigram language models perplexities were computed using 1,000 randomly selected sentences and are shown in Table 1.

Table 1. LM perplexities for different vocabulary sizes

Vocabulary size (thousand words)	1.5	3	6	10	15	20	30
Bigram perplexity		47	76	113	136	149	156	165

4   Front-End and Acoustic Modeling

The initial acoustic models for the 33 phones (32 monophones and a silence model) used 3-state left-to-right HMMs. After that, triphone models were built from the monophone models and a decision tree was designed for tying triphones with similar characteristics [4]. After each step, the models were reestimated using the Baum-Welch algorithm via HTK tools.

5   OGI-22 Corpus

The 22 Language Telephone Speech Corpus [5], which includes Brazilian Portuguese, is a spontaneous speech and telephone recordings corpus. In this work the original orthographic transcriptions were corrected, and the nonexistent created. For the experiments, the training set was composed of 2,017 files, corresponding to 184.5 minutes, and the test set had 209 files with 14 minutes.

6   Spoltech Corpus

The utterances from Spoltech corpus [6] consist of both read speech and responses to questions from a variety of regions in Brazil. The acoustic environment was not controlled, in order to allow for background conditions that would occur in application environments. In the experiments, the phonetic alphabet used was the same as the one used in the OGI-22 corpus and a pre-processing stage removed files that have poor recording quality. The training set was composed by 5,246 files that corresponding to 180 minutes and the test set used the remaining 2,000 files corresponding to 40 minutes.

7   Baseline Results

The Spoltech and OGI-22 baseline systems share the same front-end. In addition, the HMM-based acoustic models of both systems were estimated using the same procedure described in Section 4.

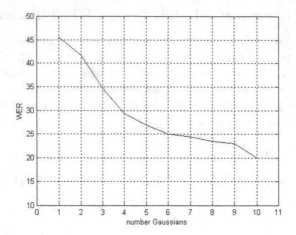

Fig. 1. Decrease in WER (%) with the number of Gaussians in each mixture for OGI-22 using a simplified bigram LM with perplexity 43

7.1 Results for Bigram LM Obtained from the Corpora Transcriptions

The first experiment used a OGI-22 bigram LM with perplexity equal to 43. The number of component mixture distributions was gradually increased from one to ten. The word error rate (WER) reduction can be observed in Fig. 1. Similarly, a bigram LM with 793 words and perplexity 7 was designed using only the Spoltech corpus. The respective WER results are shown in Fig. 2, where the number of Gaussians per mixture was also varied from 1 to 10. The WER with 10-component Gaussian mixtures is 18.6% and 19.92% for Spoltech

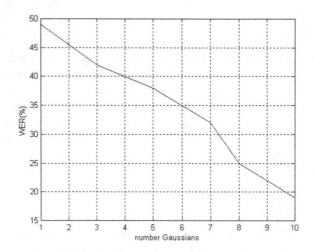

Fig. 2. WER (%) for Spoltech using a simplified bigram LM with perplexity 7

and OGI-22, respectively. The experiments finished with 10-component Gaussian mixtures, because the WER stopped to decline.

7.2 Results with Language Models Including Text from CETENFolha

Using the bigram language models mentioned in Section 3, simulations were performed setting the acoustic model created with the OGI-22 corpus and the number of Gaussians per mixture equal to ten. The WER for the system with 30,000 words is 35.87%. It can be noticed that increasing the complexity of the LM does not improve the results given that there is a mismatch between the CETENFolha text and the OGI-22 sentences.

8 Conclusions

This paper presented some baseline results for ASR in BP. The resources were made publicly available and allow for reproducing results across different sites. Future work should concentrate efforts in collecting a larger corpus with broadcast news.

Acknowledgements

This work was partially supported by CNPq, Brazil, project 478022/2006-9 *Reconhecimento de Voz com Suporte a Grandes Vocabulários para o Português Brasileiro: Desenvolvimento de Recursos e Sistemas de Referência.*

References

1. http://www.laps.ufpa.br/falabrasil (Visited in April, 2008)
2. Hosn, C., Baptista, L.A.N., Imbiriba, T., Klautau, A.: New resources for brazilian portuguese: Results for grapheme-to-phoneme and phone classification. In: VI International Telecommunications Symposium, Fortaleza (2006)
3. http://acdc.linguateca.pt/cetenfolha/ (Visited in January, 2008)
4. Young, S., Ollason, D., Valtchev, V., Woodland, P.: The HTK Book (for HTK Version 3.4). Cambridge University Engineering Department (2006)
5. Lander, T., Cole, R., Oshika, B., Noel, M.: The ogi 22 language telephone speech corpus. In: Proc. Eurospeech 1995, Madrid (1995)
6. Advancing human language technology in Brazil and the United states through collaborative research on portuguese spoken language systems (2001)

Development of a Speech Recognizer
with the Tecnovoz Database

José Lopes[1], Cláudio Neves[1], Arlindo Veiga[1], Alexandre Maciel[1],
Carla Lopes[1], Fernando Perdigão[1,2], and Luís Sá[1,2]

[1] Instituto de Telecomunicações – Pólo de Coimbra, 3030-290 Coimbra, Portugal
[2] Dep. Eng.ª Electrotécnica e de Computadores, FCTUC, 3030-290 Coimbra, Portugal
{zedavid,claudiorneves,aveiga,amam,fp,luis}@co.it.pt

Abstract. This paper describes the development of a robust speech recognition using a database collected in the scope of the Tecnovoz project. The speech recognition system is speaker independent, robust to noise and operates in a small footprint embedded hardware platform. Some issues about the database, the training of the acoustic models, the noise suppression front-end and the recognizer's confidence measure are addressed in the paper. Although the database was especially designed for specific small-vocabulary tasks, the best system performance was obtained using triphone models rather than whole-word models.

Keywords: Speech recognition, acoustic models.

1 Introduction

Tecnovoz is a cooperation project funded by the Portuguese government [1] aiming to create a body of knowledge on voice technologies and to materialize this knowledge in a series of products for the market. The authors were responsible, in the framework of the project, for the development of a speech independent connected word recognizer. As the recognizer should operate under noise adverse environments, such as factories and vehicles, it has to incorporate advanced noise reduction techniques. In addition, it should run in an embedded hardware platform.

The acoustic models are based on Hidden Markov Models (HMM). HMM have proved to be an effective basis for modelling time-varying sequences of speech spectra. However, in order to accurately capture de variation in real speech spectra (both inter-speaker and intra-speaker), it is necessary to have a large amount of speech data and to use relative complex output probability distributions [2]. Three approaches were experimented for the acoustic model units: whole-word, context-free phones and triphone models.

The paper is organized as follows. Section 2 is dedicated to a description of the database collected in the Tecnovoz project. Section 3 focuses on the noise suppression front-end, section 4 describes the training experiments, section 5 the decoder implementation and section 6 the obtained recognition results. Finally, in section 7 a discussion of the results is done and conclusions are drawn.

A. Teixeira et al. (Eds.): PROPOR 2008, LNAI 5190, pp. 260–263, 2008.

2 Speech Database

The Tecnovoz speech database (DB) was collected and annotated by a project partner. The collected speech includes about 250 commands and several phonetically rich sentences. About 30 minutes of spoken content was recorded from each speaker. There were a total of 368 speakers. Three acoustical environments were considered: Clean (TVFL), Vehicle (TVV) and Factory environment (TVF).

3 Feature Extraction

An earlier decision, concerning noise reduction, was to use a front-end which performs noise suppression or speech enhancement. Recently, ETSI standardized an Advanced Front-End (AFE) algorithm [3] based on a two-stage Mel-warped Wiener filtering system [4], for systems performing Distributed Speech Recognition. We developed an algorithm similar to the one proposed by Jin-Yu Li et al, [5]. The main differences are the following. Firstly, the waveform processing module is not performed in our system. Secondly, we found even a more efficient way to compute the smoothed Wiener filter coefficients, using a single pre-computed matrix. Thirdly, we have found that the gain factorization algorithm on the second stage is not valuable in this simplified model and so we decided not to use it. Finally, the blind equalization module was replaced by a Cepstral Mean Normalization (CMN) algorithm [6]. The overall system works in real-time with a voice activity detector different from the one recommended in the standard.

All speech files were parameterized using this front-end system which produces 12 MFCC coefficients, log energy, and their first and second derivatives, leading to a feature vector with 39 components.

4 Model Training

For training proposes only files with SNR above 15 dB were used. The command database has a total of 137,237 files (119,975 from TVFL, 8,633 from TVF and 8,629 from TVV). From these files, 75% were picked up for training, 20% for testing and 5% for development.

The training is done in several steps by applying the Baum-Welch embedded re-estimation using the HTK toolkit [7], `HERest`.

As it was referred to in section 1, three different approaches were used to find the acoustic models which best fit to the task of command recognition. Tests were done using word models, context-free phone models, and context-dependent triphone models.

In the case of whole-word models, each word is represented by an HMM with left-to-right topology. The number of states of each HMM depends upon the number of phones of the word. Three states per phone were used in the word models.

The phone set has 42 monophones, including 3 pause/sil models. All models have three states. Up to 16 Gaussians per state were employed for training the monophone HMMs.

Triphone modelling allows the parameters of a phone model to depend on the two adjacent phones and so gives considerable robustness to variations in pronunciations. The initial triphone model set was obtained from monophone models. For our vocabulary of 254 words 870 triphones are required. To perform a more efficient training, parameter tying was used, reducing the number of physical models to 846.

5 The Decoder

The decoder is based on the Viterbi algorithm applied to a grammar task. It uses the "token passing" paradigm [7]. Several optimizations were included using the floating-point extensions – Intel SSE and AMD 3DNow!.

One main characteristic of the recognizer is the inclusion of a module that measures the confidence of the recognition results. Confidence measures can be used for spotting and rejecting possible errors as well as to detect out-of-vocabulary words. To detect out-of-vocabulary (OOV) words, we used a so called "filler model" [8]. In order to calculate the confidence of a recognizer result, a "super model" was used, which is formed by taking all phone models in parallel. The aim of this model is to give a score for a sequence of phones, no matter their order or number. For well pronounced words, both the "super model" and the result's model should give almost identical scores. The scores will be very different in the case of misrecognized words. If the recognizer result has a score below the "filler model" by a given threshold, it is considered an OOV and it is rejected. All the vocabulary words have their own threshold that has been calculated using the test database. If the result is not an OOV word, then a confidence measure is computed using the "super model". The difference between result's model score and the "super model" score is normalized by number of frames. This value is then applied to a sigmoid function in order to obtain a normalized confidence measure between 0 and 100%.

6 Results

Several recognition tests were carried out using a task grammar that consists in taking all the 254 commands in parallel.

With the whole-word model set we obtained a recognition rate of 96.76% with 8 mixtures. This model set has 46.7k Gaussians (about 3.6M parameters).

For the phone model set we used a multiple pronunciation dictionary, but despite of this, we obtained a recognition rate of only 91.41% with 16 mixtures. This low performance value is obviously due to the lack of parameters: only 1888 Gaussians (150k parameters).

Table 1. Whole-word, monophone and triphone recognition rates for 8 mixtures

Word Correction (%)	Number of Gaussians	Acoustic Model
96.76	37,344	Whole-word
89.28	952	Monophone
97.03	16,104	Triphone

For the triphone model set, no multiple pronunciation dictionary was used and the result for 8 mixtures was 97.03% with 16,104 Gaussians. The better result achieved 97.5%, and was obtained with 16 mixtures (32,208 Gaussians, about 2.5M parameters). The recognition rates are shown in Table 1 for all model types with 8 mixtures.

7 Discussion and Conclusions

According to the results presented in the last section of this paper, the best score was achieved when using context dependent triphones models. More occurrences of context-dependent triphones lead to better model's parameters estimations.

When comparing the number of Gaussians in every test, the whole-word model set is by far the one with the biggest number of parameters. When performing recognition in real time this implies more memory. The use of context-dependent triphones seems the most likely solution to be adopted in this case, as it combines fewer parameters with higher recognition rate.

The training experiments prove that the database is big enough to estimate such a big number of parameters. However, as the number of speakers in this database is quite low, model adaptation for specific final users will be an important system improvement.

In terms of ongoing work, we are trying to improve the system performance using discriminative training and feature variance normalization.

References

1. Tecnovoz website (2007), http://www.tecnovoz.pt/web/home_english.asp
2. Young, S., Odell, J., Woodland, P.: Tree-Based State Tying for High Accuracy Acoustic Modelling. In: ARPA Workshop on Human Language Technology, pp. 307–312 (1994)
3. ETSI ES 202 050 v1.1.3: Speech Processing, Transmission and Quality Aspects (STQ); Distributed Speech Recognition; Advanced Front-end Feature Extraction Algorithm; Compression Algorithms. ETSI standard (2002)
4. Agarwal, A., Cheng, Y.: Two-stage Mel-warped Wiener Filter for Robust Speech Recognition. In: IEEE ASRU, Keystone, Colorado, USA, pp. 67–70 (1999)
5. Li, J.-Y., Liu, B., et al.: A Complexity Reduction of ETSI Advanced Front-end for DSR. In: IEEE ICASSP, Montreal, Canada, vol. I, pp. 61–64 (2004)
6. Peinado, A., Segura, J.: Speech Recognition over Digital Channels: Robustness and Standards. John Wiley & Sons, Ltd., England (2006)
7. Young, S., Evermann, G., et al.: The HTK Book (For Version 3.4). University of Cambridge, England (2006)
8. Yu, D., Ju, Y., Wang, Y.-Y., Acero, A.: N-Gram Based Filler Model for Robust Grammar Authoring. In: IEEE ICASSP, Toulouse, France (2006)

Dynamic Language Modeling for the European Portuguese

Ciro Martins[1,2], António Teixeira[1], and João Neto[2]

[1] Department Electronics, Telecommunications & Informatics/IEETA – Aveiro University
[2] L2F – Spoken Language Systems Lab – INESC-ID/IST, Lisbon
Ciro.Martins@l2f.inesc-id.pt,
ajst@det.ua.pt, Joao.Neto@inesc-id.pt

Abstract. Up-to-date language modeling is recognized to be a critical aspect of maintaining the level of performance for a speech recognizer over time for most applications. In particular for applications such as transcription of broadcast news and conversations where the occurrence of new words is very frequent, especially for highly inflected languages like the European Portuguese. An unsupervised adaptation approach, which dynamically adapts the active vocabulary and language model during a multi-pass speech recognition process, is presented. Experimental results confirmed the adequacy of the proposed approaches. Experiments were carried out for a European Portuguese Broadcast News transcription system with the best preliminary results yielding a relative reduction of 65.2% in OOV word rate and 6.6% in WER.

1 Introduction

Up-to-date language modeling is recognized to be a critical aspect of maintaining the level of performance for a speech recognizer over time for most applications. In particular for applications such as transcription of broadcast news (BN) and conversations where the occurrence of new words is very frequent, especially for highly inflected languages. This is the case of the European Portuguese language, where new names contain great deal of information and occur frequently in many domains as the BN one. Additionally, due to their inflectional structure, the verbs class represents another problem to overcome [1]. For a BN transcription system like the one used in this work, the ability to correctly address new words appearing in a daily basis, is an important factor to take in consideration for its performance.

In this paper, we present and compare two daily and unsupervised adaptation frameworks, which dynamically adapt the active system vocabulary and LM. Based on texts daily available on the Web, we defined two morpho-syntatic approaches to dynamically select the target vocabulary by trading off between the OOV word rate and vocabulary size [1][2]. Using an IR engine [3] and the ASR hypotheses as query material, relevant documents are extracted from a dynamic and large-size dataset to generate a story-based LM to the multi-pass speech recognition framework.

In section 2 we provide a brief description of the proposed vocabulary selection algorithms, LM adaptation procedures, and their integration into a multi-pass speech recognition framework. Section 3 describes some evaluation results.

A. Teixeira et al. (Eds.): PROPOR 2008, LNAI 5190, pp. 264–267, 2008.

2 Vocabulary Selection and Language Model Adaptation

Even though the use of very large vocabularies in recognition systems can reduce the OOV word rates, in highly inflected languages or those with a high rate of word compounding, those rates still tend to be high. In addition, just generically increasing the vocabulary size can improve the accuracy for many common words but degrades the recognition rate for less common words. Thus, defining a more rational approach to select/adapt the system vocabulary other than by simple word frequency is need.

In [1] we derived a procedure for dealing with the OOV problem by dynamically increasing the baseline system vocabulary. From the experiments derived, we observed that verbs make up for the largest portion of OOV words types, accounting for 56.2% of the OOV word types in a BN test dataset. Our approach to compensate and reduce the OOV word rate related with verbs was supported by the fact that almost all the OOV verb tokens were inflections of verbs whose lemmas were already among the lemmas set (L) of the words found in contemporary written news. Thus, the baseline vocabulary is automatically extended with all the words observed in the language model training texts and whose lemmas belong to L. Applying this adaptation approach, the baseline system vocabulary of 57K was expanded by an average of 43K new words each day. To apply this selection process, both training and adaptation word lists were morpho-syntactically classified and lemmatized using a morphological analysis tool developed for the European Portuguese [4].

In [2] we proposed another approach. It takes in consideration the differences in style across the various training corpora, especially in case of written versus spoken style. Using the same morphological analysis tool as before, we annotated both in-domain corpus and out-of-domain corpus, observing a significant difference in part-of-speech (POS) tags distribution, especially in terms of names and verbs. Hence, instead of simply adding new words to the fixed baseline system vocabulary, as the previously proposed approach, we use now the statistical information related to the distribution of POS word classes on the in-domain corpus to dynamically select words from the various training corpora available.

For LM adaptation we proposed and implemented a multi-pass speech recognition approach which creates from scratch both vocabulary and LM components in a daily basis [5]. The first-pass is being used to produce online captions for a closed-captioning system of live TV broadcasts. Based on texts daily available on the Web and static training corpora, a new vocabulary V_0 is selected for each day d using the POS-based technique described in section 2. To construct a more homogeneous adaptation dataset, we merge Web data from the current day and the 6 preceding days ($O_7(d)$). Finally, with V_0, three LMs are estimated and linearly combined. The mixture coefficients are estimated using the Expectation-Maximization (EM) algorithm to maximize the likelihood of T_{21} dataset. This T_{21} held-out dataset consists of ASR transcriptions generated by the BN transcription system itself for the 21 preceding days. A confidence measure is used to select only the most accurately recognized transcription segments.

In this multi-pass adaptation framework, a second-pass is being used to produce offline transcripts for each day using the initial set of ASR hypotheses generated during the live version and automatically segmented into individual stories, with each story ideally concerning a single topic. Using an Information Retrieval engine [3] and the text of each story segment as query material, relevant documents are extracted from a dynamic and large-size database to generate a story-based vocabulary and LM. Since those text story segments can be quite small and may contain recognition errors, a relevance feedback method for automatic query expansion was used [6]. Thus, for each story S a topic-related dataset D_S is extracted from the IR dynamic database and all words found in D_S are added to the vocabulary V_0 selected on the first-pass, generating this way a story-specific vocabulary V_S. Note that for each word added, the vocabulary size is kept constant by removing the word with the lowest frequency. With V_S, an adaptation LM trained on D_S is estimated and linearly combined with the first-pass LM to generate a story-specific LM (MIX_S). Using V_S and MIX_S in a second decoding pass the final set of ASR hypotheses is generated for each story S.

3   Evaluation Results

All experiments reported in this work were done with the AUDIMUS.media ASR system [7]. This system is part of a closed-captioning system of live TV broadcasts in European Portuguese that is daily producing online captions for the main news show of one Portuguese Broadcaster - RTP.

To evaluate the proposed framework we selected a BN dataset (RTP-07) consisting of BN shows collected from the 8 o'clock pm (prime time) news from the main public Portuguese channel, RTP. The RTP-07 BN shows were collected on May 24th and 31st of 2007, having a total duration of about 2 hours of speech and 16.1K words.

Table 1. Comparison of OOV word rates for the **RTP-07** dataset

Approach	%OOV	%reduction
BASELINE	1.40	-
1-PASS-POS	0.74	47.0
2-PASS-POS-IR	0.49	65.2

As one can observe from table 1, the proposed second-pass speech recognition approach (2-PASS-POS-IR) using the POS-based algorithm for vocabulary adaptation and the Information Retrieval Engine (IR) for LM adaptation, yields a relative reduction of 65.2% in OOV word rate (from 1.40% to 0.49%), when compared to the results obtained for the baseline system with a vocabulary of 57K words. Moreover, this approach outperformed the one based on one single-pass (1-PASS-POS).

In terms of WER (figure 1), the new approach (2-PASS-POS-IR) resulted in a 6.6% relative gain. Even using a vocabulary with only 30K we were able to get a WER better than the one obtained for the baseline system with a 57K words vocabulary. Thus, implementing the proposed multi-pass adaptation approach and increasing the vocabulary size to 100K words we could obtain a relative gain of 8.5% in terms of WER.

Analysis on the OOV words, which were found by our IR-based framework, showed that almost all the relevant terms like proper and common names were correctly recognized. This makes the proposed framework especially useful, since these words contain a great deal of information for systems where the use of automatic transcriptions is a major attribute, as is the case of our BN transcription system.

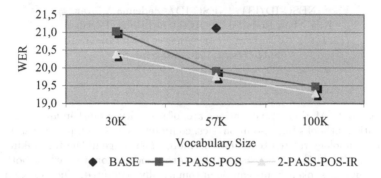

Fig. 1. WER comparison for 3 different vocabulary sizes (30K, 57K and 100K words)

Acknowledgments

This work was partially funded by PRIME National Project TECNOVOZ number 03/165 and by the FCT project POSC/PLP/58697/2004. Ciro Martins is sponsored by a FCT scholarship (SFRH/BD/23360/2005).

References

1. Martins, C., Teixeira, A., Neto, J.: Dynamic Vocabulary Adaptation for a daily and real-time Broadcast News Transcription System. In: IEEE/ACL Workshop on Spoken Language Technology (December 2006)
2. Martins, C., Teixeira, A., Neto, J.: Vocabulary Selection for a Broadcast News Transcription System using a Morpho-syntatic Approach. In: Proc. of Interspeech 2007 (2007)
3. Strohman, T., Metzler, D., Turtle, H., Croft, W.B.: Indri: A language-model based search engine for complex queries (extended version). CIIR Technical Report (2005)
4. Ribeiro, R., Mamede, N., Trancoso, I.: Morpho-syntactic Tagging: a Case Study of Linguistic Resources Reuse. In: Language Technology for Portuguese: shallow processing tools and resources, Edições Colibri, Lisbon, Portugal (2004)
5. Martins, C., Teixeira, A., Neto, J.: Dynamic Language Modeling for a daily Broadcast News Transcription System. In: Proc. of ASRU (2007)
6. Lavrenko, V., Croft, W.: Relevance-Based Language Models. In: Proc. of SIGIR 2001 (2001)
7. Meinedo, H., Caseiro, D., Neto, J., Trancoso, I.: AUDIMUS. MEDIA: A Broadcast News Speech Recognition System for the European Portuguese Language. In: Mamede, N.J., Baptista, J., Trancoso, I., Nunes, M.d.G.V. (eds.) PROPOR 2003. LNCS, vol. 2721, Springer, Heidelberg (2003)

An Approach to Natural Language Equation Reading in Digital Talking Books*

Carlos Juzarte Rolo and António Joaquim Serralheiro

INESC-ID/IST, INESC-ID/Academia Militar
Rua Alves Redol, nº 9, 1000-029 Lisboa, Portugal
{carlos.rolo,antonio.serralheiro}@l2f.inesc-id.pt
http://www.l2f.inesc-id.pt

Abstract. Mathematic equations are, of necessity, a must in any math-ematic textbooks but also in physics, communications and, in general, in any technology related texts. Furthermore, their usage in Digital Talking Books (DTB)[1] can be eased if its corresponding counterpart in both text and/or spoken forms can be automatically generated. Therefore, an automatic system to translate or convert them into text and latter to speech is needed to broaden the scope of the DTBs.

DTBs are based on different types of data, structured according to some standard. They also require a player or browser that allows users to navigate, to index and to retrieve information (text, sound, images, etc.). The player was developed using a model based framework for adaptive multi-modal environments [2]. Besides supporting the features described in the DTB standard[1], the player introduces features complementing the synchronized presentation of text and audio, such as: addition of content related images; variable synchronization units, ranging from word to paragraph; annotation controlled navigation; definition of new reading paths; adaptation of the visual elements; behavioral adaptation reflecting user interaction, amongst others.

In this paper we address the implementation of a "translation" system that converts mathematical equations into text in such a way that it resembles as much as possible the "natural" reading of those entities. The system we implemented is more than just a translator from some form of mathematical notation into text, since reading heuristics were included.

The Mathematical Markup Language (MathML) was chosen in this work over other existing alternatives, such as LaTeX[3] and Microsoft Word that, besides being able to display mathematical formulas in a correct way, are also widely used. LaTeX was not meant for integration (other than with documents of the same type), or to be parsed by exterior applications (except for its own compilers). Microsoft Word is also a *de facto* standard for documents. The MathML [4] is is a open standard and is a XML derived format, and is easily integrated in applications.

---

* This work was done under the RiCoBA project, partially funded by Fundação de Ciência e Tecnologia, Programa POSC, n.º 61042/2004.

[1] www.niso.org/standards/resources/Z39-86-2002.html

A. Teixeira et al. (Eds.): PROPOR 2008, LNAI 5190, pp. 268–271, 2008.

Furthermore, it also can be displayed inside browsers and others applications that can display XML documents. It is easily read from external applications, since a XML parser will parse it correctly. The information is well structured and, therefore, easy to be read and to process. As with XML and other derived formats, MathML files can grow in size faster than the data or the information it contains, since it requires multiple tags [5]. MathML can be used to both display the mathematical content or to represent the content of the mathematical formulas.

The easiest way to implement an equation translator is to convert the MathML tags directly into their mathematical counterparts. Let, for instance a simple equation such as $x^2 + 4x + 4 = 0$ (whose MathML description is show in fig.1) that, in most currently available "translators", is converted into text to: "*eks* to the power of begin exponent two end exponent plus four times *eks* plus four equals zero".

```
<mrow>
  <mrow>
    <msup> <mi>x</mi> <mn>2</mn> </msup> <mo>+</mo>
      <mrow>
        <mn>4</mn>
        <mo>&InvisibleTimes;</mo>
        <mi>x</mi>
      </mrow>
    <mo>+</mo>
    <mn>4</mn>
  </mrow>
  <mo>=</mo>
  <mn>0</mn>
</mrow>
```

Fig. 1. Example of MathML

But, thankfully, no one speaks or reads such simple equations this way. Easier forms, such as: "the square of *eks* plus four *eks* plus four equals zero" are commonly used either by teachers or by students. However, this sentence holds different meanings which leads to a conflict of notation and, therefore, mathematical meaning. To avoid ambiguity, punctuation marks should be included: "the square of *eks*, plus four {times} *eks*, plus four, equals {to} zero" to explicitly delineate the scope of mathematical operators (curl brackets represent optional terms).

Most of the mathematical operators are directly translated into written text. However, some of the most used ones need further processing, namely power, fraction, derivative and matrices operators. This processing avoids the direct translation of the operators, since no one reads a *simple* fraction like "fraction, *begin numerator ... end numerator* divided by *begin denominator ... end denominator* end of fraction" but rather like "*numerator* over *denominator*". The goal of this work is to mimic as much as possible the "usual" way someone reads equations so, heuristics for the above cases were employed. Of course this may lead, in some cases, to ambiguity but, the gained naturalness may compensate this problem. In the following, some reading heuristics that were implemented to increase the naturalness of the conversion will be presented: Fractions pose the problem of identifying what is the numerator and what is the denominator. To address this problem, fractions were divided in two categories, *short* or *simple* and *long* fractions. *Short* fractions are defined by having a reduced number of symbols in the denominator and/or in the numerator (a configurable threshold of three was chosen); fractions that are not *short* by this criteria are considered as *long*.

Powers have a powerful impact on the reading of an equation. For example, x^2 should be converted to "eks square" instead of "eks to the power of begin exponent two end exponent".

Derivatives such as $\frac{df(x)}{dx}$ must be identified as such and clearly separated from fractions. Otherwise, it would be converted as "fraction begin numerator derivative of $f(x)$ end numerator, begin denominator derivative of x, end denominator" instead of "derivative of $f(x)$ in order to x".

Matrices can be small or very large, full of content, sparse, etc. So, creating a heuristic to process all possibilities would result in a disproportionate effort. Informal experiences with people reading matrices showed that people always say the size of the matrix in the first place. After that, no common reading methodology was found and, as a consequence, we decided to implement the same procedure for reading matrices. The heuristic starts to evaluate the size of the matrix and writes it in the beginning of the conversion text. After that, all the lines of the matrices are read one after the other, after being numbered. Although this procedure generates lots of text for small matrices, readability and comprehension are kept.

Two sets of experiments were undertaken, since the initial informal tests showed that ambiguity could arise due to the implemented heuristics. Since it is our intent to include technical books in DTBs, any ambiguity issues can be overcome by a mere visual inspection of the equations or the formulas. However, visually impaired users may not have the capability to visualize them. So, we decided to check the translation system without any visual support of the original equation. Furthermore, DTB users can have different technical backgrounds, ranging from elementary mathematics to more advanced calculus so, the translation tests comprised an "easy test set" (ETS) and a "Difficult Test Set", (DTS). Therefore, the former test set was given to 11 persons while the latter was solved by 15 persons, according to their skills. Examinees were only given the output of the translator, without any clue of what the original equation could be and they had to write it down. No blank answers were allowed, so they had to opt for an answer even if they were unsure of its correctness. The same procedure was followed for the DTS but, in this case, a total of 13 questions were given.

To summarize the results, 316 of questions were collected and we got a total number of 38 erroneous (Table 1). Although it would be tempting to say that a comprehensibility[2] of 88% was achieved, one should bear in mind that it would be very easy to design a test with 100% of right answers. However, we knew beforehand that some of the implemented heuristics would rise ambiguity issues, namely in fractions / divisions or even in exponents. From the answers, two different situations are evident:

---

[2] Herein defined as the ratio of correct answers versus total number of answers.

Table 1. Test Results

	ETS	DTS	Total
No. Questions	121	195	316
ERROR	14	24	38
Error %	11.6	12.3	12.0

if the examined can identify from the text some well-known formula, he/she writes down the correct answer, no matter the ambiguity! By the opposite, if the text is not recognized, answers are not correct.

Overall, test results were considered very good although some errors were reported. These errors were identified as a result of ambiguity in the output text as previously expected. In either case, results showed that although ambiguity was present in some situations, if the reader could identify the equations content, he/she could immediately overcame that problem and produce the correct answers. Some of the heuristics, namely the derivatives and matrices heuristics, need improvements to cope with a broader set of formulas.

Keywords: MathML, Digital Talking Books, Speech Alignment.

References

1. Serralheiro, A., Trancoso, I., Caseiro, D., Chambel, T., Carriço, L., Guimarães, N.: Towards a Repository of Digital Talking Books. In: Proc. Eurospeech 2003, Geneva, Switzerland (September 2003)
2. Duarte, C., Carriço, L.: Users and Usage Driven Adaptation of Digital Talking Books. In: Proc. 11th International Conference on Human-Computer Interaction (HCII 2005), July 2005, Las Vegas, Nevada (2005)
3. http://www.latex-project.org/
4. W3C, MathML Standart, http://www.w3.org/TR/2003/REC-MathML2-20031021
5. Megginson, D.: Imperfect XML: Rants, Raves, Tips, and Tricks.. from an Insider. Addison Wesley Professional, Reading (2004)

Topic Segmentation in a Media Watch System

Rui Amaral[1,2,3] and Isabel Trancoso[1,3]

[1]Instituto Superior Técnico
[2]Instituto Politécnico de Setúbal
[3] L^2F - Spoken Language Systems Lab, INESC-ID
{Rui.Amaral,Isabel.Trancoso}@l2f.inesc-id.pt
https://www.l2f.inesc-id.pt

Abstract. This paper describes our on-going work on the topic segmentation module of a media watch system. The current version explores not only the typical structure of a broadcast news show, but also its contents, which are automatically produced by the speech recognition module, and the topic indexation module. The performance of the automatic topic segmentation module was compared with the manual segmentation done by a professional media watch company, yielding quite satisfactory results.

1 Introduction

Topic segmentation plays an important role in the prototype system for selective dissemination of Broadcast News (BN) in European Portuguese, developed at INESC-ID. The media watch system was initially built in the context of the ALERT European project [1] and is the object of continuous improvement in the framework of national project TECNOVOZ. The topic segmentation module (TS) described in this paper is one of the modules of the complex system and is performed off-line, exploring only audio-derived cues, for the time being. This paper starts with a brief description of our BN corpus in Section 2. The bulk of the paper is devoted to the topic segmentation module (section 3). Section 4 compares the automatic with the manual topic segmentation performed by a media watch company, and discusses the importance of video-derived cues. The final Section concludes and presents directions for future research.

2 The European Portuguese BN Corpus

The European Portuguese BN corpus includes different types of news shows, national and regional, generic and specific domains, from morning to late evening. In this work, we used 4 subsets, all manually segmented into stories, covering a wide range of scenarios. The SR (Speech Recognition) corpus contains 57h of BN shows, where 45% is presented by the lead anchor and the remaining shows also have a sports anchor. The JE (Joint Evaluation) corpus contains 13h, half of which contain only a lead anchor and the other half also include a sports anchor. To expand the segmentation scenarios, an extra BN corpus (EB) with 4h was

A. Teixeira et al. (Eds.): PROPOR 2008, LNAI 5190, pp. 272–275, 2008.

collected from a different TV station. One of the shows is presented by the lead anchor, but includes a local news commentator. The other two shows have two lead anchors, and one of them also includes the local news commentator. The need for the comparison with a professional media watch company motivated the collection of a very recent corpus (RTP07). This corpus contains around 6h, segmented by the media watch company. All the 6 shows have one lead anchor, without thematic anchors.

3   Topic Segmentation

The goal of TS module is to split the BN show into its constituent stories, exploring their characteristic structure [2]. All stories start with a segment spoken by the anchor, and are typically further developed by out-of-studio reports and/or interviews. The analysis of the typical structure of a BN show led us to train a CART (Classification and Regression Tree) with potential characteristics for each segment boundary [3]. The CART performed reasonably well for BN shows with one lead anchor, but failed with shows involving 2 lead anchors. This led us to adopt a two-stage supervised approach: in a first stage of re-clustering, the two speaker ids with the most frequent turns are clustered into a single label. After this pre-processing stage, the CART is applied.

3.1   Exploring the Topic Related Structure

To deal with a more complex structure, such as a BN show with a thematic anchor, a multi-stage approach was adopted where topic segmentation and indexation are interleaved. The first stage identifies potential story boundaries in every non-speech/anchor transitions. The second stage uses the topic indexation to isolate the thematic portion of the BN show (sports). This stage allows potential story boundaries to appear within the given theme. A third stage of boundary removal is applied using the same rules adopted by the CART. The knowledge of the topic was also used to remove false alarms in the weather forecast topic, which was typically split into multiple stories, due to the relatively long pauses made by the anchor between the forecasts for each part of the country.

3.2   Exploring Non-news Information

One recent improvement of our system is the inclusion of a non-news detector which detects the jingles that delimit the BN show, the publicity segments, and the headlines/teasers. The performance of the previous version of the TS module was seriously degraded by the presence of headlines [3], causing false alarms inside headlines, and miss boundaries after the headlines. The inclusion of the non-news information in the TS module allowed us to define another story boundary detection rule which avoided these problems.

3.3 Exploring the Contents of BN Segments

The main remaining problem was the false alarm rate due to the long anchor interventions in the middle or at the end of a story. In order to decrease these false alarms, we used the automatic transcriptions of the BN shows. The merging of short stories with either their left or right neighbors was dictated by a CART trained with the following features: the acoustic background conditions of the left and right stories, the word rate (computed at the first 7s of the short story, which is the minimal time required for a story introduction), the duration of the anchor segment, and the normalized count of matches of unigrams, bigrams and trigrams between the short story and the two neighbors. The matches are computed over the automatic transcripts and the purpose is to detect text similarities between the short story and its neighbors, to help the merge decision.

4 Results and Discussion

The results of the different versions of the segmentation algorithm are presented in Table 1. The performance of the 3-stage approach only took the sports topic splitting into account (third line). The next two lines used the single BN show of RTP07 which had weather forecast news (RTP07-1). The fourth line taked only the sports topic splitting into account, and the fifth line was obtained also taking the weather forecast merging into account. The next two lines used 3 shows of the RTP07 corpus (RTP07-3) and show the improvements achieved with the integration of non-news information (without and with, respectively). The following two lines used 6 shows of the RTP07 corpus, and show the improvements achieved with the integration of the ASR results (without and with, respectively). The last two lines of the Table show the results that would be achieved if the evaluation window is extended to 2s.

Our collaboration with video segmentation experts in the framework of European project VIDI-VIDEO and a preliminary experiment with a single recent

Table 1. Topic segmentation results

Approach	%Recall	%Precision	F-measure	corpus
Single-Stage	79.6	69.8	0.74	JE
Two-Stage	81.2	91.6	0.85	EB
Multi-Stage	88.8	56.9	0.69	JE
Multi-Stage	97.1	86.8	0.92	RTP07-1
Multi-Stage (+meteo)	97.1	89.2	0.93	RTP07-1
Multi-Stage	98.9	71.7	0.83	RTP07-3
Multi-Stage + non-news info	96.8	73.9	0.84	RTP07-3
w/o ASR (eval=1s)	88.0	81.7	0.85	RTP07
with ASR (eval=1s)	91.2	83.0	0.87	RTP07
w/o ASR (eval=2s)	93.8	87.1	0.90	RTP07
with ASR (eval=2s)	97.0	88.2	0.92	RTP07

BN allows us to discuss the feasibility of using video derived cues for the task of TS. The fusion of our topic segmentation boundaries derived only from the audio signal with the ones provided by a shot segmentation module may contribute towards a higher precision of the automatically computed boundaries. In terms of video shot representation, semantic concepts such as single news anchor, double news-anchor, news studio, etc. may contribute towards making the overall topic segmentation system more robust and autonomous. The detection of a split screen showing both the lead anchor and the field reporter might also be useful since it never happens at the very begining of a story. These are the type of video derived cues we are currently studying for the potential integration with our audio-based TS module.

5   Conclusions

This paper described our on-going work on the TS module for broadcast news. It summarized our first experiments with a single-stage CART based approach, which explored only the typical structure of BN shows. This approach evolved into a multi-stage approach, which allowed more complex structures with thematic anchors and commentators, and later also explored the topic related structure, the non-news information and the automatically produced transcripts of the BN shows.

The performance of the automatic topic segmentation module was compared with the manual segmentation done by a professional media watch company, yielding quite satisfactory results. The paper also discussed how these could be improved by merging with video derived cues, which is part of our current plans.

Acknowledgments

The present work is part of Rui Amaral's PhD thesis, initially sponsored by a FCT scholarship. This work was partially funded by PRIME National Project TECNOVOZ number 03/165, and by the European project Vidi-Video. The authors would like to acknowledge the continuing support of our colleagues J. Neto, H. Meinedo, and V. Mezaris.

References

1. Neto, J., Meinedo, H., Amaral, R., Trancoso, I.: A system for selective dissemination of multimedia information resulting from the alert project. In: Proc. MSDR 2003, Hong Kong (April 2003)
2. Barzilay, R., Collins, M., Hirschberg, J., Whittaker, S.: The rules behind roles: Identifying speaker role in radio broadcast. In: Proc. AAAI 2000, Austin, USA (July 2000)
3. Amaral, R., Trancoso, I.: Exploring the structure of broadcast news for topic segmentation. In: Proc. LTC 2007, Poznan, Poland (October 2007)

Author Index

Lecture Notes in Artificial Intelligence (LNAI)

Vol. 4930: I. Wachsmuth, G. Knoblich (Eds.), Modeling Communication with Robots and Virtual Humans. X, 337 pages. 2008.

Vol. 4929: M. Helmert, Understanding Planning Tasks. XIV, 270 pages. 2008.

Vol. 4924: D. Riaño (Ed.), Knowledge Management for Health Care Procedures. X, 161 pages. 2008.

Vol. 4923: S.B. Yahia, E.M. Nguifo, R. Belohlavek (Eds.), Concept Lattices and Their Applications. XII, 283 pages. 2008.

Vol. 4914: K. Satoh, A. Inokuchi, K. Nagao, T. Kawamura (Eds.), New Frontiers in Artificial Intelligence. X, 404 pages. 2008.

Vol. 4911: L. De Raedt, P. Frasconi, K. Kersting, S. Muggleton (Eds.), Probabilistic Inductive Logic Programming. VIII, 341 pages. 2008.

Vol. 4908: M. Dastani, A. El Fallah Seghrouchni, A. Ricci, M. Winikoff (Eds.), Programming Multi-Agent Systems. XII, 267 pages. 2008.

Vol. 4898: M. Kolp, B. Henderson-Sellers, H. Mouratidis, A. Garcia, A.K. Ghose, P. Bresciani (Eds.), Agent-Oriented Information Systems IV. X, 292 pages. 2008.

Vol. 4897: M. Baldoni, T.C. Son, M.B. van Riemsdijk, M. Winikoff (Eds.), Declarative Agent Languages and Technologies V. X, 245 pages. 2008.

Vol. 4894: H. Blockeel, J. Ramon, J. Shavlik, P. Tadepalli (Eds.), Inductive Logic Programming. XI, 307 pages. 2008.

Vol. 4885: M. Chetouani, A. Hussain, B. Gas, M. Milgram, J.-L. Zarader (Eds.), Advances in Nonlinear Speech Processing. XI, 284 pages. 2007.

Vol. 4874: J. Neves, M.F. Santos, J.M. Machado (Eds.), Progress in Artificial Intelligence. XVIII, 704 pages. 2007.

Vol. 4870: J.S. Sichman, J. Padget, S. Ossowski, P. Noriega (Eds.), Coordination, Organizations, Institutions, and Norms in Agent Systems III. XII, 331 pages. 2008.

Vol. 4869: F. Botana, T. Recio (Eds.), Automated Deduction in Geometry. X, 213 pages. 2007.

Vol. 4865: K. Tuyls, A. Nowe, Z. Guessoum, D. Kudenko (Eds.), Adaptive Agents and Multi-Agent Systems III. VIII, 255 pages. 2008.

Vol. 4850: M. Lungarella, F. Iida, J.C. Bongard, R. Pfeifer (Eds.), 50 Years of Artificial Intelligence. X, 399 pages. 2007.

Vol. 4845: N. Zhong, J. Liu, Y. Yao, J. Wu, S. Lu, K. Li (Eds.), Web Intelligence Meets Brain Informatics. XI, 516 pages. 2007.

Vol. 4840: L. Paletta, E. Rome (Eds.), Attention in Cognitive Systems. XI, 497 pages. 2007.

Vol. 4830: M.A. Orgun, J. Thornton (Eds.), AI 2007: Advances in Artificial Intelligence. XIX, 841 pages. 2007.

Vol. 4828: M. Randall, H.A. Abbass, J. Wiles (Eds.), Progress in Artificial Life. XII, 402 pages. 2007.

Vol. 4827: A. Gelbukh, Á.F. Kuri Morales (Eds.), MICAI 2007: Advances in Artificial Intelligence. XXIV, 1234 pages. 2007.

Vol. 4826: P. Perner, O. Salvetti (Eds.), Advances in Mass Data Analysis of Signals and Images in Medicine, Biotechnology and Chemistry. X, 183 pages. 2007.

Vol. 4819: T. Washio, Z.-H. Zhou, J.Z. Huang, X. Hu, J. Li, C. Xie, J. He, D. Zou, K.-C. Li, M.M. Freire (Eds.), Emerging Technologies in Knowledge Discovery and Data Mining. XIV, 675 pages. 2007.

Vol. 4811: O. Nasraoui, M. Spiliopoulou, J. Srivastava, B. Mobasher, B. Masand (Eds.), Advances in Web Mining and Web Usage Analysis. XII, 247 pages. 2007.

Vol. 4798: Z. Zhang, J.H. Siekmann (Eds.), Knowledge Science, Engineering and Management. XVI, 669 pages. 2007.

Vol. 4795: F. Schilder, G. Katz, J. Pustejovsky (Eds.), Annotating, Extracting and Reasoning about Time and Events. VII, 141 pages. 2007.

Vol. 4790: N. Dershowitz, A. Voronkov (Eds.), Logic for Programming, Artificial Intelligence, and Reasoning. XIII, 562 pages. 2007.

Vol. 4788: D. Borrajo, L. Castillo, J.M. Corchado (Eds.), Current Topics in Artificial Intelligence. XI, 280 pages. 2007.

Vol. 4775: A. Esposito, M. Faundez-Zanuy, E. Keller, M. Marinaro (Eds.), Verbal and Nonverbal Communication Behaviours. XII, 325 pages. 2007.

Vol. 4772: H. Prade, V.S. Subrahmanian (Eds.), Scalable Uncertainty Management. X, 277 pages. 2007.

Vol. 4766: N. Maudet, S. Parsons, I. Rahwan (Eds.), Argumentation in Multi-Agent Systems. XII, 211 pages. 2007.

Vol. 4760: E. Rome, J. Hertzberg, G. Dorffner (Eds.), Towards Affordance-Based Robot Control. IX, 211 pages. 2008.

Vol. 4755: V. Corruble, M. Takeda, E. Suzuki (Eds.), Discovery Science. XI, 298 pages. 2007.

Vol. 4754: M. Hutter, R.A. Servedio, E. Takimoto (Eds.), Algorithmic Learning Theory. XI, 403 pages. 2007.

Vol. 4737: B. Berendt, A. Hotho, D. Mladenic, G. Semeraro (Eds.), From Web to Social Web: Discovering and Deploying User and Content Profiles. XI, 161 pages. 2007.

Vol. 4733: R. Basili, M.T. Pazienza (Eds.), AI*IA 2007: Artificial Intelligence and Human-Oriented Computing. XVII, 858 pages. 2007.

Vol. 4724: K. Mellouli (Ed.), Symbolic and Quantitative Approaches to Reasoning with Uncertainty. XV, 914 pages. 2007.

Vol. 4722: C. Pelachaud, J.-C. Martin, E. André, G. Chollet, K. Karpouzis, D. Pelé (Eds.), Intelligent Virtual Agents. XV, 425 pages. 2007.

Vol. 4720: B. Konev, F. Wolter (Eds.), Frontiers of Combining Systems. X, 283 pages. 2007.

Vol. 4702: J.N. Kok, J. Koronacki, R. Lopez de Mantaras, S. Matwin, D. Mladenič, A. Skowron (Eds.), Knowledge Discovery in Databases: PKDD 2007. XXIV, 640 pages. 2007.

Vol. 4701: J.N. Kok, J. Koronacki, R. Lopez de Mantaras, S. Matwin, D. Mladenič, A. Skowron (Eds.), Machine Learning: ECML 2007. XXII, 809 pages. 2007.